BASIC NETWORK THEORY
WITH COMPUTER APPLICATIONS

BASIC NETWORK THEORY
WITH COMPUTER APPLICATIONS

JIŘÍ VLACH

Department of Electrical and Computer Engineering
University of Waterloo
Waterloo, Ontario, Canada

VNR VAN NOSTRAND REINHOLD
New York

Library of Congress Catalog Card Number 91-25099
ISBN 0-442-00900-3

Manufactured in the United States of America

Published by Van Nostrand Reinhold
115 Fifth Avenue
New York, New York 10003

Chapman and Hall
2–6 Boundary Row
London, SE1 8HN, England

Thomas Nelson Australia
102 Dodds Street
South Melbourne 3205
Victoria, Australia

Nelson Canada
1120 Birchmount Road
Scarborough, Ontario M1K 5G4, Canada

16 15 14 13 12 11 10 9 8 7 6 5 4 3 2 1

Library of Congress Cataloging-in-Publication Data
Vlach, Jiří.
    Basic network theory : with computer applications / Jiří Vlach.
      p. cm.
    Includes index.
    ISBN 0-442-00900-3
    1. Electric networks. I. Title.
TK454.2.V54   1991
621.319′24—dc20   91-25099
                                 CIP

PREFACE

This book is intended for a two-semester introductory course in network theory for electrical or computer engineering students. Usually such a course is taught in second year and a certain mathematical background, like trigonometric functions, standard calculus and basic algebra, are already available. Nevertheless, the book treats all needed subjects in detail, even if they have been already covered elsewhere. The writing of equations for frequency domain as well as time domain is unified, with emphasis on their solution by matrix methods. This prepares the student for computerized applications in later studies. The book differs slightly from similar introductory texts by including, at the end, more recent subjects in network theory. All theoretical steps are explained with the help of numerous solved problems which are placed into the text.

Presentation of the material is independent of any use of computers and in this respect the book does not differ from other texts. However, these days many students already have personal computers and one can assume that all have access to personal computers at their schools. For this reason the book is accompanied by three programs for IBM PC (or equivalent) computers. The programs can help the student in confirming correctness of his solutions. More about these programs is in the section which follows this preface.

The first chapter deals with fundamental definitions and presents, on an informal basis, all linear elements. Two elements are next treated in detail: independent sources and resistors. The chapter explains Kirchhoff's laws, concepts of power and energy and connections of resistors.

The following four chapters are devoted to methods of writing and solving equations for resistive networks. We start from the simplest cases and develop them to the extent that any network can be transformed into either the nodal or the mesh formulation. Chapter 2 is brief and only two elements are used here: current sources with conductances in nodal formulation, and voltage sources with resistors in mesh formulation.

Because the students may not have sufficient practical knowledge of matrix algebra, the third chapter explains in detail how to solve systems of linear equations by matrix methods. Emphasis is on *practical* methods for *hand calculations* and the theory is demonstrated on several simple resistive networks.

Chapter 4 is devoted to the study of dependent sources. The voltage-controlled current source is ideally suited for nodal formulation while the current-controlled voltage source is directly applicable in mesh formulation. The other two dependent sources need preprocessing, which is explained in detail. In this chapter we deal with networks which have only one type of source. Those with current sources, dependent or independent, are analyzed by the nodal formulation. If the network has only voltage sources, dependent or independent, then mesh formulation is applied.

Chapter 5 finally collects all information needed for analysis of arbitrary resistive networks, in either nodal or mesh formulation. The sources can be transformed by the Thévenin and Norton theorems. It is shown how to handle situations when we need transformation of the controlling terminals of controlled sources. Splitting of current or voltage sources helps in modifying the network into a form suitable for nodal or mesh formulations. At the end of this chapter the student will have enough knowledge to write either nodal or mesh equations for any resistive network. If he/she understands the material of the first five chapters, he/she will have no difficulties in setting up equations for arbitrary networks later. Recommendations are also given: from now on, the nodal formulation is preferred for all networks. This is in agreement with the more modern approaches and prepares the student for a general method, the nodal formulation, covered in Chapter 17.

Chapter 6 exposes the student for the first time to the concepts of capacitors and inductors, still without initial conditions. One section demonstrates how one could write the differential equations for first-order networks, but the book does not solve time domain problems by preparing differential equations. In order to unify frequency and time domain methods, we prefer to use the Laplace transform.

Instead of going directly to time domain solutions, as is often done at this point, we introduce the s variable and expose the student to the concept of impedances and admittances. Chapter 7 covers analysis of networks which contain capacitors, inductors, or both elements. Their impedances or admittances are used, as necessary by the particular formulation method. The networks are solved by keeping the varible s and/or some of the elements as variables. As a result of this chapter, the student will be able to set up the network equations for either frequency or time domain.

Frequency domain methods are explained in Chapter 8. It starts with the definition of phasors and gives all details how to handle complex numbers. Preference is given to algebraic methods instead of graphical solutions, because the latter are less accurate and these days all students have calculators capable of handling complex numbers in rectangular or polar form. Frequency domain analysis is done by means of phasors. The student is given the choice of either inserting $j\omega$ from the beginning or keeping the s variable in all derivations. Frequency domain analysis is applied to simple RC, RL and RLC networks. The last section of the chapter shows how to use phasors in power considerations.

Since the student is, at this point, familiar with all the necessary network analysis methods, we turn in Chapter 9 to the theory of the Laplace transform. It is traditionally taught in other courses, and for this reason we limit the scope to subjects which are needed in the analysis of linear networks with lumped elements. The chapter provides enough mathematical background to express initial conditions on capacitors and inductors.

Time domain analysis is finally covered in Chapter 10. Since it is based on the

Laplace transform, it is possible to use exactly the same analysis steps as in previous chapters. If the network has zero initial conditions, then such a situation was, in fact, already handled in Chapter 7. If there are initial conditions, then it is shown that they are represented by special independent sources and as such are mathematically expressed by changes in the right-hand side vector of the matrix equation. Simple first- and second-order networks are also explained in detail. Networks with switches show how the initial conditions are obtained practically.

Chapter 11 deals with network functions, the concepts of poles, zeros and stability. It is shown how one can get frequency domain responses (amplitude, phase, group delay) graphically from the positions of poles and zeros. The chapter is concluded with a section on Bode plots.

Chapter 12 is devoted to the analysis of active networks. The method is based on nodal formulation with special handling of voltage sources; as a consequence, the student automatically gets the minimal set of equations. Operational amplifiers can be ideal, with infinite gain, or nonideal, with frequency dependent characteristics. In the second case the number of equations grows by one for every nonideal OPAMP. A number of practical examples is discussed at the end of the chapter.

Two-port theory is presented in Chapter 13. The theory is still important as a theoretical tool, although it has lost its importance in practical applications. The chapter introduces some specialized elements, like gyrators and impedance converters. The instructor should decide how much of this material is needed for his particular needs.

Chapter 14 explains analysis methods for networks with ideal, perfect, technical and general transformers, the last linear elements considered in the book. Equivalent transformer networks are derived and applied in various situations, including coupled tuned circuits.

The material of the three following chapters is not traditionally taught in introductory network theory courses. Chapter 15 is an introduction to modeling and has two parts, linear and nonlinear. The linear part should be covered: modeling of sources, modeling of linear amplifiers and linear transistor models. The second, nonlinear part, is an introducton about how to handle general nonlinear systems. It shows that nonlinearities create new frequencies so that frequency domain methods and the superposition principle do not apply. It demonstrates how to set up network equations in the presence of nonlinearities and how to solve them iteratively by the Newton-Raphson method. Important here is the fact that the Jacobian of the Newton–Raphson equation has the same form as if the elements were linear. The value of the derivative appears in the same positions of the matrix as if the element were linear, and the value changes from iteration to iteration. Linear elements stay in exactly the same places of the Jacobian as they would if the network was linear.

One of the most important and rarely covered areas of linear network theory are sensitivities with respect to element changes. The subject is usually left to courses dealing with filters. Chapter 16 covers all aspects of linear network sensitivities, under the assumption that a transfer function in symbolic form has been derived. There is no need to go into all the mathematical details, but a proper understanding should be achieved. Sensitivity calculations, with plotting, are available in the program NETOPT and demonstrations may be more helpful than theoretical details.

Network transformations, which proved so useful in writing the minimal set of network equations in nodal or mesh form, are suitable for hand analysis but not for general applications. Chapter 17 presents a modern method for setting up network equations. It is called the modified nodal formulation and is used in all modern programs, like the well-known SPICE. It is not necessary to cover this section in an introductory course, but its presentation will help the student in his/her later studies.

Chapter 18 explains how to use the program NETOPT. The program is based on recent theoretical developments and was specially written for analysis and design of linear networks. The instructor can find theoretical foundations in the book by J. Vlach and K. Singhal: *Computer Methods for Circuit Analysis and Design*, Van Nostrand Reinhold, 1983. The program itself was described in E. Christen and J. Vlach: 'NETOPT—A program for multiobjective design of linear networks', *IEEE Transactions on Computer-Aided Design*, vol. 7, no. 5, pp. 567–577, May 1988. Description in Chapter 18 covers only those parts which are useful for the course. Omitted are the design features of the program; they can be found in a manual, available separately.

Fourier series and transform are traditionally included into basic texts on network theory. They are not directly related to the subject, but we have included the material in Chapter 19. Only basic information, with emphasis on understanding, is given.

PROGRAMS

Three programs are available (see order form in back of book) to help the student in checking his results and his understanding. They are prepared for IBM PC (or equivalent) computers, preferably with plotting facility (CGA, EGA, VGA, or Hercules). A math coprocessor is not needed but use of it is made, if present.

The first program, to be used with the study of Chapter 3, solves systems of linear equations. Input is either interactive or from a file. The program differs from similar programs by giving the results in the same form as if they were obtained with Cramer's rule: it prints the system determinant (denominator), the numerators for all variables, and finally the results as ratios of the numerators and the determinant. Appendix A gives information about its use.

The second program can calculate and plot time domain and frequency domain responses of network functions. It is interactive and when loaded comes first with a menu for the choice of input:

a. Network function in terms of poles and zeros, given interactively
b. Network function in terms of poles and zeros, from a file
c. Network function as a rational function, given interactively
d. Network function as a rational function, from a file.

The second menu gives the choice of calculating and plotting time domain responses to a Dirac impulse, unit step, sine or cosine. In this case it also displays the necessary residues. In the frequency domain the program can plot amplitude responses in magnitude or dB, group delay and phase. The results appear as plots on the screen and numerical values are printed in files. The program can be used for problems starting with Chapter 8. Information about its use is in Appendix B.

The third program, NETOPT, is an educational version of a general interactive program for analysis and design of linear networks. NETOPT has many more features than needed in the first two courses on network theory. Its use for this particular course is described in Chapter 18. NETOPT can be applied to any linear network problem, starting from Chapter 2, and used in all chapters of the book. It has many features which are generally not available in other programs, like symbolic functions, sensitivities to network element changes, poles, zeros and their sensitivities, etc. The program plots responses on screen (if the necessary hardware is available) and writes files of numerical results.

Appendix C gives derivation of a practical solution method for numerical systems of linear equations. The theory, with full pivoting, is coded in FORTRAN and is available on the diskettes accompanying the book. The subroutines are suitable for systems with complex coefficients, as needed for frequency domain solutions, and supplement the theory developed in Chapter 17.

CONTENTS

BASIC NETWORK THEORY
WITH COMPUTER APPLICATIONS

CHAPTER 1

Basic Concepts

1.0. INTRODUCTION

This chapter starts with the concept of electric current and voltage, introduces basic terminology and shows how networks are drawn. It gives a brief overview of all linear elements which are used in network theory, plus a few of the most important practical devices. After this informal explanation we give definitions of the independent current and voltage sources, of resistors, of power delivered or consumed. Three fundamental laws are stated: Ohm's law and Kirchhoff's current and voltage laws. Based on this knowledge we study the voltage and current dividers and also explain how to obtain equivalent resistances for combinations of resistors. The chapter is concluded by a brief description of voltmeters and ammeters and of their desired properties.

1.1. VOLTAGES AND CURRENTS

Network theory deals with voltages, currents and network elements; in this section we will consider in detail the first two, and also introduce some basic concepts.

Electric *current* is a flow of electrons and is measured in *amperes*, denoted by the letter A. Voltage is a force which causes electrons to flow and is measured in *volts*, denoted by V. The concept of current is usually somewhat easier to grasp because we can visualize it as a current of water.

FIGURE 1.1.1   A two-terminal element.

When we talk about the flow of water, we associate it with a certain direction and we know, from daily experience, that the direction is from a higher point to a lower point. We also accept without difficulties that there must be some force which determines the direction of the flow. A similar situation exists in electrical networks and the force that pushes the current through the elements is the voltage. The word *potential* can be used as well.

In the early days of electricity, only chemical batteries were available and the concept of electrons was unknown. One of the connecting points of the battery was arbitrarily marked with a + sign, the other with a − sign and it was assumed that the current flowed into the external network from the + terminal to the − terminal. The current direction is usually marked by an arrow, like in Fig. 1.1.1 where the box denotes some not yet specified element. It is a standard practice to use letters i or I for the current and letters v or V for the voltage.

Since voltage is a force and force must always be referred to some point with respect to which it acts, we must indicate a reference point when talking about voltage. Such point is usually *ground*. In electrical instruments ground is understood to be the metal base of the equipment and not earth, to which the instrument may or may not be connected. We will use the term ground even if the equipment is in a flying aircraft.

The concept of a positive direction of current is fundamental to network theory. If the arrow denoting the current is directed from a point with higher voltage to a point with lower voltage, e.g. from + to −, then we say that the current has *positive* direction. If the arrow points from − to +, a negative sign is associated with the current. Positive direction of current is indicated in Fig. 1.1.1. We will say more about the current direction and its implications later in this chapter.

Each idealized element has its own symbol, but for a start we will represent the simplest one by a small box with leads as shown in Fig. 1.1.1. The lines coming out of the box have no special electrical properties; their only purpose is to connect the element to other elements. The small circles at the ends of the lines are called *terminals*. The element has two terminals and is thus called a *two-terminal* element. We will see later that we may have three- and four-terminal elements as well.

Several elements may be connected together to form a network, for instance as shown in Fig. 1.1.2a. The drawing, although showing correctly what we intended to do, is not very pleasing to the eye and is also not simple to draw. In order to avoid difficulties in drawing schematics of networks and also in understanding them, we have a convention that a line which connects elements has no special

Fig. 1.1.2   Fig. 1.1.3

FIGURE 1.1.2   Drawing networks: (a) Possible way. (b) Accepted method.

FIGURE 1.1.3   Conventions for connections: (a) Electrically connected. (b) Electrically not connected.

meaning; it just indicates a connection. Using this rule, we can redraw the same network as shown in Fig. 1.1.2b. This figure is clearly much easier to produce and is also easier to interpret. The full dots indicate that there is a true connection of the lines (we could think that they are soldered together). A point connecting two or more elements is called the *node*. Connection of two lines is indicated in Fig. 1.1.3a.

In more complicated situations we may not be able to draw the lines without some intersections, although there is no electrical connection at the point where they intersect. In order to clearly indicate that the lines are *not* connected, there are two conventions. Intersection *without* a full dot means that the lines are not connected and can be thought of as running one above the other. This is shown in Fig. 1.1.3b. Another convention draws a small semicircle, shown in the same figure. True connection of *three or more* wires will *always* be marked by a full dot. We may use any of the above two conventions to indicate that the lines are not connected.

Another common expression says that a certain point is *grounded. Ground* is marked by the symbol at the bottom of both parts of Fig. 1.1.2. It is a somewhat misleading concept because we do not ground the network in the way a lightning rod is grounded. The symbol only indicates this reference point. Every network must always have one such reference point. In many network drawings, the bottom long line is automatically assumed to represent the reference point.

We still need a few more fundamental concepts. A voltage at a node, measured with respect to ground, is called *nodal voltage*. Two nodes are sketched in Fig. 1.1.4a. It is a common practice to give the node a name, usually a positive integer number, and the voltage may be indicated at the node, as shown in Fig. 1.1.4a. The two nodes are not connected and we use the expression *open circuit* for such a situation. Rather obviously, no current can flow through an open circuit and arbitrary voltages may exist at the two nodes and thus across the open circuit. The opposite situation would be a *short circuit*, shown in Fig. 1.1.4b. It is

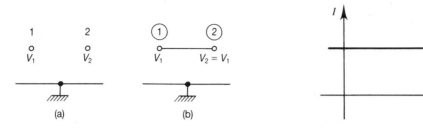

Fig. 1.1.4

Fig. 1.1.5

FIGURE 1.1.4   Two nodes: (a) Open circuit. (b) Short circuit.

FIGURE 1.1.5   Direct current (dc) as function of time.

represented by a connecting line, the same line as used in Fig. 1.1.2. Two nodes connected by a short circuit have the same voltage (zero voltage difference between the two nodes) and an arbitrary current can flow through a short circuit in any direction. The node numbers may be in circles, as in Fig. 1.1.4b.

The current can have a value which does not change with time, or it can vary with time. We can draw a diagram in which time is measured on the horizontal axis and current is marked on the vertical axis. If the current value does not change with time, such diagram will be a horizontal line, shown in Fig. 1.1.5. We speak about *direct current* and denote it by *dc*. A dc current is supplied, for instance, by a car battery. In other situations we may have currents whose instant values change. Figure 1.1.6 shows another typical case called *alternating current*, denoted as *ac*. This type of current is supplied by the outlet at your home. It is a periodic function and the current repeatedly changes its value. An example of still another type of current is given in Fig. 1.1.7, where the current flows in short pulses, with periods of time when the current does not flow at all. The three examples belong to more typical cases. There are many ways in which current can change with time.

Fig. 1.1.6

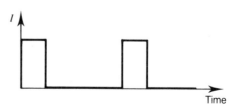

Fig. 1.1.7

FIGURE 1.1.6   Alternating current (ac) as function of time.

FIGURE 1.1.7   Graph of current in pulses.

Whatever we said about time dependence of current equally applies to voltages, with the only difference that the units measured on the vertical axis will be volts. Even if voltages and currents are changing values, in our schematics we will not consider time functions. A plus sign may still be given at some node and a minus sign at some other node.

1.2. NETWORK ELEMENTS

Our first section introduced some of the most important conventions and explained the concepts of currents and voltages. In this section we will give a brief, informal overview of elements which are used in network theory. All the elements will be discussed in detail later.

1.2.1. Independent Voltage and Current Sources

An electric network can function only if a source of a voltage or current is attached to it. Network theory uses two sources: the *independent voltage*, and the *independent current* source. Their symbols are in Fig. 1.2.1. The same symbols are usually used for a dc, ac or, in fact, any actual type of source. Sometimes, in order to clearly indicate that there is a special source, another symbol may be used. For instance, it is common to use the symbol in Fig. 1.2.2 for a battery.

1.2.2. Resistor

A resistor is an element produced from some material which "resists" the flow of electrons. The property is called *resistance*. At normal temperatures, all materials exhibit some resistance. Silver and copper have small resistances and copper is commonly used in wires for connections. Some alloys or carbon mixed with other materials have larger resistances. The letter R is used for the resistance; it is measured in Ohms and the symbol is Ω. As an example, if a resistor has the resistance of ten Ohms, we would write next to its symbol $R = 10\ \Omega$, $R = 10$, or just $10\ \Omega$. The symbol for a resistor is shown in Fig. 1.2.3. The voltage across the

Fig. 1.2.1   Fig. 1.2.2

FIGURE 1.2.1   Symbols for independent voltage (a) and current (b) sources.

FIGURE 1.2.2   Symbol for battery.

Fig. 1.2.3   Fig. 1.2.4

FIGURE 1.2.3   Symbol for a resistor.

FIGURE 1.2.4   Symbol for a capacitor.

resistor and the current flowing through the resistor are coupled by Ohm's law

$$V = RI$$

Sometimes we use the inverted value of the resistance,

$$G = \frac{1}{R}$$

It is called *conductance*, is measured in Siemens and the symbol for the unit is S. We will learn more about resistors later in this chapter.

1.2.3. Capacitor

The symbol for a capacitor is in Fig. 1.2.4 and in schematics is denoted by the letter C. It is an element formed by two layers of conducting material, separated by a material which does not conduct, or as we say by an insulator. Many materials can serve as insulators and even waxed paper is sometimes used. Electrons can be stored on the capacitor; in such case we say that there is a charge on the capacitor. Charge is usually denoted by the letter Q or q, and is measured in Coulombs, C. If the charge is changing with time, a current flows through the capacitor according to

$$i(t) = \frac{dq(t)}{dt}$$

The equation indicates that no current can flow through the capacitor if the charge does not change with time (derivative of a constant is zero). If the capacitor is independent of any external influences, the equation simplifies to

$$i(t) = C \frac{dv(t)}{dt}$$

where C is the symbol for the *capacitance*, the property of the capacitor to store electrons. *Capacitance* is measured in Farads, F. If the capacitor has a capacitance of two Farads, we would write next to the symbol either $C = 2$ F, or $C = 2$ or just 2 F. The element will be introduced later in this book, with all details postponed until then.

1.2.4. Inductor

The symbol for an inductor is shown in Fig. 1.2.5. The classical form of an inductor is a coil of insulated wire, possibly placed into a core of ferromagnetic material, such as iron. A current flowing through the inductor creates a magnetic field, or flux. Flux is usually denoted by the letter Φ and is measured in Webers, Wb. There is a relationship between the flux and the voltage across the inductor,

$$v(t) = \frac{d\Phi(t)}{dt}$$

This equation indicates that if the flux is constant (created by a dc current), then there is no voltage across the inductor, because the derivative of a constant is zero. If the inductor is independent on external influences, then the above equation simplifies to

$$v(t) = L\frac{di(t)}{dt}$$

where L is the property of the inductor, called the *inductance*. Inductance is measured in Henries, H. In schematics, the element is denoted by the letter L. For instance, if an inductor has an inductance of three Henries we would write next to its symbol $L = 3$ H or $L = 3$ or just 3 H. This element will also be discussed in greater detail later in the book.

1.2.5. Dependent Sources

Unlike the independent sources introduced above, dependent sources deliver voltages or currents whose values depend on a voltage or current somewhere else in the network. This means that the elements have two controlling terminals, where

L

FIGURE 1.2.5   Symbol for an inductor.

they do the sensing, and two controlled terminals, where they deliver a voltage or current. There are four possible types of dependent sources:

1. *Voltage-controlled voltage source, VV*, Fig. 1.2.6a. The voltage difference at the terminals on the left, V_1, is multiplied by a coefficient, μ, and delivered by the source on the right:

$$V_2 = \mu V_1$$

The coefficient μ is often called the *gain* of the VV.

2. *Voltage-controlled current source, VC*, Fig. 1.2.6b. The voltage difference at the terminals on the left, V_1, is multiplied by a conversion constant, g, and is delivered as a current at the terminals on the right:

$$I_2 = g V_1$$

3. *Current-controlled voltage source, CV*, Fig. 1.2.6c. The current through the short circuit on the left, I_1, is multiplied by a conversion constant, r, and applied as a voltage to the terminals on the right:

$$V_2 = r I_1$$

4. *Current-controlled current source, CC*, Fig. 1.2.6d. The current flowing through the short circuit on the left, I_1, is multiplied by a constant, α, and applied as a current on the terminals on the right:

$$I_2 = \alpha I_1$$

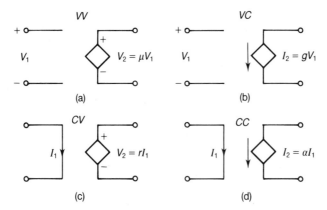

FIGURE 1.2.6   Symbols for dependent sources: (a) Voltage-controlled voltage source, VV. (b) Voltage-controlled current source, VC. (c) Current-controlled voltage source, CV. (d) Current-controlled current source, CC.

The elements will be discussed in detail in Chapter 4. They are very important because they are used to model properties of useful devices in real-life situations.

1.2.6. Transformer

If two (or more) coils are near each other, or are placed on one ferromagnetic core, then they form a transformer. There are many forms of transformers, but the most common one has two coils and its symbol is given in Fig. 1.2.7. The element will be discussed in detail in one of the later chapters.

1.2.7. Operational Amplifier

An operational amplifier, OPAMP, is a complicated electronic device whose simplified properties can be summarized as follows: it acts as a voltage-controlled voltage source, with one fundamental difference. In the ideal case, the amplification factor, μ, approaches infinity. The most common symbol for the operational amplifier is given in Fig. 1.2.8 and we will use it in our studies of active networks.

1.2.8. Other Elements

The above elements are all the elements used in network theory. In various combinations, they are used to describe the properties of actual, physically existing elements. We will not go into details now, but we will mention two more elements, because they are widely used. They are the field-effect transistors and bipolar transistors. In Fig. 1.2.9a is the symbol for a bipolar transistor, and in Fig. 1.2.9b is the symbol for a field-effect transistor, FET. Their properties are modeled by combinations of dependent sources, resistors, capacitors, and possibly inductors. Transistors are also used in the design of operational amplifiers.

1.2.9. Units and Their Prefixes

To complete this section we summarize the terminology introduced so far in Table 1.2.1. Theoretical units like Ohms for a resistor, or Henries for the inductor, may

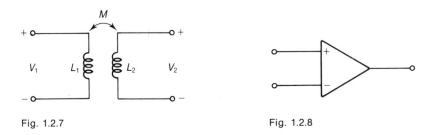

Fig. 1.2.7   Fig. 1.2.8

FIGURE 1.2.7   Symbol for a transformer.

FIGURE 1.2.8   Symbol for an operational amplifier.

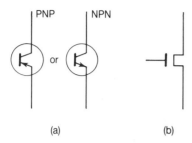

FIGURE 1.2.9   Symbols for transistors: (a) Bipolar transistor. (b) Field-effect transistor, FET.

in some cases be either too large or too small in practical applications. For instance, a unit to measure capacitance is one Farad, 1 F. This happens to be an impractically large unit. To express that a practical value is, for instance, 10^{-12} times smaller, we write the letter p before the symbol F and have one picofarad, $1\,\mathrm{pF} = 10^{-12}$ F. The prefixes for all units are summarized in Table 1.2.2.

1.2.10. Examples of Networks

Since we already know the elements which we will be using in this book, we now give some examples. They will show how we draw the networks and what information we usually write into the schematics.

Figure 1.2.10 shows a network with one independent voltage source, $E = 5$ V, two resistors, two capacitors, one inductor and a current-controlled current source. If there are several elements of the same type, we either give them different names or use subscripts, like in the figure. Subscripts are not needed here for the inductor, L, and the current-controlled current source, CC, but can be given. The controlling current is denoted by I and its direction is indicated by an arrow. It is the current

TABLE 1.2.1   Units and Symbols

Name	Notation	Unit Name	Unit Symbol
Resistance	R	Ohm	Ω
Conductance	G	Siemens	S
Capacitance	C	Farad	F
Inductance	L	Henry	H
Current	I, i	Ampere	A
Voltage	V, v	Volt	V
Energy	W, w	Joule	J
Power	P, p	Watt	W
Charge	Q	Coulomb	C
Flux	Φ	Weber	Wb
Time	t	Second	s

TABLE 1.2.2   **Prefixes for Electrical Quantities**

Prefix	Name	Value
f	femto	10^{-15}
p	pico	10^{-12}
n	nano	10^{-9}
μ	micro	10^{-6}
m	milli	10^{-3}
k	kilo	10^{3}
M	mega	10^{6}
G	giga	10^{9}
T	tera	10^{12}

flowing through the resistor R_2. The source part of the CC, connected between ground and node three, has a current gain $\alpha = 3$. We have marked the nodes of the network by numbers in circles; this is not a standard way, but is quite convenient. We have also written nodal voltages to these nodes. The ground node, or reference node, is at the bottom of the picture. Note especially the short circuit between nodes 3 and 4. It is the sensing part of the current-controlled current source and is indicated separately. Nodes 3 and 4 are connected by a short circuit.

The voltage source is the *input* of the network. Also marked, on the right side, is the *output*, in this case a voltage V_{out}. We will use this notation quite often. By output we mean that the variable (voltage or current) will be measured (or calculated) because we need to know it.

Another network is in Fig. 1.2.11. It has a floating current source. The word *floating* refers to elements which are not connected to ground. In this case the source current is $J = 2$ A. The network has two capacitors, an inductor, a resistor and a voltage-controlled voltage source. The controlling voltage is taken between nodes 1 and 2 and the gain of the dependent source is $\mu = 3$. The output voltage is $V_{out} = V_3$. The nodes as well as the nodal voltages are written into the figure. In some schematics the symbol for ground is omitted and the bottom line is understood to be the reference node. We have done so in this example.

FIGURE 1.2.10   First example of a network.

FIGURE 1.2.11   Second example of a network.

In general, the output can be any nodal voltage, a difference of two nodal voltages, for instance $V_{out} = V_2 - V_3$, or the current through any of the elements. A network may have several outputs.

1.3. INDEPENDENT VOLTAGE AND CURRENT SOURCES

The first set of idealized elements which we discuss in detail are the independent voltage and current sources. An ideal *independent voltage source* is an element for which we will use the symbol shown in Fig. 1.3.1a. The word *independent* means that the properties of the source will not change due to any external influence. A voltage source always has the symbol + at one of its terminals and the symbol − at the other terminal. Because of the rule we introduced in Section 1.1, positive direction of the current flowing through the voltage source will be from the + terminal to the − terminal. It has to be understood that our definition does not mean that the current will actually flow that way; more about this will be said when we have covered the concept of resistors and power.

We will *always* denote an independent voltage source by the letter *E*. The use of this letter may differ from other books, but there is a strong advantage to be gained if we give the independent source a special letter. When we get later to the point of writing the equations, we will see that the independent source always appears on the other side of the equation than the network variables, voltages and currents. By assigning special letters we eliminate many possible mistakes.

The voltage of an independent voltage source is given, but we can say nothing about the current flowing through it. An ideal voltage source can supply *any* amount of current and the current can flow in *any* direction, depending on the network to which it is connected. Since the voltage delivered by the independent voltage source is fixed and any amount of current can flow through it in any direction, we can draw a graph (sometimes also called the characteristic) describing these properties. If we measure current on the horizontal axis and the voltage on the vertical axis, then the characteristic is a horizontal line, as shown in Fig. 1.3.1b. If the voltage available from the source changes, for instance from two to four

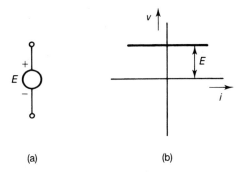

(a)   (b)

FIGURE 1.3.1   Independent voltage source: (a) Symbol. (b) Its characteristic.

volts, the horizontal line will shift upwards. Should we apply a voltage -2 V without changing the positions of the $+$ and $-$ signs, the horizontal line would shift below the horizontal axis. One very important case is what happens if we have a voltage source which is delivering zero voltage. In such case the horizontal line in Fig. 1.3.1 drops down on the horizontal axis. Such element will have zero voltage across it but a current of arbitrary magnitude can flow through it in any direction. If we recall the definition of the short circuit from our previous discussion, then we see that *a voltage source with zero voltage is equivalent to a short circuit.*

An ideal *independent current source* is an element for which we will use the symbol shown in Fig. 1.3.2a. Similarly as in the above case, there exists no possibility of changing the amount of the current delivered by some external influence. The direction of the current is indicated by an arrow. We will *always* denote an independent current source by the letter J, for the same reasons as discussed above. The current of an independent current source is given, but we can say nothing about the voltages at its terminals, because they depend on the network to which the source is connected. Since the current delivered by the

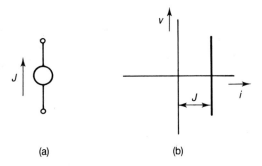

(a)   (b)

FIGURE 1.3.2   Independent current source: (a) Symbol. (b) Its characteristic.

independent current source is fixed, but any voltage difference can be measured between its terminals, we can draw a graph (characteristic) describing these properties. It is shown in Fig. 1.3.2b. If the amount of current delivered by the source changes from, say, 2 A to 4 A, the straight line will shift to the right. Should we apply, say, -2 A without changing the direction of the arrow in the figure, the vertical line will shift to the left of the vertical axis. If we consider a current source which delivers zero current, then the vertical line will coincide with the vertical axis. In such case no current can flow through the element but arbitrary voltage can appear across its terminals; we conclude that *a current source with zero current is equivalent to an open circuit.*

A battery is an example of a nonideal *dc* source, because the voltage at the terminals depends on the amount of current we draw from it. Nevertheless, in many cases we take it as an ideal source and denote it by the letter *E*. The outlet at your home is another source of voltage, again nonideal, but it delivers an alternating current, *ac*. This voltage changes with time but we will still use the symbol *E*.

Independent sources cannot be connected together arbitrarily and we now discuss the acceptable and the prohibited cases. If we connect two voltage sources *in series*, as shown in Fig. 1.3.3a, then the voltage between the external terminals will be $E_1 + E_2$. Should one of them have reversed direction, as shown in Fig. 1.3.3b, then the voltage between the external terminals will be $E_1 - E_2$. We can *never* connect two *different* voltage sources *in parallel*, something like in Fig. 1.3.4, because it is not possible to have simultaneously two different voltages between two nodes. There is, of course, the possibility of connecting in parallel two voltage sources with *exactly the same voltage* and this is sometimes used in theoretical considerations, as we will learn later. From a practical point of view, this is a meaningless situation, since an ideal voltage source can deliver any current and can thus do the same job as two or more such sources in parallel.

Independent current sources can be connected in parallel, as shown in Fig. 1.3.5a. In this case the current delivered from this combination into the external

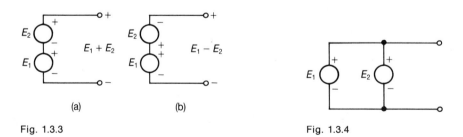

Fig. 1.3.3   Fig. 1.3.4

FIGURE 1.3.3   Permitted connections of voltage sources: (a) The voltages add. (b) The voltages subtract.

FIGURE 1.3.4   Prohibited connection for $E_1 \neq E_2$.

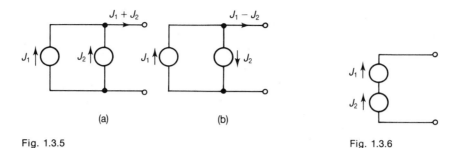

Fig. 1.3.5

Fig. 1.3.6

FIGURE 1.3.5   Permitted connections of current sources: (a) The currents add. (b) The currents subtract.

FIGURE 1.3.6   Prohibited connection for $J_1 \neq J_2$.

network will be $J_1 + J_2$. Should they be connected as shown in Fig. 1.3.5b, the current flowing into the external network will be $J_1 - J_2$. We cannot connect two independent current sources in series, as shown in Fig. 1.3.6, because the combination cannot simultaneously deliver two different currents into the external network. We could connect two such sources in series if they had *exactly* the same current values. Similarly as above, this is sometimes used in theoretical considerations, but is meaningless practically, because one ideal source can do exactly the same job as two equal sources in series.

Connecting one voltage and one current source either in series or in parallel is not prohibited theoretically, but is not useful practically. Finally, we should mention that in a network schematic a voltage source should never be shortcircuited. By our definition, it should still maintain the same voltage across its terminals, but the current flowing into a short circuit would be infinitely large. Similarly, a current source should never be left without a connection to some circuit which can return the current back to the other terminal of the source.

Example 1.

Consider the combination of voltage sources in Fig. 1.3.7a. What is the voltage we obtain at the terminals of the network? The voltage sources E_1 and E_3 have one orientation, the other sources act in the opposite direction. The resulting voltage is

$$E = E_1 - E_2 + E_3 - E_4 - E_5 = -7$$

as shown in Fig. 1.3.7b. The same is achieved by reversing the $+$ and $-$ signs, as shown on the right of Fig. 1.3.7c, and making the voltage positive 7 V.

Example 2.

Consider the parallel connection of five current sources, Fig. 1.3.8a. The sum of currents flowing upwards is 8, the sum flowing downwards is 7. Thus the equivalent current source

FIGURE 1.3.7   Equivalent voltage of five voltage sources: (a) Original connection. (b) and (c) Two equivalent results.

FIGURE 1.3.8   Equivalent current of five current sources: (a) Original connection. (b) and (c) Two equivalent results.

will be 1 A upwards, as shown in Fig. 1.3.8b. Equivalently, this could also be drawn as a negative current of 1 A pointing downwards, as shown in Fig. 1.3.8c.

1.4. RESISTORS AND OHM'S LAW

In this section we consider the simplest of all elements, the resistor, and the way it relates the voltage across it to the current flowing through it.

The electrical property of the resistor, called the *resistance*, can be best understood by considering a small experiment, at least in our mind, if not in reality. Take several materials (silver, copper, iron, wood, china, glass etc.), make thin rods with exactly equal dimensions, connect them to the same source of voltage and measure the current flowing through them. We will discover that a large current will flow through the rod made of silver, less current through copper, still less through iron, very little through wood and next to nothing through china or glass. The properties of the materials are such that they offer different resistance to the flow of current. More detailed measurements would lead us to the conclusion that the voltage, the current, and the properties of the material are related by Ohm's law which states

$$V = IR$$ (1.4.1)

Here V is the voltage across the element, measured in *Volts*, V, I is the current flowing through the element, measured in *Amperes*, A, and R is the resistance, measured in *Ohms*, Ω. A resistor will have a resistance of $1\,\Omega$ if the voltage across it is $1\,V$ and the current flowing through it is $1\,A$. For the resistor we will use the left symbol shown in Fig. 1.4.1a.

If we draw a graph with perpendicular axes, measure the current on the horizontal axis and the voltage on the vertical axis, then Eq. (1.4.1) represents a straight line passing through the origin. We call it the *characteristic* of the resistor, see the graph in Fig. 1.4.1b. The tangent of the angle, ϕ, is the value of the resistance. The equation describes a *linear* resistor, a resistor whose value is independent of any external influences. In true life, nothing is as simple as that. The resistor will change its value with time or temperature or with the current which flows through it. Figure 1.4.2 shows several cases of nonlinear resistors. Note that the straight line which *does not* pass through the origin would also represent a nonlinear resistor.

Since the resistance of a linear resistor is a constant and Eq. (1.4.1) represents a straight line, we can invert the equation and write

$$I = \frac{1}{R} V = GV \tag{1.4.2}$$

The inverted value

$$G = \frac{1}{R} \tag{1.4.3}$$

is called the *conductance* and is measured in *Siemens*, S.

Should the resistor change with time, t, we could denote it by the symbol $R(t)$. Its resistance can also depend on the voltage across it. In such case the resistor

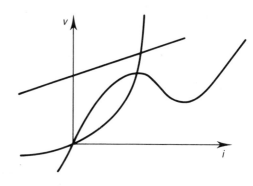

(a)   (b)

Fig. 1.4.1   Fig. 1.4.2

FIGURE 1.4.1   Linear resistor: (a) Symbol. (b) Its characteristic.

FIGURE 1.4.2   Characteristics of nonlinear resistors.

depends on v and we could express the dependence by writing $R(v)$. Whatever the situation may be, Ohm's law is always valid. In practical life we talk mostly about resistors and their values in ohms (Ω), kiloohms ($1\,k\Omega = 10^3\,\Omega$) or megaohms ($1\,M\Omega = 10^6\,\Omega$). The notation is standardized and is summarized in Tables 1.2.1 and 1.2.2.

Example 1.

A voltage source delivering 100 V is connected to a resistor having the value $R = 10\,k\Omega$. What is the current through the resistor? The answer is

$$I = \frac{10^2}{10^4} = 10^{-2}\,A = 10 \times 10^{-3}\,A = 10\,mA$$

Example 2.

Let the resistor be described by the voltage–current relationship $V = I^3$. If the current through the element is 0.5 A, what is the voltage across it? The answer is $V = 0.5^3 = 0.125$ V.

1.5. POWER AND ENERGY

Power and energy are the next two fundamental concepts to be covered in this chapter. Energy, usually denoted by $w(t)$ or W, is measured in joules, J; it is the amount of work which a source of energy can deliver. In network theory, energy is delivered by the voltage or current sources and is consumed by resistors. The notation $w(t)$ indicates that the energy is changing with time while W indicates that it is constant.

Closely related to energy is power, denoted by $p(t)$ or P and measured in watts, W. The two definitions are related by

$$p(t) = \frac{dw(t)}{dt}$$

$$w(t) = \int_{t_1}^{t_2} p(t)\,dt$$

(1.5.1)

The relation of their units is $1\,W = 1\,J/s$. In electrical terms, power is expressed by the product of current and voltage,

$$p(t) = v(t)i(t)$$

(1.5.2)

and the expression is valid at any instant of time. Substituting from Ohm's law we can also write

$$p(t) = \frac{v^2(t)}{R} = i^2(t)R$$

(1.5.3)

When dealing with direct current, then

$$P = VI = \frac{V^2}{R} = I^2 R \qquad (1.5.4)$$

or replacing $R = 1/G$,

$$P = VI = V^2 G = \frac{I^2}{G} \qquad (1.5.5)$$

In applications, we must distinguish between power which is delivered by the source and power consumed by the network or element. A resistor *always* consumes power and the current flows through it from $+$ to $-$. If the network has only one voltage source, then the source delivers power into the network and the current will flow from the $+$ terminal into the network and return through the network to the $-$ terminal. If the network has a single current source, then the terminal at the arrow will be the positive one and the other will be negative.

Example 1.

Consider the network in Fig. 1.5.1 and calculate the currents and powers delivered and consumed in the network.

Applying Ohm's law we get the current $I = 3/10 = 0.3$ A. It will flow in the indicated direction. The power *consumed* by the resistor will be $P = VI = 0.3 \times 3$ W and the same amount of power will be *delivered* by the source.

Example 2.

In the network Fig. 1.5.2 the current source delivers 3 A. Calculate the voltages and powers delivered and consumed.

Fig. 1.5.1

Fig. 1.5.2

FIGURE 1.5.1   Power from a voltage source.

FIGURE 1.5.2   Power from a current source.

The voltage across the resistor is $V = 3 \times 10 = 30$ V. The power consumed by the resistor (and converted into heat) is $P = V \times I = 3 \times 30 = 90$ W. The same power is delivered by the current source.

The situation becomes more complicated if there is more than one independent source in the network. In a general case, the question of how the powers are distributed can be answered only by setting up the appropriate equations and by solving them. We will learn how to do that later. Here we will take a simple network where the answer can be found by the theory covered until now.

Example 3.

Consider the network in Fig. 1.5.3 and calculate the current and powers delivered or consumed by the elements.

The voltage across the resistor is $V = 5 - 2 = 3$ V and the current in the network is $I = 3/2 = 1.5$ A. The power consumed by the resistor is $P_R = IV = 3 \times 1.5 = 4.5$ W. The source on the left is delivering the power $P = 5 \times 1.5 = 7.5$ W. On the right, the current flows through the source in the direction from $+$ to $-$ and thus this source *consumes* power. The amount is $P = 2 \times 1.5 = 3$ W. The power delivered by the left source is exactly equal to the power consumed by the resistor and the other source.

Example 4.

A parallel connection of a voltage and current source is in Fig. 1.5.4. Calculate the powers delivered and consumed by the respective elements.

Due to the voltage source, the voltage across the resistor *must* be 5 V and the current through the resistor must be, due to Ohm's law, $I_R = 6/2 = 3$ A. The current source is supplying 10 A. Three amperes will be forced through the resistor, the rest must flow through the voltage source, from $+$ to $-$. The powers are distributed as follows. The current source delivers $P = 6 \times 10 = 60$ W. Of this the resistor consumes $P = 3 \times 6 = 18$ W while the voltage source consumes $P = 6 \times 7 = 42$ W. As always, the power delivered is equal to the power consumed.

Fig. 1.5.3   Fig. 1.5.4

FIGURE 1.5.3   Network with two voltage sources.

FIGURE 1.5.4   Network with current and voltage source.

1.6. KIRCHHOFF'S LAWS

The laws discovered by Kirchhoff are fundamental to the theory of electricity. They are valid in *any* situation, even if the elements are nonlinear or time varying.

1.6.1. Kirchhoff Current Law

The Kirchhoff current law (KCL) states that **the sum of currents flowing away from a node is zero**. The definition remains valid if we replace the words *away from* by the word *to*. A *node* is a point where we connect several elements and one such case is shown in Fig. 1.6.1a. Since the node is only a connection of wires, it is easily understood that it cannot store electrons and thus any amount of electrons flowing into the node through some of the connecting lines must also leave through some other lines. In the statement of the law we used the word *away* because in this book we will consider a current flowing *away* from a node as having *positive* direction.

In Fig. 1.6.1a we have five elements connected to the central node. *At this node*, the currents i_1, i_2 and i_3 have positive directions while currents i_4 and i_5 have negative directions. The Kirchhoff current law can be written as

$$i_1 + i_2 + i_3 - i_4 - i_5 = 0 \qquad (1.6.1)$$

We drew the directions of the currents in the figure arbitrarily. There is, however, one situation where we do not have this freedom of choice, namely when there are independent current sources connected to the node. Figure 1.6.1b shows a node with two independent current sources and three elements which are not sources. We arbitrarily select directions of currents for the resistors. Following the

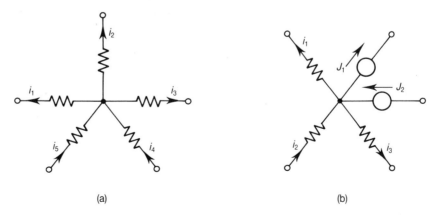

(a)   (b)

FIGURE 1.6.1   A node with five elements: (a) All current directions are arbitrary. (b) Directions of current sources given, rest arbitrary.

KCL it must be true that

$$i_1 - i_2 + i_3 + J_1 - J_2 = 0 \qquad (1.6.2)$$

In mathematics, it is common to write unknown variables on the left side and known quantities on the right side of the equation. Because the currents and directions of independent current sources are known, we can follow the usual mathematical strategy and rewrite the above equation in the form

$$i_1 - i_2 + i_3 = -J_1 + J_2 \qquad (1.6.3)$$

We now see the reason why we used different letters to describe properties of independent sources: without much thinking about what is dependent and what is independent, we know immediately that the letter J must go to the right side of the equation *with a change in sign.*

1.6.2. Kirchhoff Voltage Law

The Kirchhoff voltage law (KVL) states that **the sum of voltages around a loop is zero.** A loop is such a connection of elements that we can walk around it and reach the same point from which we started. An example of a loop with five elements is shown in Fig. 1.6.2a. We know nothing about the voltages and currents in this loop, because they will be influenced by the connections to other parts of the network (indicated in the figure by the open-ended lines out of the loop). We can assign, completely arbitrarily, the signs + and − to the elements, as was done in the figure. However, once we have done that, there is nothing arbitrary any more, except for the direction of our walk around the loop. Let us choose the one

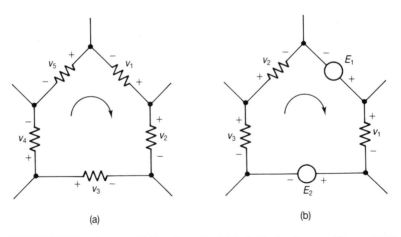

(a)   (b)

FIGURE 1.6.2   A loop with five elements: (a) Individual voltages arbitrary. (b) Voltages of voltage sources given, rest arbitrary.

indicated in the figure by the arrow. If this direction goes through the element from $+$ to $-$, then we add the voltage of this element. If the direction of our walk is such that we go through the element from $-$ to $+$, we subtract the voltage. For the network in Fig. 1.6.2a this results in the equation

$$-v_1 + v_2 - v_3 + v_4 - v_5 = 0 \qquad (1.6.4)$$

We have the freedom to assign the directions of voltages to all elements except the independent voltage sources where the $+$ and $-$ signs are given. The network in Fig. 1.6.2b has two voltage sources with their $+$ and $-$ specified. For the remaining elements we choose arbitrary positions of $+$ and $-$, for instance as shown. Once we have done this, the only arbitrary thing is the direction of the walk around the loop. If we decide to walk as shown, the equation must be written as

$$-E_1 + v_1 + E_2 - v_3 + v_2 = 0 \qquad (1.6.5)$$

Since the sources are known, we follow the usual mathematical practice and write their voltages on the right side of the equation. The sum of voltages around the loop can be rewritten as

$$v_1 + v_2 - v_3 = +E_1 - E_2 \qquad (1.6.6)$$

Here we take advantage of our rule that independent voltage sources are marked by the letter E. Since we use different letters for the unknowns, v, and for the knowns, E, we do not need much thinking to determine what should go to the right side of the equation *with opposite sign* and what should stay on the left side. The same applies to general currents, i, and current sources, J.

Example 1.

The sketch in Fig. 1.6.3 represents part of some larger network. The nodal voltage at the center is V, the others have subscripts 1 through 4. Current directions are also marked. Applying the KCL we can write for the node in the center

$$-I_1 + I_2 - I_3 + I_4 = 0 \qquad (1.6.7)$$

From these directions of the currents we conclude that $V_1 > V$, $V_2 < V$, $V_3 > V$ and $V_4 < V$. Express the currents in terms of the voltages across the conductances:

$$
\begin{aligned}
I_1 &= (V_1 - V)G_1 \\
I_2 &= (V - V_2)G_2 \\
I_3 &= (V_3 - V)G_3 \\
I_4 &= (V - V_4)G_4
\end{aligned}
\qquad (1.6.8)
$$

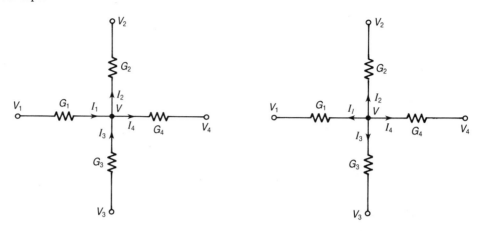

Fig. 1.6.3   Fig. 1.6.4

FIGURE 1.6.3   A node with specified current directions.

FIGURE 1.6.4   A node with other current directions.

These equations can be inserted into Eq. (1.6.7) to get

$$-(V_1 - V)G_1 + (V - V_2)G_2 - (V_3 - V)G_3 + (V - V_4)G_4 = 0$$

We can open the brackets and rewrite in the form

$$V(G_1 + G_2 + G_3 + G_4) - V_1G_1 - V_2G_2 - V_3G_3 - V_4G_4 = 0 \qquad (1.6.9)$$

Example 2.

In this example we take the same network, but assume that the voltage V is larger than all the other indicated voltages. Due to this assumption, the directions of the currents must change. They will be as indicated in Fig. 1.6.4. KCL for the central node provides

$$I_1 + I_2 + I_3 + I_4 = 0 \qquad (1.6.10)$$

The currents through the branches are

$$I_1 = (V - V_1)G_1$$
$$I_2 = (V - V_2)G_2$$
$$I_3 = (V - V_3)G_3 \qquad (1.6.11)$$
$$I_4 = (V - V_4)G_4$$

Inserting these currents into Eq. (1.6.10) results in

$$(V - V_1)G_1 + (V - V_2)G_2 + (V - V_3)G_3 + (V - V_4)G_4 = 0$$

Rearranging terms we get

$$V(G_1 + G_2 + G_3 + G_4) - V_1 G_1 - V_2 G_2 - V_3 G_3 - V_4 G_4 = 0 \qquad (1.6.12)$$

The result is interesting, in both cases we get *exactly* the same equation. This is an important result, because it clearly does not matter which voltage we think is higher and which is lower. All we have to do is correctly relate the currents in terms of the voltages and conductances. We will make use of this fact in the next chapter.

1.7. CONNECTIONS OF RESISTORS

The first application of Ohm's and Kirchhoff's laws comes in the study of series and parallel connections of resistors. The simplest case is one loop of resistors, for instance as shown in Fig. 1.7.1. The arrow indicates the direction of our walk around the loop and in this case also indicates the direction of the current, I, which will flow through the combination. Since the same current flows through all resistors, the + and − signs of the voltages across the individual resistors are as shown. Let the source be a dc source, $E = 22$ V. From Kirchhoff's voltage law we have

$$V_1 + V_2 + V_3 + V_4 - E = 0$$

the minus sign before E coming from the fact that the circulating current enters the voltage source at its minus sign and thus has negative direction according to our rules. The voltage of the source is known and we transfer it to the other side of the equation:

$$V_1 + V_2 + V_3 + V_4 = E \qquad (1.7.1)$$

FIGURE 1.7.1   Voltage divider with four resistors.

Each of the resistors follows Ohm's law:

$$V_1 = IR_1$$
$$V_2 = IR_2$$
$$V_3 = IR_3 \qquad (1.7.2)$$
$$V_4 = IR_4$$

Substituting into Eq. (1.7.1)

$$IR_1 + IR_2 + IR_3 + IR_4 = E$$

or

$$I(R_1 + R_2 + R_3 + R_4) = E \qquad (1.7.3)$$

We can introduce the notation

$$R_{tot} = R_1 + R_2 + R_3 + R_4 \qquad (1.7.4)$$

and obtain the overall resistance of the combination, in our case $R_{tot} = 11\,\Omega$. The voltage of the source is 22 V and the equation is solved for the current

$$I = \frac{22}{11} = 2\ \text{A}$$

Returning to Eq. (1.7.2), we can now obtain the voltage across each resistor: $V_1 = 4$ V, $V_2 = 10$ V, $V_3 = 2$ V and $V_4 = 6$ V. The sum of these individual voltages must be equal to the voltage of the voltage source.

Since the voltage of the source was divided into several partial voltages, the network in Fig. 1.7.1 is also called a *voltage divider*. We say that the resistors are connected in *series*. We will also remember that the overall resistance of the resistors connected in series is equal to the sum of the individual resistances. In other words, we do not have to go through all the steps derived above. All we have to do is add the resistances of resistors connected in series.

A different situation is shown in Fig. 1.7.2 where we have the one current source and several resistors connected *in parallel*. We would like to know the voltage common to all these resistors and the currents flowing through them. It is obvious that the same voltage, V, appears across all the resistors and the sum of the currents flowing through them must be equal to the current delivered by the current source. The Kirchhoff current law states that the sum of currents at the upper node must be equal to zero:

$$I_1 + I_2 + I_3 + I_4 - J = 0 \qquad (1.7.5)$$

FIGURE 1.7.2   Current divider with four branches.

where the negative sign for the current source comes from the fact that its current flows *into* the node. As always, known values are transferred to the right-hand side of the equation with a change of sign:

$$I_1 + I_2 + I_3 + I_4 = J \qquad (1.7.6)$$

The current through each resistor can be expressed in terms of its conductance:

$$I_1 = VG_1$$
$$I_2 = VG_2$$
$$I_3 = VG_3 \qquad (1.7.7)$$
$$I_4 = VG_4$$

Inserting these values into Eq. (1.7.6)

$$VG_1 + VG_2 + VG_3 + VG_4 = J$$

or

$$V(G_1 + G_2 + G_3 + G_4) = J \qquad (1.7.8)$$

The total conductance is

$$G_{tot} = G_1 + G_2 + G_3 + G_4$$

and the equation simplifies into

$$V = \frac{J}{G_{tot}} = \frac{7}{14} = 0.5 \text{ V}$$

Going back to the individual equations of Eq. (1.7.6), we calculate the currents through all resistors as $I_1 = 1.5$ A, $I_2 = 2.5$ A, $I_3 = 1$ A and $I_4 = 2$ A. The sum is 7 A. The current of the source was divided into individual currents of the resistors and for this reason the network is also called the *current divider*.

The steps explained above can be summarized into two simple rules:

a. If all resistors are in series, add their resistances to obtain the overall resistance.
b. If all resistors are in parallel, add their conductances to obtain the overall conductance.

The rules help us to find the overall resistance of more complicated networks without the need to calculate the currents. We give several examples with increasing complexity.

Example 1.

In the network Fig. 1.7.3 find the current and power delivered into the network by the source.

On the right, the two resistors are in parallel and thus their conductances are added to get $G = 3/4$, or $R = 4/3$. This is redrawn in the middle of the figure. Now we have two resistors in series and we can add their resistances to obtain the overall resistance $R_{tot} = 1 + 4/3 = 7/3$. The network is again redrawn on the right. We now have a source $E = 14$ V, a resistance $R_{tot} = 7/3$ and thus the current flowing from the source is $I = E/R = 6$ A. The power delivered into the combination is $P = E \times I = 14 \times 6 = 84$ W.

Example 2.

Find the voltage across the current source in Fig. 1.7.4. Also find the power delivered by the source.

In the solution we first add the resistances of the two resistors on the right, $R = 3 + 9 = 12\,\Omega$. This is redrawn in the middle. Now we have two resistors in parallel and we must add their conductances, $G = 1/2 + 1/12 = 7/12$. The network can be redrawn again, shown on the right, where we have a resistance $R = 12/7$. The voltage across the source (and also across the equivalent resistor) is $V = R \times J = 12$ V and the power delivered into the network is $P = J \times V = 7 \times 12 = 84$ W.

FIGURE 1.7.3   Calculating equivalent resistance.

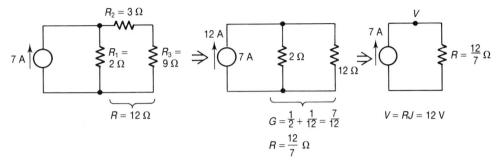

FIGURE 1.7.4   Calculating equivalent resistance.

Example 3.

Find the equivalent resistance of the combination in Fig. 1.7.5.

Starting from the right we first add $R_3 + R_4 = R' = 3\,\Omega$. We now have two resistors in parallel and we must add their conductances. This is done in the middle of the figure. Their resistance is $R'' = 6/5$. Finally, the resistance of R_1 is added to R'' to obtain the total resistance $R_{tot} = 11/5\,\Omega$.

Example 4.

Find the equivalent resistance of the combination in Fig. 1.7.6.

The steps are all indicated in the figure, with the result $R_{tot} = 292/141\,\Omega$.

Example 5.

We will use the voltage divider to introduce the concept of an internal resistance of a *nonideal* voltage source. In Fig. 1.7.7 is one ideal voltage source and two resistors. This is a simple voltage divider but consider the voltage source with the resistor R_1 as a model of

FIGURE 1.7.5   Calculating equivalent resistance.

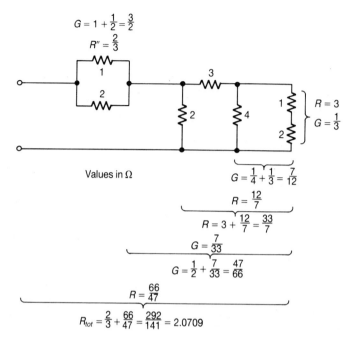

$$G = 1 + \tfrac{1}{2} = \tfrac{3}{2}$$
$$R'' = \tfrac{2}{3}$$

Values in Ω

$$G = \tfrac{1}{4} + \tfrac{1}{3} = \tfrac{7}{12}$$
$$R = \tfrac{12}{7}$$
$$R = 3 + \tfrac{12}{7} = \tfrac{33}{7}$$
$$G = \tfrac{7}{33}$$
$$G = \tfrac{1}{2} + \tfrac{7}{33} = \tfrac{47}{66}$$
$$R = \tfrac{66}{47}$$
$$R_{tot} = \tfrac{2}{3} + \tfrac{66}{47} = \tfrac{292}{141} = 2.0709$$

FIGURE 1.7.6   Calculating equivalent resistance.

a nonideal source, for instance a battery. The current flowing from the source will be

$$I = \frac{E}{R_1 + R_2}$$

and the voltage across R_2 will be

$$V = IR_2 = \frac{ER_2}{R_1 + R_2}$$

If R_2 is very large, $R_2 \to \infty$, then there will be no current and the voltage V will

FIGURE 1.7.7   Voltage source with internal resistance.

be equal to E. If we replace R_2 by a short circuit, then the voltage will be zero. Any finite value of R_2 will make the output voltage $0 < V < E$. We can also calculate the power delivered by the source into the resistor R_2. It will be equal to

$$P = I \times V = \frac{E}{R_1 + R_2} \times \frac{ER_2}{R_1 + R_2} = E^2 \frac{R_2}{(R_1 + R_2)^2}$$

If we select $R_2 = 0$, then the power consumed in it is zero. If we select $R_2 \to \infty$ (the same as if we disconnect R_2 from the network), then the power will also be zero. For all other cases there will be some power consumed in R_2. We could plot such curve for a number of values. We would discover that the largest power is delivered when $R_2 = R_1$, in which case the voltage $V = E/2$.

Example 6.

A nonideal current source can be modeled as in Fig. 1.7.8. Using the KCL, we have

$$V(G_1 + G_2) = J$$

or

$$V = \frac{J}{G_1 + G_2}$$

The current flowing through G_2 is

$$I = VG_2 = \frac{JG_2}{G_1 + G_2}$$

and the power consumed (or as we say dissipated) in G_2 is

$$P = V \times I = J^2 \frac{G_2}{(G_1 + G_2)^2}$$

For the two extreme values of G_2, namely zero and infinity, there will be no power consumed in G_2. In all other cases the power will be nonzero and its maximum will occur when $G_1 = G_2$.

FIGURE 1.7.8   Current source with internal conductance.

1.8. VOLTMETER AND AMMETER

It is not the purpose of this book to discuss the construction of voltmeters (instruments which measure voltage between two points) and ammeters (which measure the current flowing through some element). What we need to know are the properties which such instruments should have ideally. Although we will rarely draw these elements, their symbols are in Fig. 1.8.1a and b.

Figure 1.8.2a shows a simple voltage divider for which we find

$$I = \frac{E}{R_1 + R_2} \tag{1.8.1}$$

$$V_2 = \frac{ER_2}{R_1 + R_2} \tag{1.8.2}$$

Figure 1.8.2b shows the use of a voltmeter. To measure the voltage across the resistor R_2, we connect the instrument in parallel. An actual voltmeter will have some resistance, R_V, and we would like to see how such resistance influences the result. Two parallel resistors are equivalent to

$$R_e = \frac{R_2 R_V}{R_2 + R_V} \tag{1.8.3}$$

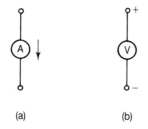

(a)   (b)

FIGURE 1.8.1   Symbols for measuring instruments: (a) Ammeter. (b) Voltmeter.

(a)   (b)   (c)

FIGURE 1.8.2   Nonideal measuring instruments: (a) Simple voltage divider. (b) Nonideal voltmeter. (c) Nonideal ammeter.

and the voltage across these two resistors is

$$V_{out} = \frac{ER_e}{R_1 + R_e} \qquad (1.8.4)$$

If we insert for R_e from Eq. (1.8.3), we get after a few steps

$$V_2 = \frac{ER_2}{R_1 + R_2 + \dfrac{R_1 R_2}{R_V}} \qquad (1.8.5)$$

An ideal measuring instrument should not influence the network behavior and Eq. (1.8.5.) should be equal to Eq. (1.8.2). This is possible only if R_V is very large, ideally approaching infinity, $R_V \to \infty$. We conclude an ideal voltmeter must have infinitely large internal resistance and is thus equivalent to an open circuit.

If we wish to measure the current through some element, we must modify the network by breaking the connection and inserting the instrument as shown in Fig. 1.8.2c. R_A denotes the resistance of the ammeter. The current through this combination is

$$I = \frac{E}{R_1 + R_2 + R_A} \qquad (1.8.6)$$

The instrument should not influence the current flowing in the original network; this means that Eq. (1.8.6) should be equal to Eq. (1.8.1). It is obvious that this is possible only if the ammeter has zero resistance, $R_A = 0$, and an ideal ammeter is equivalent to a short circuit.

The use of ammeters is more complicated than the use of voltmeters. Connection of an ammeter requires modification of the network, creation of a new node and insertion of the instrument in series.

Example 1.

Figure 1.8.3a is a network in which we wish to measure the voltage across R_4 and the current through R_2. Indicate how the instruments are connected.

The connection is shown in Fig. 1.8.3b in which we had to create an additional node to insert the ammeter. It was also necessary to renumber the nodes.

FIGURE 1.8.3   Connections of measuring instruments: (a) Original network. (b) The same network with ammeter and voltmeter.

PROBLEMS CHAPTER 1

P.1.1a through i. Find equivalent resistances of the networks.

(h)

(i)

P.1.2a through e. Find equivalent resistances.

(a)

(b)

(c)

All R = 1 Ω

(d)

(e)

P.1.3a through e. Find equivalent resistances of the networks.

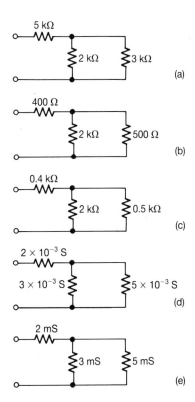

(a)

(b)

(c)

(d)

(e)

P.1.4a through d. Find one equivalent source for the given combinations. Are a and b possible?

(a)

(b)

(c)

(d)

P.1.5 Find the voltage across the resistor R_2.

P.1.6a and b. Find the voltages across the resistors and currents flowing through them.

(a)

(b)

P.1.7 Find the current flowing through the resistor.

P.1.8 Find the nodal voltage V.

P.1.9 Find the current flowing through the resistor.

P.1.10 What current is flowing through the resistor? Which of the sources is delivering power and which does consume power?

P.1.11a and b. Find the powers delivered or consumed in the three elements.

(a)   (b)

P.1.12 Find the powers delivered or consumed by the three elements.

P.1.13 Find the power consumed in the resistor.

P.1.14a and b. Find the voltages, currents and powers delivered and consumed.

(a)

(b)

P.1.15 Find the voltage across the combination and powers delivered or consumed by all elements.

P.1.16 Find the voltages across the elements and powers delivered or consumed by each element.

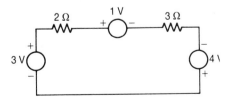

P.1.17 Find the voltages across the resistors, currents flowing through them and power consumed in them.

P.1.18 If voltages are given in volts and resistances in ohms, the currents are in amperes. Similarly, if voltages are given in volts and resistances in kiloohms, then the currents are in milliamperes. This simplifies solutions because we do not have to write all the zeros or multiplying factors. Use this fact to calculate the currents.

P.1.19 A voltmeter with internal resistance is connected to the voltage divider. Calculate the voltage across the resistor R_2 for the case when the voltmeter is not connected and then when it is connected.

P.1.20a and b. We can also model the voltmeter as an ideal device with infinite resistance, in parallel to a resistor, as shown in a. Calculate the voltage V for the case when the voltmeter is not connected and then when it is connected. Do a similar calculation for the case b where the resistor is thought to be in series with a device which ideally does not have any resistance.

(a)

(b)

P.1.21a and b. Calculate the currents and what errors will be caused by the measuring instruments. The dotted lines indicate a short circuit in which we obtain the current without the instrument.

(a)

(b)

P.1.22 Calculate the current and the error caused by connecting the ammeter as shown. The dotted line indicates a short circuit before connecting the instrument. What are the powers dissipated in all three resistors?

P.1.23 Design a voltage divider such that you have four taps with indicated voltages. The power dissipated in any element must not exceed 1 W.

P.1.24 Let the voltage source P.1.23 have an internal resistance 10 Ω. Design the resistors to get the same voltages at the taps.

CHAPTER 2

Nodal and Mesh Analysis

2.0. INTRODUCTION

The voltage and current dividers explained in the previous chapter are the first useful networks, but they are needed only rarely. In most situations we need voltages at several (or all) nodes or currents through the elements. Such problems lead to systems of simultaneous equations which must be solved.

We will explain two classical methods to write equations for networks: the nodal and the mesh formulations. The nodal formulation is based on the use of Kirchhoff's current law and is suitable for networks with current sources. The mesh method uses Kirchhoff's voltage law and is easy to apply for networks with voltage sources.

The equations which we set up have to be solved and practical methods to do it will be given in the next chapter. Our main interest in this brief chapter is to learn how to write the two types of equations and master enough skills so that we will be able to write them by inspection.

2.1. NODAL ANALYSIS

Nodal analysis is a way of writing network equations by considering the Kirchhoff current law, KCL, at all ungrounded nodes of the network. In this chapter we will deal with only two types of elements: independent current sources and resistors, given in terms of their conductance values.

FIGURE 2.1.1   Network with estimated current directions.

We will start with two simple situations. Consider first the network in Fig. 2.1.1, with one source and five resistors. We can make a good estimate of which way the currents will flow through the resistors; in fact we have marked them in the network.

Consider next the same network, but attach another source on the right, Fig. 2.1.2. Suddenly we have a dilemma: we cannot guess which way we should mark directions of the currents in the resistors. The currents will obviously depend on the current values of the sources as well as on the values of the resistors. If $J_2 \gg J_1$ it is likely that in the floating resistors (those not connected to ground) the currents will flow from right to left. If we have a larger network with several sources, nobody will be able to guess the directions. To avoid problems in assigning the directions, we must develop a unified method which works for any values of resistors and any number of sources. Such method is the subject of this section.

Let us now turn our attention to Fig. 2.1.3. The bottom line is the reference node (ground), always assumed to have zero voltage. Assign voltages to the nodes which are not grounded. Normally we use subscripted variables, V_1 and V_2.

The KCL states that the sum of currents leaving a node is zero. Consider the first node. We do not know the voltage V_1, but if we *assume* that it is the largest voltage in the network, then the currents through all resistors connected to the node must flow away from the node. Since the first resistor is connected to ground, its current will be, as follows from Ohm's law, $I_{G1} = G_1 V_1$. The voltage across the second resistor is equal to $V_1 - V_2$ and using Ohm's law, the current will be $I_{G2} = (V_1 - V_2)G_2$. Note the way of writing. By giving first V_1 and by subtracting from it V_2 we indicate that we consider V_1 to be more positive than V_2. Currents flowing away from the node are considered to be positive by our definition. The

FIGURE 2.1.2   Network with estimated current directions.

FIGURE 2.1.3   Network with one current source.

current from the source must be taken negatively, because it flows into the node. The KCL equation is

$$G_1V_1 + (V_1 - V_2)G_2 - J = 0 \qquad (2.1.1)$$

Let us now move to the second node. As before, we know nothing about the voltages and we can still make the assumption that now it is V_2 which is the highest voltage in the network. Once we make this assumption, all currents through the resistors must flow *away* from node 2. The voltage across the second resistor is now $V_2 - V_1$ (because of our assumption we must put first the voltage of the second node and subtract the voltage of the first node). The current through this resistor will be $I_{G2} = (V_2 - V_1)G_2$. The current through the third resistor must flow down and $I_{G3} = G_3V_2$. Since there is no source at node two, the KCL for the second node is

$$G_3V_2 + (V_2 - V_1)G_2 = 0 \qquad (2.1.2)$$

Rearrange the terms and transfer the independent source to the right, with a change of sign. We get a set of two equations

$$(G_1 + G_2)V_1 - G_2V_2 = J$$
$$-G_2V_1 + (G_2 + G_3)V_2 = 0 \qquad (2.1.3)$$

Numerical values simplify them to

$$3V_1 - 2V_2 = 1$$
$$-2V_1 + 5V_2 = 0 \qquad (2.1.4)$$

At this moment we do not know yet a good method for the solution of systems of equations, but the reader certainly knows elimination. The result of such solution is $V_1 = 5/11$ and $V_2 = 2/11$, as can be easily checked by inserting these values into the two equations. We can also calculate the currents through the elements and their directions:

$$I_{G1} = \frac{5}{11} \text{ A}$$

$$I_{G2} = \left(\frac{5}{11} - \frac{2}{11} \right) 2 = \frac{6}{11} \text{ A}$$

$$I_{G3} = \frac{2}{11} 3 = \frac{6}{11} \text{ A}$$

The network of Fig. 2.1.4 has two current sources and we wish to calculate the nodal voltages, the currents and their directions. As before, consider the first node as the most positive one; the KCL is

$$G_1 V_1 + G_2(V_1 - V_2) - J_1 = 0$$

or, rearranging the terms

$$(G_1 + G_2)V_1 - G_2 V_2 = J_1$$

and numerically

$$3V_1 - 1V_2 = 1 \tag{2.1.5}$$

Moving to the next node and considering it as the most positive one we obtain the KCL equation

$$G_3 V_2 + G_2(V_2 - V_1) - J_2 = 0$$

or, rearranging the terms

$$-G_2 V_1 + (G_2 + G_3)V_2 = J_2$$

and numerically

$$-V_1 + 4V_2 = 5 \tag{2.1.6}$$

In this case we will use elimination to solve. From Eq. (2.1.5) get

$$V_2 = 3V_1 - 1 \tag{2.1.7}$$

FIGURE 2.1.4   Network with two current sources.

and insert into Eq. (2.1.6) with the result

$$V_1 = \frac{9}{11} \text{ V} \qquad\qquad (2.1.8)$$

Inserting this result into Eq. (2.1.7)

$$V_2 = \frac{16}{11} \text{ V} \qquad\qquad (2.1.9)$$

The currents are $I_{G1} = V_1 G_1 = 18/11$ A and $I_{G2} = (V_1 - V_2)G_2 = -7/11$ A. The minus sign in the last case indicates that the current actually flows from node 2 to node 1. For the third resistor we get $I_{G3} = V_2 G_3 = 48/11$ A. The resulting current directions are indicated in the network by arrows. Taking absolute values of the currents *and* the directions indicated in the figure we confirm that the sum of currents at each node is indeed zero.

Example 1.

Consider Fig. 2.1.5, write the KCL equations for the two nodes and collect terms multiplied by the same voltages.

Keeping first the general notation we get for the first node

$$(G_1 + G_2)V_1 - G_2 V_2 = J_1 - J_3$$

and for the second node

$$-G_2 V_1 + (G_2 + G_3)V_2 = J_2 + J_3$$

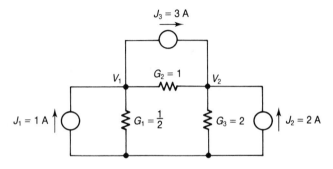

FIGURE 2.1.5   Network with three current sources.

Using numerical values, the equations become

$$\frac{3}{2}V_1 - V_2 = -2$$

$$-V_1 + 3V_2 = 5$$

We will use one more network, Fig. 2.1.6, to write the KCL equations. This network is already so complicated that even a skilled person cannot see which way the currents will flow. Following the rules stated above, we first assume that node 1 is the most positive one. All currents through the conductances must leave this node: $I_{G1} = G_1 V_1$, $I_{G2} = G_2(V_1 - V_2)$, $I_{G5} = G_5(V_1 - V_3)$, and the KCL for the first node is

$$G_1 V_1 + G_2(V_1 - V_2) + G_5(V_1 - V_3) - J_1 + J_3 = 0 \qquad (2.1.10)$$

Going to node 2, we assume that it is now this node which is most positive. The currents leaving the second node will be $I_{G2} = G_2(V_2 - V_1)$, $I_{G3} = G_3 V_2$, $I_{G4} = G_4(V_2 - V_3)$, and the sum of currents for the second node is

$$G_2(V_2 - V_1) + G_3 V_2 + G_4(V_2 - V_3) = 0 \qquad (2.1.11)$$

Going to node 3 we make the assumption that now this node is the most positive one. The currents of the conductances leaving the third node will be: $I_{G4} = G_4(V_3 - V_2)$, $I_{G5} = G_5(V_3 - V_1)$, and the KCL at node 3 is

$$G_4(V_3 - V_2) + G_5(V_3 - V_1) - J_2 - J_3 = 0 \qquad (2.1.12)$$

If we rearrange the terms and transfer the independent sources on the right side,

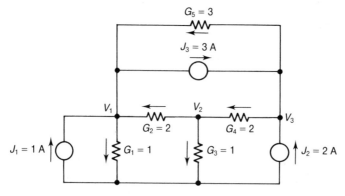

FIGURE 2.1.6   Network with three current sources.

we get the system

$$(G_1 + G_2 + G_5)V_1 - G_2V_2 - G_5V_3 = J_1 - J_3$$
$$-G_2V_1 + (G_2 + G_3 + G_4)V_2 - G_4V_3 = 0 \qquad (2.1.13)$$
$$-G_5V_1 - G_4V_2 + (G_4 + G_5)V_3 = J_2 + J_3$$

Insertion of numerical values leads to

$$6V_1 - 2V_2 - 3V_3 = -2$$
$$-2V_1 + 5V_2 - 2V_3 = 0 \qquad (2.1.14)$$
$$-3V_1 - 2V_2 + 5V_3 = 5$$

We state the result of the solution of this system: $V_1 = 53/37$, $V_2 = 58/37$ and $V_3 = 92/37$. The currents through the resistors are

$$I_{G1} = V_1G_1 = \frac{53}{57} \text{ A}$$

$$I_{G2} = (V_1 - V_2)G_2 = -\frac{10}{37} \text{ A}$$

$$I_{G3} = V_2G_3 = \frac{58}{37} \text{ A}$$

$$I_{G4} = (V_2 - V_3)G_4 = -\frac{68}{7} \text{ A}$$

$$I_{G5} = (V_1 - V_3)G_5 = -\frac{117}{37} \text{ A}$$

In this case the currents through all floating resistors actually flow from right to left. The directions of the currents are drawn in the network. Taking absolute values of the currents and the marked directions we confirm that the sum of currents at each node is indeed equal to zero.

Let us return to the set of Eq. (2.1.13) because they give us the opportunity to find a general rule how to write the equations. Consider the first node: The sum of conductances connected to the first node is multiplied by the nodal voltage of the first node. Subtracted from this sum is the product of the same conductances, multiplied by the voltage of the other node to which they are connected. Check the other equations to convince yourself that the same applies for the other nodes as well.

We are now in a position to formulate a general rule for writing KCL at any node, i:

1. Multiply the sum of all conductances connected to node i by V_i.
2. Subtract from this sum the same conductances, each multiplied by the voltage of the other node where the element is connected.
3. Since the voltage of ground is defined to be zero, no subtraction occurs for grounded resistors.
4. The current of an independent current source flowing *into* node i is taken positive *on the right side of the equation*. The sign changes if the current flows in the opposite direction.

Using the above rules and the numerical values of the above examples, set up the equations directly, without first defining the currents.

We have not attempted to solve the systems of three equations at this moment. It is a fairly laborious process and elimination is not a very good way to proceed. Our next chapter will give the most convenient methods to solve systems of linear equations; here we take one more example.

Example 2.

Consider the network of Fig. 2.1.7. Using the rules stated above write the nodal equations by inspection.

The set of equations is

$$(G_1 + G_2 + G_6)V_1 - G_2 V_2 - G_6 V_3 = J_1 - J_4$$

$$-G_2 V_1 + (G_2 + G_3 + G_4)V_2 - G_4 V_3 = J_2 - J_3$$

$$-G_6 V_1 - G_4 V_2 + (G_4 + G_5 + G_6)V_3 = J_3 + J_4 + J_5$$

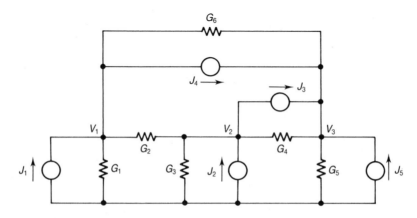

FIGURE 2.1.7   Network with five current sources.

2.2. MESH ANALYSIS

Mesh analysis is a way of writing network equations by applying the Kirchhoff voltage law, KVL: the sum of voltages around a loop is zero. In this section we will deal with the principle of mesh analysis and only two types of elements: independent voltage sources and resistors, this time given by their resistance values.

We start with the network in Fig. 2.2.1. The voltage source delivers currents into the network and in this simple case we can easily guess which sides of the elements will be marked with the + sign and which with the − sign; this is shown in the figure. The way we walk around a loop is indicated by an arrow. We have two such arrows in the "windows", or "meshes", marked as 1 and 2. These two cases are not the only ones we can select. Another loop can be closed around the whole network, also indicated in the figure. Using the KVL we can write the following equations:

Mesh one:

$$V_{R1} + V_{R2} - E = 0$$

Here we must subtract the voltage of the voltage source because we are walking through the source from − to +.

Mesh two:

$$-V_{R2} + V_{R3} + V_{R4} = 0$$

The negative sign comes from the fact that the current I_2 flows through R_2 from − to +. The loop around the network:

$$V_{R1} + V_{R3} + V_{R4} - E = 0$$

We can think of the arrows in the meshes or around the whole network as directions of currents. Such circulating currents do not actually exist but the notion helps; they are the first of several abstractions we will make.

FIGURE 2.2.1   Possible loops on a simple network.

FIGURE 2.2.2   Possible loops on another network.

In larger networks we can find many loops around which we could walk. In Fig. 2.2.2 we have marked six possible loops: three in the meshes, two more circling two meshes and one around the whole network. In still larger networks this could become very complicated and we must accept certain simplifications. The first and most important one is that the network must be *planar*. A planar network is such that we can draw its diagram on paper without any connection crossing some other connection. The two networks we considered so far were clearly planar. The network in Fig. 2.2.3a seems not to be planar, but we can redraw it as shown in Fig. 2.2.3b and discover that it actually is planar. The network in Fig. 2.2.4 is not planar because some line will cross another line no matter how we redraw the network.

Why did we use so much space to discuss planar networks? Because only planar networks can be analyzed by the method we will discuss here. In every planar network we can identify the meshes and draw directions for the voltage summations. The directions around the meshes are arbitrary but it is advisable to select one direction and use it everywhere. We will use the clockwise direction.

Consider the network in Fig. 2.2.5; it has two meshes, one voltage source and four resistors. The arrows in the meshes indicate the direction in which we will collect the voltages. If we take the directions as currents, then the product of a current and a resistance gives the voltage across the resistor, which is exactly what we need for the summations required by the KVL. There will also be cases where two currents in opposite directions will flow through a resistor; in our case it is

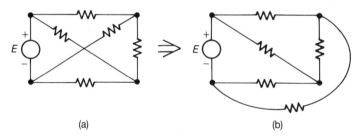

(a)   (b)

FIGURE 2.2.3   Is the network planar? (a) Seemingly nonplanar. (b) Redrawn for planarity.

Fig. 2.2.4   Fig. 2.2.5

FIGURE 2.2.4   Nonplanar network.

FIGURE 2.2.5   Network with one voltage source.

resistor R_2. When we are summing the voltages around the first mesh, the current I_1 has positive direction while the current I_2 has negative direction. Thus the voltage which contributes to the first mesh will be $R_2(I_1 - I_2)$. When we move to the second loop and walk around it, it will be the current I_2 which will have positive direction and the current I_1 will have to be subtracted. This is similar to the situation we had in section 1 where we always considered the node for which we were writing the equations as the most positive one.

In Fig. 2.2.5 I_1 flows through the resistor R_1 and the voltage across it must be $V_{R1} = I_1 R_1$, with the $+$ sign on the left side of R_1. As explained above, the voltage across the second resistor will be $V_{R2} = R_2(I_1 - I_2)$. The voltage of the voltage source must be subtracted *on the left side of the equation* because we are walking through it from $-$ to $+$. Thus the sum of voltages around the first mesh is

$$R_1 I_1 + R_2(I_1 - I_2) - E = 0 \qquad (2.2.1)$$

We proceed similarly for the second mesh where we are summing the voltages in the direction of the second arrow. The voltage across R_2 is now $V_{R2} = R_2(I_2 - I_1)$, the voltage across resistor R_3 is $V_{R3} = R_3 I_2$ and the voltage across the fourth resistor is $V_{R4} = R_4 I_2$. The second equation becomes

$$R_2(I_2 - I_1) + R_3 I_2 + R_4 I_3 = 0 \qquad (2.2.2)$$

Transferring the known E in the first equation to the right and collecting terms multiplied by the same currents we get

$$(R_1 + R_2)I_1 - R_2 I_2 = E$$
$$-R_2 I_1 + (R_2 + R_3 + R_4)I_2 = 0$$

$$(2.2.3)$$

Substituting the resistor values we simplify to the form

$$3I_1 - 2I_2 = 1$$
$$-2I_1 + 9I_2 = 0$$

This is a system of two equations in two unknowns, I_1 and I_2, and we give its solution: $I_1 = 9/23$ and $I_2 = 2/23$.

The second network we consider is in Fig. 2.2.6. In the first mesh the voltage across R_1 is $V_{R1} = 2I_1$, the voltage across the second resistor is $V_{R2} = 4(I_1 - I_2)$, similarly as in the above problem. The first mesh equation is $2I_1 + 4(I_1 - I_2) - 2 = 0$. The second equation is written similarly as above. The contributions are $V_{R2} = 4(I_2 - I_1)$, $V_{R3} = I_2$ and the voltage of the voltage source is added, because the current flows through it from + to −: $4(I_2 - I_1) + 1I_2 + 4 = 0$. Rearranging terms multiplied by the same currents and transferring the constant terms on the right side of the equations we get the final form

$$6I_1 - 4I_2 = 2$$
$$-4I_1 + 5I_2 = -4$$

Similarly as in the nodal formulation, we can now try to find some general rule how to write these equations by inspection. Refer to Eq. (2.2.3). In the first equation, when we consider mesh 1, we multiply by I_1 the sum of all resistors in mesh 1. If a resistor is in two meshes, then we subtract its resistance, multiplied by the current of the other mesh. When we move to mesh 2, a similar pattern appears. The current I_2 multiplies the sum of the resistors in the second mesh. If a resistor is shared by two meshes, its resistance, multiplied by the current of the other mesh, is subtracted. For the voltage source we see that if the circulating current flows into the mesh from the + terminal then its voltage appears *on the right side of the equation* with a positive sign.

We summarize the steps in the following rule for any mesh i:

1. Draw all mesh currents with clockwise directions.
2. Multiply the sum of all resistances in mesh i by I_i.

FIGURE 2.2.6   Network with two voltage sources.

FIGURE 2.2.7   Network with four voltage sources.

3. If an opposite current of mesh j also flows through the resistor, subtract the product of its resistance and I_j.
4. The voltage of an independent voltage source appears on the *right side of the equation* with positive sign if the ith equation current flows into the mesh from the + terminal of the source.

Apply the above rules for the network in Fig. 2.2.7. For the first mesh

$$(R_1 + R_2)I_1 - R_2 I_2 = E_1 + E_2 + E_3$$

For the second mesh

$$-R_2 I_1 + (R_2 + R_3 + R_4)I_2 = -E_3 + E_4 - E_5$$

Inserting numerical values

$$5I_1 - 3I_2 = 4$$
$$-3I_1 + 6I_2 = 1$$

Example 1.

Use the above rule to write the equations for the network in Fig. 2.2.8.

$$(R_1 + R_2)I_1 - R_2 I_2 - R_1 I_3 = E_1 - E_2$$
$$-R_2 I_1 + (R_2 + R_3)I_2 - R_3 I_3 = E_2 - E_3 \qquad (2.2.4)$$
$$-R_1 I_1 - R_3 I_2 + (R_1 + R_3 + R_4)I_3 = -E_4$$

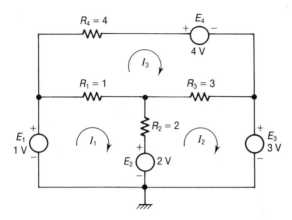

FIGURE 2.2.8   Network with four voltage sources.

Inserting numerical values

$$3I_1 - 2I_2 - I_3 = -1$$

$$-2I_1 + 5I_2 - 3I_3 = -1$$

$$-I_1 - 3I_2 + 8I_3 = -4$$

For those who wish to solve the system the result is $I_1 = -94/44$ A, $I_2 = -86/44$ A and $I_3 = -66/44$ A. The negative signs mean that all three circulating currents actually flow in opposite directions. The currents can be used to find the voltage across each resistor.

A word of caution: Our rule for writing the mesh equations assumes that the circulating currents all have the same directions, either clockwise or counterclockwise. Only then is it possible to perform the steps as easily as above, because the current of the neighboring mesh flows always in opposite direction. If we select arbitrary directions, we will have to check where we must add the currents and where we must subtract them.

Example 2.

Apply the mesh method to the network in Fig. 2.2.9 by writing the equations directly.
    The equations are:

$$(R_1 + R_2 + R_3)I_1 - R_3I_2 - R_1I_4 = E_1 + E_2 - E_3$$

$$-R_3I_1 + (R_3 + R_4 + R_6)I_2 - R_6I_3 - R_4I_4 = -E_6 + E_7$$

$$-R_6I_2 + (R_6 + R_7 + R_8)I_3 - R_7I_4 = -E_7 + E_8$$

$$-R_1I_1 - R_4I_2 - R_7I_3 + (R_1 + R_4 + R_5 + R_7)I_4 = E_3 - E_4 + E_5 + E_6 - E_8$$

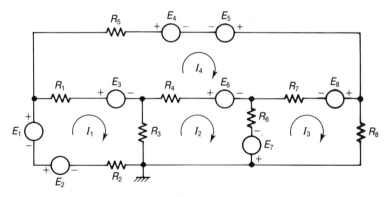

FIGURE 2.2.9   Network with eight voltage sources.

PROBLEMS CHAPTER 2

All problems can be solved by the program NET-OPT; a description on how to use it is in Chapter 18.

P.2.1 Find the indicated voltage V using one mesh formulation.

P.2.2 Find the current I using one mesh equation.

P.2.3 Find the voltage V using one nodal equation.

P.2.4 Find the voltage V using one nodal equation.

P.2.5 Write the mesh equations and find the voltage V.

P.2.6 Connecting an element in series with a current source is not typical because the current forced through the element is known and for the rest of the network the voltages are the same as if the element was missing. The only difference is the voltage formed across the current source. Use this information, apply nodal formulation and calculate the current I. Also calculate the voltage across the current source.

P.2.7 Find the indicated V and I. Use simplifications of the network to get to the result more easily.

P.2.8 Calculate the indicated voltages V_1 and V_2 using any method or simplifications.

P.2.9 Use mesh formulation to obtain the indicated currents.

P.2.10 Use mesh formulation to calculate the currents.

P.2.11 Calculate the voltage V using nodal formulation.

P.2.12 Calculate the nodal voltages V_1 and V_2.

P.2.13 Use mesh formulation to get the currents. Also calculate the nodal voltages V_1 and V_2.

P.2.14 Calculate V_1 and V_2.

P.2.18 Obtain the voltages across the resistors.

P.2.15 Calculate V_1 and V_2.

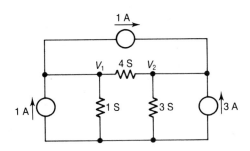

P.2.19 Calculate the currents and the voltage V.

P.2.16 Calculate the nodal voltages V_1, V_2 and V_3.

P.2.20 Calculate the currents and the voltage V.

P.2.17 Obtain the voltages across the resistors.

P.2.21 Calculate the currents and the voltage V.

P.2.22 Calculate the currents, the voltages and the powers either delivered or consumed in all elements.

P.2.23 Calculate the nodal voltages V_1 and V_2.

P.2.24 Obtain the indicated currents and the voltage V.

P.2.25 It is generally meaningless to connect an element in series with a current source because the element can be removed without influencing the balance of currents and voltages in the remaining network. Show that this is true by analyzing the network (a), then do the same with network (b).

The only difference will be the voltage across the current source in (a). Find this voltage.

(a)

(b)

P.2.26 It is generally meaningless to connect an element in parallel with a voltage source because the element can be removed without influencing the balance of currents and voltages in the remaining network. Show that this is true by analyzing the network (a), then do the same with network (b). The only difference will be the total current delivered by the source, increased by the amount flowing through the removed element. Show that this is true and find the total current delivered by the source in (a).

(a)

(b)

CHAPTER 3

Matrix Methods

3.0. INTRODUCTION

Solutions of linear networks lead to solutions of systems of linear equations and the reader probably knows the elimination method: a variable is separated in one of the equations, substituted into the remaining ones and the process is repeated. Although conceptually simple, this method is often confusing.

The reader may also know the Gaussian elimination: the right-hand side vector is written next to the matrix and row operations are performed on both the matrix and the vector until the equations are reduced to a triangular system. This is a good numerical method but is not particularly suitable if we need to solve the same system with another right-hand side. It is also not suitable for solutions of systems in which we must keep element names because we do not know their values. Another method which the reader probably knows is Cramer's rule. It is an excellent theoretical tool and one of the best methods for hand solution of small systems. On the other hand, it turns out to be unsuitable for computer programming.

All these choices may be quite confusing. What we need are a few selected methods which are good, reliable and easy to apply. Selection of such methods *for hand calculations* is the subject of this chapter. It is assumed that the reader already has some background in matrix algebra, knows multiplication of matrices and vectors and has heard about matrix inversion. In some places we just give the necessary rules, without deriving them. Where needed, we derive all the necessary steps.

3.1. LINEAR EQUATIONS IN MATRIX FORM

In mathematics, the unknowns are usually denoted by x_1, x_2 etc. As an example, we may have a system

$$3x_1 - 2x_2 = 1$$
$$-2x_1 + 9x_2 - 4x_3 = 0 \qquad (3.1.1)$$
$$-4x_2 + 9x_3 = 3$$

Suppose that we apply elimination and separate x_1 from the first equation

$$x_1 = \frac{1}{3} + \frac{2}{3}x_2 \qquad (3.1.2)$$

We can now replace x_1 in the remaining two equations by the right-hand side of Eq. (3.1.2) but here we have encountered something unpleasant: *the method invariably leads to fractions.* If we use a calculator to divide, we will almost certainly lose some digits in repunching the numbers and the final result will not be exact. It is usually better not to use the calculator and keep the fractions as long as possible, because they are *exact*. Division should be done only in the final result.

The above Eq. (3.2.1) can also be conveniently written in matrix form

$$\begin{bmatrix} 3 & -2 & 0 \\ -2 & 9 & -4 \\ 0 & -4 & 9 \end{bmatrix} \begin{bmatrix} x_1 \\ x_2 \\ x_3 \end{bmatrix} = \begin{bmatrix} 1 \\ 0 \\ 3 \end{bmatrix}$$

Standard mathematical notation uses bold letters for vectors and matrices and thus the equation could be expressed by

$$\mathbf{Ax} = \mathbf{b}$$

Here $\mathbf{A}$ is the system matrix (known), $\mathbf{x}$ is the unknown vector and $\mathbf{b}$ is the right-hand side vector (known). We are looking for the best practical methods to solve such systems.

The above problem is still relatively simple because all entries of the matrix are numbers. In the solution of networks we often do not know the numerical values of the elements. All we know are the types of elements (resistors, capacitors etc.) and we wish to determine the values later so that the network behaves in a certain desired manner. This means that we must first get the solution in terms of the element *variables* and only later we try to select some particular values in order to achieve what is required.

This so-called *symbolic* solution is a much more difficult problem. As an example, we give the simple resistive network in Fig. 3.1.1. In nodal formulation the equations are

$$(G_1 + G_2)V_1 - G_2V_2 = 1$$

$$-G_2V_1 + (G_2 + G_3 + G_4)V_2 - G_4V_3 = 0$$

$$-G_4V_2 + (G_4 + G_5)V_3 = 3$$

where V_1, V_2 and V_3 are the nodal voltages with respect to ground. If we insert the element values written in Fig. 3.1.1, we get the numerical example given above.

Consider this set of equations and the elimination method: separate V_1 from the first equation

$$V_1 = \frac{1}{G_1 + G_2} + \frac{G_2}{G_1 + G_2} V_2$$

and substitute for V_1 in the remaining equation: The fractions will stay with us through the whole solution process.

Such division can be avoided if we use matrix algebra and rewrite the equations in matrix form:

$$\begin{bmatrix} (G_1 + G_2) & -G_2 & 0 \\ -G_2 & (G_2 + G_3 + G_4) & -G_4 \\ 0 & -G_4 & (G_4 + G_5) \end{bmatrix} \begin{bmatrix} V_1 \\ V_2 \\ V_3 \end{bmatrix} = \begin{bmatrix} 1 \\ 0 \\ 3 \end{bmatrix}$$

Summarizing, we have presented two typical problems encountered in the solution of networks. The first and easier one has numbers in all entries of the matrix. Such case is usually solved by numerical methods, possibly by a calculator or a computer. The second, theoretically a much more important case, keeps either all or some elements as variables (letters). Such problem can be solved only for relatively small networks and the solution must be done by hand. In the following we select appropriate methods for both these cases.

FIGURE 3.1.1   Network with three nodes.

Example 1.

Write Eq. (2.1.14) in matrix form.

$$
\begin{bmatrix}
6 & -2 & -3 \\
-2 & 5 & -2 \\
-3 & -2 & 5
\end{bmatrix}
\begin{bmatrix}
V_1 \\
V_2 \\
V_3
\end{bmatrix}
=
\begin{bmatrix}
-2 \\
0 \\
5
\end{bmatrix}
$$

Example 2.

Write Eq. (2.1.13) in matrix form.

$$
\begin{bmatrix}
G_1 + G_2 + G_5 & -G_2 & -G_5 \\
-G_2 & G_2 + G_3 + G_4 & -G_4 \\
-G_5 & -G_4 & G_4 + G_5
\end{bmatrix}
\begin{bmatrix}
V_1 \\
V_2 \\
V_3
\end{bmatrix}
=
\begin{bmatrix}
J_1 - J_3 \\
0 \\
J_2 + J_3
\end{bmatrix}
$$

3.2. DETERMINANTS

This section will summarize some rules for the evaluation of determinants. You will learn how to find determinants, but all theoretical details are left to specialized mathematical books.

Matrices may be rectangular or square but determinants are defined only for square matrices. Using standard notation with two subscripts, a 3×3 matrix **A** is written as

$$
\mathbf{A} =
\begin{bmatrix}
a_{11} & a_{12} & a_{13} \\
a_{21} & a_{22} & a_{23} \\
a_{31} & a_{32} & a_{33}
\end{bmatrix}
\tag{3.2.1}
$$

The first subscript refers to rows and increases going down from the top, the second subscript refers to columns and increases from left to right.

The concept of a transpose matrix is also important. Columns of the transpose matrix, $\mathbf{A}^t$, are the rows of the original matrix, **A**, and vice versa. Sometimes it is easier to think about the process of transposition as if the matrix was rotated along an axis formed by its main diagonal (a line connecting the top left and bottom right entry). The transpose of the matrix in Eq. (3.2.1) is

$$
\mathbf{A}^t =
\begin{bmatrix}
a_{11} & a_{21} & a_{31} \\
a_{12} & a_{22} & a_{32} \\
a_{13} & a_{23} & a_{33}
\end{bmatrix}
\tag{3.2.2}
$$

Many possible ways of calculating the value of a determinant exist; some are advantageous in certain situations, some are not. We will concentrate on those best suited *for solutions by hand.* First we will give methods for the evaluation of 2×2 and 3×3 determinants. In order to simplify things, let us use for each entry a different letter without a subscript. The determinant of a 2×2 matrix is

$$\det \mathbf{A} = |\mathbf{A}| = \begin{vmatrix} a & b \\ c & d \end{vmatrix} = ad - bc \qquad (3.2.3)$$

The determinant of a 3×3 matrix is

$$|\mathbf{A}| = \begin{vmatrix} a & b & c \\ d & e & f \\ g & h & i \end{vmatrix} = (aei + dhc + gbf) - (gec + dbi + ahf) \qquad (3.2.4)$$

The easiest way to remember the steps for the 3×3 determinant is to copy the first two lines under the determinant as shown in Fig. 3.2.1. Products along the lines pointing down are added, products along the lines pointing up are subtracted. Study carefully which entries are involved in the products because it is advantageous to memorize the method without copying the additional lines. We feel that this method is the *best* one for hand evaluation of a 3×3 determinant. Its main advantage is that *no fractions are involved.* Note that the method applies only to 3×3 determinants. It cannot be extended to determinants of other sizes. Hand evaluations of larger determinants will be discussed in Section 5.

Having mastered the first two cases, we can now turn our attention to some

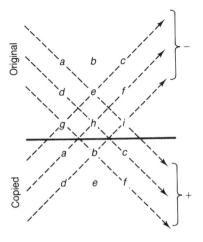

FIGURE 3.2.1   Evaluation of a 3×3 determinant.

useful rules:

1. If all entries of one row (column) are zero, then $|\mathbf{A}| = 0$.
2. If two rows (columns) are equal, then $|\mathbf{A}| = 0$.
3. If two rows (columns) are interchanged, then the determinant changes its sign.
4. If any row (column) is multiplied by a nonzero coefficient α, then the whole determinant is multiplied by this coefficient. In other words,

$$\alpha \begin{vmatrix} a_{11} & a_{12} \\ a_{12} & a_{22} \end{vmatrix} = \begin{vmatrix} \alpha a_{11} & a_{12} \\ \alpha a_{12} & a_{22} \end{vmatrix} = \begin{vmatrix} \alpha a_{11} & \alpha a_{12} \\ a_{12} & a_{22} \end{vmatrix} \qquad (3.2.5)$$

5. If we add to any row (column) some other row (column) multiplied by a nonzero coefficient α, the determinant does not change. For instance,

$$\begin{vmatrix} a_{11} & a_{12} \\ a_{21} & a_{22} \end{vmatrix} = \begin{vmatrix} a_{11} & a_{12} + \alpha a_{11} \\ a_{21} & a_{22} + \alpha a_{21} \end{vmatrix}$$

6. The determinant of the transpose matrix is equal to the determinant of the original matrix, $|\mathbf{A}| = |\mathbf{A}^t|$.

It is good to remember the above rules because sometimes we may obtain the determinant by simply inspecting the matrix or by performing some simple operation.

Let us now summarize our recommendations. For symbolic calculations (with letters instead of numbers) we recommend the above two rules for determinants of matrices 2×2 and 3×3. For larger matrices we recommend applying the Laplace expansion, covered in Section 5. The main advantage of these steps lies in the fact that no fractions are involved.

It is more difficult to give unified recommendations for numerical solutions. Sometimes, inspection and application of the above rules may help. For 2×2 and 3×3 cases, it may still be easiest to use the above rules. Larger numerical problems are usually solved by special numerical methods. Some modern calculators for scientific applications can also solve smaller numerical systems of equations.

Example 1.

Find the determinant of the matrix

$$\mathbf{A} = \begin{bmatrix} 3 & -2 \\ -2 & 5 \end{bmatrix}$$

Using Eq. (3.2.3) we find $|\mathbf{A}| = 15 - 4 = 11$.

Example 2.

Find the determinant of the matrix

$$\mathbf{A} = \begin{bmatrix} G_1 + G_2 & -G_2 \\ -G_2 & G_2 + G_3 \end{bmatrix}$$

Using Eq. (3.2.3) we get

$$|\mathbf{A}| = (G_1 + G_2)(G_2 + G_3) - G_2^2 = G_1 G_2 + G_1 G_3 + G_2 G_3$$

Example 3.

Find the determinant of the matrix

$$\mathbf{A} = \begin{bmatrix} 6 & -2 & -3 \\ -2 & 5 & -2 \\ -3 & -2 & 5 \end{bmatrix}$$

Using Eq. (3.2.4) or Fig. 3.2.1

$$|\mathbf{A}| = [(6)(5)(5) + (-2)(-2)(-3) + (-3)(-2)(-2)]$$
$$- [(-3)5(-3) + (-2)(-2)5 + 6(-2)(-2)] = 126 - 89 = 37$$

3.3. MATRIX INVERSION

Although we will not need matrix inversion to solve network problems, it is important to know some of the definitions and how to obtain the inverse theoretically.

First of all, matrix inversion is defined for square matrices only. Let us have an invertible square matrix $\mathbf{A}$; then its inverse, $\mathbf{A}^{-1}$, is a matrix satisfying

$$\mathbf{A}\mathbf{A}^{-1} = \mathbf{A}^{-1}\mathbf{A} = \mathbf{1} \tag{3.3.1}$$

Here $\mathbf{1}$ is the *unit matrix*, having ones along the main diagonal and zeros everywhere else, like in the 3×3 matrix below:

$$\mathbf{1} = \begin{bmatrix} 1 & 0 & 0 \\ 0 & 1 & 0 \\ 0 & 0 & 1 \end{bmatrix}$$

In Eq. (3.3.1), the sequence of the matrices is not important; *in all other situations it is.*

Inversion of a matrix requires evaluation of its determinant and calculation of a matrix which is called adjoint, adj **A**. Then

$$\mathbf{A}^{-1} = \frac{\text{adj } \mathbf{A}}{|\mathbf{A}|} \tag{3.3.2}$$

We already know how to calculate the determinant; all we need are the steps to obtain the adjoint matrix. We give these steps as a set of rules by considering a 3×3 example. Let the matrix be

$$\mathbf{A} = \begin{bmatrix} a_{11} & a_{12} & a_{13} \\ a_{21} & a_{22} & a_{23} \\ a_{31} & a_{32} & a_{33} \end{bmatrix} \tag{3.3.3}$$

Step 1. Write the transpose of this matrix

$$\mathbf{A}^t = \begin{bmatrix} a_{11} & a_{21} & a_{31} \\ a_{12} & a_{22} & a_{32} \\ a_{13} & a_{23} & a_{33} \end{bmatrix} \tag{3.3.4}$$

Step 2. Cross out one row and one column. The lines intersect at a given entry of a matrix; to be specific, let us cross out the second row and third column of $\mathbf{A}^t$. Copy the remaining entries of the matrix into a matrix which has the size one less than the original matrix and calculate the determinant. In our case we must evaluate

$$\begin{vmatrix} a_{11} & a_{21} \\ a_{13} & a_{23} \end{vmatrix} = a_{11}a_{23} - a_{13}a_{21}$$

Step 3. Attach to the result a sign which is equal to $(-1)^{i+j}$ where i is the subscript of the row and j is the subscript of the column which were crossed out. In our case, the determinant will be multiplied by $(-1)^{2+3} = -1$ and the result is placed as one entry of the inverse matrix into the very same position we crossed out on the transpose matrix. In our example, this is the entry in second row and third column.

Step 4. Repeat steps 2 and 3 for all other entries of the matrix.

We will take one example of inversion using

$$\mathbf{A} = \begin{bmatrix} 1 & 1 & 0 \\ 1 & 0 & 1 \\ 1 & 2 & 2 \end{bmatrix}$$

In step 1 we transpose the matrix

$$\mathbf{A}^t = \begin{bmatrix} 1 & 1 & 1 \\ 1 & 0 & 2 \\ 0 & 1 & 2 \end{bmatrix}$$

In step 2 we cross out the first row and first column of $\mathbf{A}^t$. This asks for the evaluation of the determinant

$$\begin{vmatrix} 0 & 2 \\ 1 & 2 \end{vmatrix} = -2$$

In step 3 we find that this value is multiplied by $+1$. The result will be the $(1,1)$ entry of the adjoint matrix. We now repeat steps 2 and 3 for the other positions. Crossing out row one and column two leads to the evaluation of the determinant

$$\begin{vmatrix} 1 & 2 \\ 0 & 2 \end{vmatrix} = 2$$

and this must be multiplied by -1. Similarly, for the $(1, 3)$ position we evaluate the determinant

$$\begin{vmatrix} 1 & 0 \\ 0 & 1 \end{vmatrix} = 1$$

and this result is multiplied by $+1$. Continuing similarly for the other positions we get the adjoint matrix

$$\mathrm{adj}\ \mathbf{A} = \begin{bmatrix} -2 & -2 & 1 \\ -1 & 2 & -1 \\ 2 & -1 & -1 \end{bmatrix}$$

We must still calculate the determinant of the original (or transpose) matrix; using the method of the previous section we find its value to be -3. Each entry of the adjoint matrix is divided by this number. The resulting inverse matrix is

$$\mathbf{A}^{-1} = \begin{bmatrix} \dfrac{2}{3} & \dfrac{2}{3} & \dfrac{-1}{3} \\[2mm] \dfrac{1}{3} & \dfrac{-2}{3} & \dfrac{1}{3} \\[2mm] \dfrac{-2}{3} & \dfrac{1}{3} & \dfrac{1}{3} \end{bmatrix}$$

Check the correctness of the result by multiplying $\mathbf{A}$ and $\mathbf{A}^{-1}$; you must get a unit matrix.

As can be seen, inversion is a rather laborious process and is rarely needed. For instance, in theoretical mathematics, the solution of the system $\mathbf{Ax} = \mathbf{b}$ is often described by simply writing $\mathbf{x} = \mathbf{A}^{-1}\mathbf{b}$. Although this is formally correct, inversion in this case is *never* used. In smaller problems we may use Cramer's rule, explained in the next section. In larger numerical problems we may use the well known Gaussian elimination, or other available numerical methods.

There is still one theoretical concept we must introduce. If the determinant of the matrix $\mathbf{A}$ happens to be zero, then such a matrix *does not have an inverse* and we say that the matrix is singular. The situation is extremely rare in cases when the matrix describes a practical network. Should it happen to you, then better check twice that you have not made some mistake somewhere.

If the above steps are applied to a 2×2 matrix,

$$\mathbf{A} = \begin{bmatrix} a & b \\ c & d \end{bmatrix} \tag{3.3.5}$$

then the inverse is

$$\mathbf{A}^{-1} = \frac{1}{ad - bc} \begin{bmatrix} d & -b \\ -c & a \end{bmatrix} \tag{3.3.6}$$

The entries on the main diagonal are interchanged and a minus sign is attached to the other two entries. It is useful to memorize this formula. It is the only practical case when it is advantageous to use matrix inversion to solve a system of linear equations.

Example 1.

Solve the system

$$2x_1 + 3x_2 = 2$$
$$-1x_1 + 4x_3 = 3$$

by writing it as a matrix equation and by inverting the matrix.
   The system is

$$\begin{bmatrix} 2 & 3 \\ -1 & 4 \end{bmatrix} \begin{bmatrix} x_1 \\ x_2 \end{bmatrix} = \begin{bmatrix} 2 \\ 3 \end{bmatrix}$$

Using matrix inversion

$$\begin{bmatrix} x_1 \\ x_2 \end{bmatrix} = \frac{1}{8+3} \begin{bmatrix} 4 & -3 \\ 1 & 2 \end{bmatrix} \begin{bmatrix} 2 \\ 3 \end{bmatrix} = \frac{1}{11} \begin{bmatrix} 8-9 \\ 2+6 \end{bmatrix} = \begin{bmatrix} -1/11 \\ 8/11 \end{bmatrix}$$

3.4. CRAMER'S RULE

Cramer's rule is probably the best method for hand solutions of equations. It is especially advantageous if only one variable is needed or if we have a matrix with symbols instead of numbers. It should be stressed that Cramer's rule is not useful for programming.
   We will derive Cramer's rule by using a 3 × 3 example. Consider the linear equations

$$a_{11}x_1 + a_{12}x_2 + a_{13}x_3 = b_1$$
$$a_{21}x_1 + a_{22}x_2 + a_{23}x_3 = b_2 \qquad (3.4.1)$$
$$a_{31}x_1 + a_{32}x_2 + a_{33}x_3 = b_3$$

or, in matrix form

$$\mathbf{Ax} = \mathbf{b} \qquad (3.4.2)$$

The determinant of the system matrix is

$$|\mathbf{A}| = \begin{vmatrix} a_{11} & a_{21} & a_{31} \\ a_{21} & a_{22} & a_{32} \\ a_{31} & a_{32} & a_{33} \end{vmatrix} \qquad (3.4.3)$$

Using rule 4 from Section 3.2, we know that multiplication of any column (row) by α means that the determinant is multiplied by α. Suppose that we multiply the third column by x_3. The result will be

$$x_3|\mathbf{A}| = \begin{vmatrix} a_{11} & a_{21} & a_{31}x_3 \\ a_{21} & a_{22} & a_{32}x_3 \\ a_{31} & a_{32} & a_{33}x_3 \end{vmatrix}$$

According to rule 5 of Section 3.2, we can multiply any column by some α, add it to any other column and the determinant will not change. We choose to multiply column 1 by x_1 and column 2 by x_2 and add them to the third column:

$$x_3|\mathbf{A}| = \begin{vmatrix} a_{11} & a_{21} & a_{11}x_1 + a_{12}x_2 + a_{13}x_3 \\ a_{21} & a_{22} & a_{21}x_1 + a_{22}x_2 + a_{23}x_3 \\ a_{31} & a_{32} & a_{31}x_1 + a_{32}x_2 + a_{33}x_3 \end{vmatrix}$$

Compare the third column with the left side of Eq. (3.4.1): they are identical. We can thus replace the column by the right side of Eq. (3.4.1):

$$x_3|\mathbf{A}| = \begin{vmatrix} a_{11} & a_{21} & b_1 \\ a_{21} & a_{22} & b_2 \\ a_{31} & a_{32} & b_3 \end{vmatrix}$$

However, this is the determinant of a matrix similar to $\mathbf{A}$, but with the third column replaced by the right-hand side. Let us denote this matrix by $\mathbf{A}_3$ and obtain

$$x_3|\mathbf{A}| = |\mathbf{A}_3|$$

or

$$x_3 = \frac{|\mathbf{A}_3|}{|\mathbf{A}|}$$

In the general case of a $n \times n$ system, we find the solution of the kth variable x_k by

a. calculating the determinant of the system matrix, $|\mathbf{A}|$, and
b. calculating the determinant of a similar matrix $|\mathbf{A}_k|$ in which the kth column was replaced by the right-hand side vector.

Then

$$x_k = \frac{|\mathbf{A}_k|}{|\mathbf{A}|} \tag{3.4.4}$$

Consider

$$x_1 + 2x_2 + x_3 = 1$$
$$-x_1 + 3x_2 + 2x_3 = 0$$
$$2x_1 - 2x_2 + x_3 = 2$$

The determinant of the system, D, is

$$D = |\mathbf{A}| = \begin{vmatrix} 1 & 2 & 1 \\ -1 & 3 & 2 \\ 2 & -2 & 1 \end{vmatrix} = 13$$

Replacing the first column by the right-hand side we get the first numerator of Eq. (3.4.4)

$$N_1 = |\mathbf{A}_1| = \begin{vmatrix} 1 & 2 & 1 \\ 0 & 3 & 2 \\ 2 & -2 & 1 \end{vmatrix} = 9$$

Replacing the second column by the right-hand side we get

$$N_2 = |\mathbf{A}_2| = \begin{vmatrix} 1 & 1 & 1 \\ -1 & 0 & 2 \\ 2 & 2 & 1 \end{vmatrix} = -1$$

Replacing the third column by the right-hand side

$$N_3 = |\mathbf{A}_3| = \begin{vmatrix} 1 & 2 & 1 \\ -1 & 3 & 0 \\ 2 & -2 & 2 \end{vmatrix} = 6$$

The solutions are

$$x_1 = \frac{N_1}{D} = \frac{|\mathbf{A}_1|}{|\mathbf{A}|} = \frac{9}{13}$$

$$x_2 = \frac{N_2}{D} = \frac{|\mathbf{A}_2|}{|\mathbf{A}|} = -\frac{1}{13}$$

$$x_3 = \frac{N_3}{D} = \frac{|\mathbf{A}_3|}{|\mathbf{A}|} = \frac{6}{13}$$

Example 1.

Solve the system

$$\begin{bmatrix} 3 & 2 \\ 2 & 4 \end{bmatrix}\begin{bmatrix} x_1 \\ x_2 \end{bmatrix} = \begin{bmatrix} 2 \\ -2 \end{bmatrix}$$

Using Cramer's rule.
    The determinant of the system is $D = 12 - 4 = 8$. The first numerator is

$$N_1 = \begin{vmatrix} 2 & 2 \\ -2 & 4 \end{vmatrix} = 8 + 4 = 12$$

The second numerator is

$$N_2 = \begin{vmatrix} 3 & 2 \\ 2 & -2 \end{vmatrix} = -6 - 4 = -10$$

The solution is

$$x_1 = \frac{N_1}{D} = \frac{12}{8}$$

$$x_2 = \frac{N_2}{D} = -\frac{10}{8}$$

Example 2.

Using Cramer's rule, solve the system

$$2x_1 + x_3 = 1$$
$$3x_1 + 2x_2 - 3x_3 = 2$$
$$x_1 - x_2 = 0$$

In matrix form these equations are

$$\begin{bmatrix} 2 & 0 & 1 \\ 3 & 2 & -3 \\ 1 & -1 & 0 \end{bmatrix}\begin{bmatrix} x_1 \\ x_2 \\ x_3 \end{bmatrix} = \begin{bmatrix} 1 \\ 2 \\ 0 \end{bmatrix}$$

The denominator is

$$D = \begin{vmatrix} 2 & 0 & 1 \\ 3 & 2 & -3 \\ 1 & -1 & 0 \end{vmatrix} = -3 - 2 - 6 = -11$$

The first numerator is

$$N_1 = \begin{vmatrix} 1 & 0 & 1 \\ 2 & 2 & -3 \\ 0 & -1 & 0 \end{vmatrix} = -2 - 3 = -5$$

The second numerator is

$$N_2 = \begin{vmatrix} 2 & 1 & 1 \\ 3 & 2 & -3 \\ 1 & 0 & 0 \end{vmatrix} = -3 - 2 = -5$$

The third numerator is

$$N_3 = \begin{vmatrix} 2 & 0 & 1 \\ 3 & 2 & 2 \\ 1 & -1 & 0 \end{vmatrix} = -3 - 2 + 4 = -1$$

and the solutions are

$$x_1 = \frac{N_1}{D} = \frac{5}{11}$$

$$x_2 = \frac{N_2}{D} = \frac{5}{11}$$

$$x_3 = \frac{N_3}{D} = \frac{1}{11}$$

3.5. EXPANSION OF DETERMINANTS

Hand evaluation of determinants larger than 3×3 is best done by the Laplace expansion. This expansion is especially convenient if we work with variables (letters) because no division is involved. Although more sophisticated expansions exist, we will use only that along a row or column.

Let us start with a numerical example

$$|\mathbf{A}| = \begin{vmatrix} 2 & 3 & -2 & 4 \\ 3 & -2 & 1 & 2 \\ 3 & 2 & 3 & 4 \\ -2 & 4 & 0 & 5 \end{vmatrix} \qquad (3.5.1)$$

Expansion of the determinant can be done along *any row or column* and proper signs have to be attached to the results, as explained below. In the example we will use expansion along the first column. Apply the following steps. Cross out the first row and column; take the number at the intersection of these two lines, in our case 2, and copy the remaining determinant less the crossed out lines:

$$2 \begin{vmatrix} -2 & 1 & 2 \\ 2 & 3 & 4 \\ 4 & 0 & 5 \end{vmatrix} = 2 \cdot (-48)$$

Return to the original determinant, go down to the next entry (first column, second row) and cross out the first column and second row. Write the number at the intersection and copy the remaining determinant, less the crossed out lines:

$$-3 \begin{vmatrix} 3 & -2 & 4 \\ 2 & 3 & 4 \\ 4 & 0 & 5 \end{vmatrix} = -3 \cdot (-15)$$

We will return specially to the sign in front of this expression. The third expansion about the first column crosses out the first column and third row:

$$3 \begin{vmatrix} 3 & -2 & 4 \\ -2 & 1 & 2 \\ 4 & 0 & 5 \end{vmatrix} = 3 \cdot (-37)$$

and the last case crosses out the first column and fourth row:

$$-(-2) \begin{vmatrix} 3 & -2 & 4 \\ -2 & 1 & 2 \\ 2 & 3 & 4 \end{vmatrix} = 2 \cdot (-62)$$

Again, we will speak about the sign in front of this expression later. Taking the sum of all partial results

$$|\mathbf{A}| = 2\cdot(-48) - 3\cdot(-15) + 3\cdot(-37) + 2\cdot(-62) = -286$$

Now about the signs in front of the expressions. Theoretically, each is multiplied by $(-1)^{i+j}$ where i and j are the subscripts of the term lying on the intersection of the lines we used for crossing out. Going down in the first column, the sequence will alternate signs, $+$, $-$, $+$, $-$, and this was used above.

The following mnemonic rule gives the same result but is probably easier to remember. Start in the upper left corner with $+$ sign. Walk to the entry under consideration in any sequence of horizontal and vertical moves, in each step always changing the sign. When we reach the desired entry, the sign is correct. This is shown schematically in Fig. 3.5.1 where one way is straightforward and one quite complicated. Nevertheless, the sign is always the same one.

It is advisable to look for any row or column which has the largest number of zeros, and expand the determinant along this row or column. In such a case some partial determinants would be multiplied by zero and we would not calculate them at all. Looking at Eq. (3.5.1) we see that expansion about the third column would have saved us one determinant. Using the above rule for determining the signs, the expansion will be

$$|\mathbf{A}| = -2\begin{vmatrix} 3 & -2 & 2 \\ 3 & 2 & 4 \\ -2 & 4 & 5 \end{vmatrix} - 1\begin{vmatrix} 2 & 3 & 4 \\ 3 & 2 & 4 \\ -2 & 4 & 5 \end{vmatrix} + 3\begin{vmatrix} 2 & 3 & 4 \\ 3 & -2 & 222 \\ -2 & 4 & 5 \end{vmatrix}$$

$$= -2\cdot(60) - 1\cdot(-17) + 3\cdot(-61) = -286$$

The reader should also note that addition or subtraction of columns may

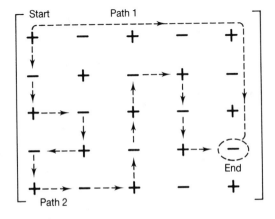

FIGURE 3.5.1   Determining sign for a subdeterminant.

considerably simplify the expansion. Using the above example and adding the first row to the last row will create a zero in the (4, 1) position. Similarly, subtracting the second row from the third row will create a zero at the (3, 1) position and the determinant will be

$$|\mathbf{A}| = \begin{vmatrix} 2 & 3 & -2 & 4 \\ 3 & -2 & 1 & 2 \\ 0 & 4 & 2 & 2 \\ 0 & 7 & -2 & 9 \end{vmatrix} = 2 \begin{vmatrix} -2 & 1 & 2 \\ 4 & 2 & 2 \\ 7 & -2 & 9 \end{vmatrix} - 3 \begin{vmatrix} 3 & -2 & 4 \\ 4 & 2 & 2 \\ 7 & -2 & 9 \end{vmatrix}$$

$$= 2 \cdot (-110) - 3 \cdot (22) = -286$$

Instead of doing the expansion about the column we can achieve the same by doing the expansion about any row.

Summarizing this section we state that Laplace expansion of determinants is best suited for evaluations of determinants having letters instead of numbers, a frequent case in network solutions. Numerical solutions can be handled by many methods, but if done by hand, then Laplace expansion and Cramer's rule still retain one big advantage: there are no intermediate fractions, only products. Only the last step of Cramer's rule requires a division and this is done without any loss of accuracy.

Example 1.

Evaluate the determinant of the matrix in which not all entries are numbers:

$$\begin{bmatrix} 1 & 2 & 0 & 0 \\ 3 & 0 & A & 1 \\ 1 & B & 0 & 2 \\ 0 & 1 & 0 & 3 \end{bmatrix}$$

Expanding about the third column this reduces to the evaluation of

$$-A \begin{vmatrix} 1 & 2 & 0 \\ 1 & B & 2 \\ 0 & 1 & 3 \end{vmatrix} = 8A - 3AB$$

Less advantageous is the expansion about the first row which leads to

$$1 \begin{vmatrix} 0 & A & 1 \\ B & 0 & 2 \\ 1 & 0 & 3 \end{vmatrix} - 2 \begin{vmatrix} 3 & A & 1 \\ 1 & 0 & 2 \\ 0 & 0 & 3 \end{vmatrix} = 1[2A - 3AB] - 2[-3A] = 8A - 3AB$$

Another possibility is to expand about the fourth row. This leads to the evaluation of

$$1\begin{vmatrix} 1 & 0 & 0 \\ 3 & A & 1 \\ 1 & 0 & 2 \end{vmatrix} + 3\begin{vmatrix} 1 & 2 & 0 \\ 3 & 0 & A \\ 1 & B & 0 \end{vmatrix} = 2A + 3(2A - 3AB) = 8A - 3AB$$

3.6. SOLUTIONS OF RESISTIVE NETWORKS

Resistive linear networks are the simplest practical problems and we will go through several examples.

Example 1.

Consider the network in Fig. 3.6.1. Using mesh formulation and matrix algebra find the two mesh currents. Also find the equivalent resistance of the combination by calculating the ratio E/I_1. The two equations are

$$4I_1 - 3I_2 = E$$
$$-3I_1 + 6I_2 = 0$$

In matrix form:

$$\begin{bmatrix} 4 & -3 \\ -3 & 6 \end{bmatrix}\begin{bmatrix} I_1 \\ I_2 \end{bmatrix} = \begin{bmatrix} E \\ 0 \end{bmatrix}$$

The determinant is

$$D = 24 - 9 = 15$$

Using Cramer's rule we find the numerator for I_1

$$N_1 = \begin{vmatrix} E & -3 \\ 0 & 6 \end{vmatrix} = 6E$$

FIGURE 3.6.1   Network with two meshes.

and for I_2

$$N_2 = \begin{vmatrix} 4 & E \\ -3 & 0 \end{vmatrix} = 3E$$

The currents are $I_1 = (6/15)E$ and $I_2 = (3/15)E$. The current I_1 is flowing from the source into the network. By taking the ratio $I_1/E = 6/15$ we will get the conductance of the combination. Check this result by finding the overall conductance of the network as a parallel/series combination of resistors (as explained in Chapter 1).

Example 2.

The network in Fig. 3.6.2 is suitable for nodal equations. Find the nodal voltages and the overall resistance of the network, as viewed from the source.

The equations are

$$4V_1 - 2V_2 - 1V_3 = J$$

$$-2V_1 + 5V_2 - 2V_3 = 0$$

$$-1V_1 - 2V_2 + 5V_3 = 0$$

In matrix form

$$\begin{bmatrix} 4 & -2 & -1 \\ -2 & 5 & -2 \\ -1 & -2 & 5 \end{bmatrix} \begin{bmatrix} V_1 \\ V_2 \\ V_3 \end{bmatrix} = \begin{bmatrix} J \\ 0 \\ 0 \end{bmatrix}$$

The determinant of the system is $D = 51$. Using Cramer's rule, we evaluate the determinants

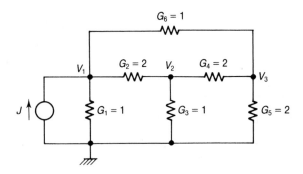

FIGURE 3.6.2   Network with three nodes.

for the numerators:

$$N_1 = \begin{vmatrix} J & -2 & -1 \\ 0 & 5 & -2 \\ 0 & -2 & 5 \end{vmatrix} = 21J$$

$$N_2 = \begin{vmatrix} 4 & J & -1 \\ -2 & 0 & -2 \\ -1 & 0 & 5 \end{vmatrix} = 12J$$

$$N_3 = \begin{vmatrix} 4 & -2 & J \\ -2 & 5 & 0 \\ -1 & -2 & 0 \end{vmatrix} = 9J$$

The results are

$$V_1 = \frac{21}{51} J$$

$$V_2 = \frac{12}{51} J$$

$$V_3 = \frac{9}{51} J$$

The overall equivalent resistance of the network, as viewed from the source, is equal to $V_1/J = 21/51$.

Example 3.

Consider the network in Fig. 3.6.3 where all conductances are $G_i = 1$. Find the voltage V_1 and the equivalent resistance of the network, seen from the terminals of the source.

FIGURE 3.6.3   Network with four nodes.

The nodal equations are

$$3V_1 - 1V_2 - 1V_3 + 0V_4 = J$$
$$-1V_1 + 4V_2 - 1V_3 - 1V_4 = 0$$
$$-1V_1 - 1V_2 + 4V_3 - 1V_4 = 0$$
$$0V_1 - 1V_2 - 1V_3 + 3V_4 = 0$$

or in matrix form

$$
\begin{bmatrix}
3 & -1 & -1 & 0 \\
-1 & 4 & -1 & -1 \\
-1 & -1 & 4 & -1 \\
0 & -1 & -1 & 3
\end{bmatrix}
\begin{bmatrix}
V_1 \\
V_2 \\
V_3 \\
V_4
\end{bmatrix}
=
\begin{bmatrix}
J \\
0 \\
0 \\
0
\end{bmatrix}
$$

In this example we find only the voltage V_1. The determinant of the system matrix is obtained by expansion about the first column:

$$
D = 3\begin{vmatrix} 4 & -1 & -1 \\ -1 & 4 & -1 \\ -1 & -1 & 3 \end{vmatrix} + 1\begin{vmatrix} -1 & -1 & 0 \\ -1 & 4 & -1 \\ -1 & -1 & 3 \end{vmatrix} - 1\begin{vmatrix} -1 & -1 & 0 \\ 4 & -1 & -1 \\ -1 & -1 & 3 \end{vmatrix} = 105 - 15 - 15 = 75
$$

The numerator is found by replacing the first column of the system matrix by the right-hand side and by expanding the determinant about the first column. Since only one entry in this column is nonzero, this expansion is much simpler:

$$
N_1 = \begin{vmatrix} J & -1 & -1 & 0 \\ 0 & 4 & -1 & -1 \\ 0 & -1 & 4 & -1 \\ 0 & -1 & -1 & 3 \end{vmatrix} = J\begin{vmatrix} 4 & -1 & -1 \\ -1 & 4 & -1 \\ -1 & -1 & 3 \end{vmatrix} = 35J
$$

The resulting voltage is $V_1 = (35/75)J$ and the input resistance of the network is

$$\frac{V_1}{J} = R_{in} = \frac{35}{75}$$

Example 4.

The network in Fig. 3.6.4 is suitable for nodal equations. Find all nodal voltages and the equivalent resistance of the network. All conductances are equal to $G_i = 1$. Write directly

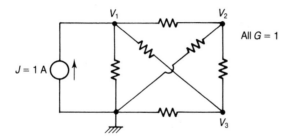

FIGURE 3.6.4   Network with three nodes.

the matrix equation.

$$
\begin{bmatrix} 3 & -1 & -1 \\ -1 & 3 & -1 \\ -1 & -1 & 3 \end{bmatrix} \begin{bmatrix} V_1 \\ V_2 \\ V_3 \end{bmatrix} = \begin{bmatrix} 1 \\ 0 \\ 0 \end{bmatrix}
$$

Determinant of the matrix is $D = 16$, $N_1 = 8$, $N_2 = 4$, $N_3 = 4$. The voltages are $V_1 = 1/2$, $V_2 = 1/4$ and $V_3 = 1/4$. The input resistance is

$$
\frac{V_1}{J} = \frac{V_1}{1} = V_1 = \frac{1}{2}
$$

Example 5.

The network in Fig. 3.6.5 has two voltage sources and is suitable for mesh equations. Write the mesh equations, the matrix equation, and calculate the voltage V_{out} indicated in the figure.

The equations are

$$
5I_1 - 4I_2 - I_3 = 2
$$

$$
-4I_1 + 8I_2 = -3
$$

$$
-1I_1 + 3I_3 = 3
$$

In matrix form:

$$
\begin{bmatrix} 5 & -4 & -1 \\ -4 & 8 & 0 \\ -1 & 0 & 3 \end{bmatrix} \begin{bmatrix} I_1 \\ I_2 \\ I_3 \end{bmatrix} = \begin{bmatrix} 2 \\ -3 \\ 3 \end{bmatrix}
$$

We need the current I_2 to be able to calculate the voltage V_{out}. The system matrix determinant is $D = 64$. For the second mesh current we must find

$$
N_2 = \begin{vmatrix} 5 & 2 & -1 \\ -4 & -3 & 0 \\ -1 & 3 & 3 \end{vmatrix} = -6
$$

FIGURE 3.6.5   Network with three meshes.

The second mesh current is $I_2 = N_2/D = -6/16$. This current flows through the resistor $R_3 = 4\,\Omega$ and thus the output voltage is

$$V_{out} = R_3 I_2 = -4 \cdot \frac{6}{16} = -\frac{3}{2}\,\text{V}$$

Example 6.

Use mesh equations for the network in Fig. 3.6.6, find the current I_1 and the equivalent resistance seen from the source. All resistors have unit values, $R_i = 1\,\Omega$.

Writing the mesh equations is left to the student, we give directly the matrix form:

$$\begin{bmatrix} 2 & -1 & -1 & 0 \\ -1 & 3 & -1 & -1 \\ -1 & -1 & 3 & -1 \\ 0 & -1 & -1 & 3 \end{bmatrix} \begin{bmatrix} I_1 \\ I_2 \\ I_3 \\ I_4 \end{bmatrix} = \begin{bmatrix} 1 \\ 0 \\ 0 \\ 0 \end{bmatrix}$$

The determinant of the system matrix is found by Laplace expansion about the first column and is equal to $D = 8$. Replacing the first column by the right-hand side we find $N_1 = 16$ and thus $I_1 = 16/8 = 2\,\text{A}$. The input resistance $R_{in} = E/I_1 = 0.5\,\Omega$.

FIGURE 3.6.6   Network with four meshes.

Numerical systems of linear equations can be solved by the program described in Appendix A. Although the numerical method is different from Cramer's rule, the results are presented in the same form as if Cramer's rule had actually been applied: for each variable the program prints the system determinant, all numerators and all solution variables. Use the program for the above examples or other problems.

PROBLEMS CHAPTER 3

Appendix A gives instructions how to use the program for solution of systems of linear equations. The systems in P.3.1 were solved by it.

Several network problems are also given. Solve them by hand using the methods explained in this chapter and use the program from Appendix A to confirm your results. In addition, all network problems can be solved by the program NETOPT, described in Chapter 18.

P.3.1 Find the solutions of the following problems by using Cramer's rule. For systems larger than 3×3 use expansion of determinants described in Section 3.5.

$$x_1 + 2x_2 + x_3 = 1$$
$$2x_1 + 6x_2 + 2x_3 = 0 \qquad \text{(a)}$$
$$3x_1 + 7x_2 + 6x_3 = -2$$

$$2x_1 + x_2 + 5x_3 + x_4 = 5$$
$$x_1 + x_2 - 3x_3 - 4x_4 = 1 \qquad \text{(b)}$$
$$3x_1 + 6x_2 - 2x_3 + x_4 = 8$$
$$2x_1 + 2x_2 + 2x_3 - 3x_4 = 2$$

$$2x_1 + 0x_2 + 3x_3 + 0x_4 = 1$$
$$0x_1 + 3x_2 + 0x_3 + 1x_4 = 2 \qquad \text{(c)}$$
$$6x_1 + 6x_2 + 10x_3 + 4x_4 = 0$$
$$2x_1 + 0x_2 + 6x_3 + 8x_4 = 3$$

$$2x_1 + 4x_2 + 0x_3 + 2x_4 + 0x_5 = 4$$
$$0x_1 + 3x_2 + 6x_3 + 0x_4 + 3x_5' = 0$$
$$1x_1 + 2x_2 + 1x_3 + 2x_4 + 0x_5 = 3 \qquad \text{(d)}$$
$$2x_1 + 4x_2 + 1x_3 + 4x_4 + 0x_5 = 6$$
$$0x_1 + 1x_2 + 2x_3 + 2x_4 + 3x_5 = 2$$

$$1x_1 + 0x_2 + 0x_3 + 2x_4 + 1x_5 = -1$$
$$3x_1 + 2x_2 + 0x_3 + 0x_4 + 2x_5 = 3$$
$$0x_1 + 1x_2 + 1x_3 + 0x_4 + 0x_5 = 0 \qquad \text{(e)}$$
$$0x_1 + 0x_2 + 1x_3 + 2x_4 + 1x_5 = 0$$
$$0x_1 + 0x_2 + 1x_3 + 1x_4 + 1x_5 = 1$$

$$11x_1 - 10x_2 + 0x_3 + 0x_4 = 10$$
$$-10x_1 + 19x_2 - 4x_3 + 0x_4 = 20$$
$$\qquad \text{(f)}$$
$$0x_1 - 4x_2 + 39x_3 - 20x_4 = -20$$
$$0x_1 + 0x_2 - 20x_3 + 22x_4 = -15$$

$$3x_1 - 1x_2 - 1x_3 + 0x_4 = 5$$
$$-1x_1 + 3x_2 - 1x_3 - 1x_4 = -10$$
$$\qquad \text{(g)}$$
$$-1x_1 - 1x_2 + 3x_3 - 1x_4 = 10$$
$$0x_1 - 1x_2 - 1x_3 + 3x_4 = -5$$

$$1x_1 + 0x_2 + 1x_3 + 0x_4 = 0$$
$$2x_1 + 2x_2 + 2x_3 + 4x_4 = 0$$
$$\qquad \text{(h)}$$
$$0x_1 + 1x_2 + 3x_3 + 11x_4 = 0$$
$$1x_1 + 0x_2 - 1x_3 - 5x_4 = 3$$

P.3.2 Find the voltage V across the last resistor.

P.3.3 Find the indicated current I.

P.3.4 Use nodal formulation to obtain the three nodal voltages.

P.3.5 Use mesh equations, obtain the currents and calculate the indicated voltages V_1 and V_2.

P.3.6 Use mesh equations to obtain the three currents.

P.3.7 Use nodal formulation to obtain the three nodal voltages.

P.3.8 Use nodal formulation to obtain the three nodal voltages.

- **P.3.9** Use nodal formulation to obtain the three nodal voltages.

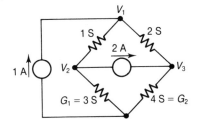

P.3.10 Find the three currents.

P.3.11 Find the three currents.

P.3.12 Set up the nodal equations and write them in matrix form. Use determinant expansion to obtain the nodal voltages. Check by the program from Appendix A.

All $G = 1$ S

P.3.13 Do the same as in P.3.12. All $G_i = 1$ S.

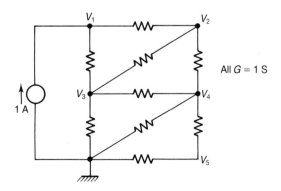

All $G = 1$ S

P.3.14 Use mesh equations to set up the matrix equation. Solve by hand and check by the program in Appendix A.

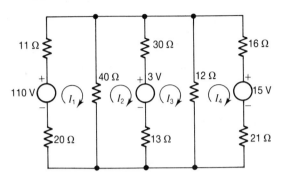

P.3.15 Use nodal equations to solve. Find the nodal voltages and also calculate the indicated current I.

P.3.16 Connection of the current source on the left is not typical but since the same current flows from the source as through the resistor on the left, one can first replace the first resistor by a short circuit. This will not influence the distribution of voltages and currents in the remaining network. Find the nodal voltages this way first. Afterwards find the voltage across the current source on the right.

P.3.17 Calculate the current I_2 and the voltage across the resistor R_4 by keeping this resistor as a variable.

P.3.18 This network was already solved in P.3.4 but now keep the conductance on the right as a variable, G. Calculate the voltage V_3. Afterwards substitute $G = 3$ S; you should get the same result as in P.3.4.

P.3.19 In problem P.3.6 keep the resistors R_1 and R_2 as variables and obtain the solution. Afterwards substitute the same values; your should obtain the same result.

P.3.20 In problem P.3.7 keep G_5 as variable and get the result. Afterwards substitute the same value as there and check that you have the same result.

P.3.21 In problem P.3.8 keep G_1 and G_2 as variables and find V_3.

P.3.22 In problem P.3.9 keep G_1 and G_2 as variables and find the voltage V_1.

P.3.23 Keep the conductance G as variable and find the voltage V_2. Afterwards get such value of G that $V_2 = 2$ V.

CHAPTER 4

Dependent Sources

4.0. INTRODUCTION

Since we now have the two fundamental equation formulation methods and also the means to solve them, we can turn our attention to dependent sources. They are extremely important elements because the properties of semiconductors are modeled by them.

An independent voltage source, as we defined it in Chapter 1, delivered a voltage of E volts and this voltage did not depend on anything inside the network. The same was true about the independent current source which delivered a current of J amperes, independent on any external influences.

If we make the value of the voltage (or current) delivered by the source dependent on some other voltage (or current) in the network, we will have a dependent source. There are altogether four possibilities:

1. Voltage-controlled current source, VC.
2. Current-controlled voltage source, CV.
3. Current-controlled current source, CC.
4. Voltage-controlled voltage source, VV.

The elements were already introduced in Section 1.6. In the abbreviations the first letter refers to the controlling variable and the second to the type of source.

We will now study all these controlled sources in more detail and show how we use them in the nodal or mesh formulations. If the student has access to an IBM-PC computer, he should start using the NETOPT program for analysis of linear networks. Instructions on how to use it are in Chapter 18.

4.1. VOLTAGE-CONTROLLED CURRENT SOURCE

The voltage-controlled current source, VC, is an element which can easily be incorporated into the nodal equations and we consider it first. The element is shown in Fig. 4.1.1. The terminals on the left are called the *input* or *controlling* terminals, those on the right the *output* or *controlled* terminals. Voltages are measured with respect to ground and *positive* directions of currents are taken as flowing *away* from the terminals. Since no current flows into the element at the input terminals

$$I_i = I_j = 0 \qquad (4.1.1)$$

The currents at the output terminals, k and l, are

$$I_k = g(V_i - V_j)$$
$$I_l = -g(V_i - V_j) \qquad (4.1.2)$$

The constant g is called *transconductance* and is measured in Siemens, S. The amount of current delivered by the source at the controlled terminals depends on the transconductance value and on the voltage difference $V_i - V_j$. If this difference is zero, the element does not deliver any current and behaves as an open circuit. Recall here that independent current source not delivering current also behaves as an open circuit.

Consider the simple network in Fig. 4.1.2. The VC in the network has its controlling terminals between nodes 1 and 0, its controlled terminals between nodes 0 and 2. The sequence must be given from 0 to 2, because of the direction of the current. All the necessary information is already in the figure. Identifying

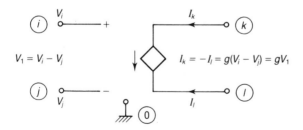

FIGURE 4.1.1   Voltage-controlled current source, VC.

FIGURE 4.1.2   Network with one VC.

the voltages with those in the diagram of the VC we have $V_i = V_1$, $V_j = 0$, $V_k = 0$ and $V_l = V_2$, and $I_l = -gV_1$.

The nodal equations are written in exactly the same way as we have done in Section 2.1. Consider node 1 and assume that it has the highest voltage with respect to ground. Having made this assumption we must accept the consequence that all currents through the resistors flow *away* from the node. The Kirchhoff current law states that the sum of these currents must be equal to zero:

$$G_1 V_1 + G_2(V_1 - V_2) - J = 0$$

When considering node 2 we assume that the voltage of the second node is now the highest one. This means that all currents through the resistors must flow away from this node. The current gV_1 flows into node 2 and has to be taken negatively in the KCL:

$$G_3 V_2 + G_2(V_2 - V_1) - gV_1 = 0$$

There is no difference in writing equations for the dependent or independent current sources. Since the transconductance is multiplied by some unknown voltage, the term must stay on the left side of the equations. The independent source, which is known, is transferred to the right side with a change of sign. Collecting terms multiplied by the same voltage

$$(G_1 + G_2)V_1 - G_2 V_2 = J$$
$$-(G_2 + g)V_1 + (G_2 + G_3)V_2 = 0$$

If we use the values indicated in Fig. 4.1.2, the system of equations will be

$$3V_1 - 2V_2 = 1$$
$$-6V_1 + 5V_2 = 0$$

Using matrix algebra

$$\begin{bmatrix} 3 & -2 \\ -6 & 5 \end{bmatrix} \begin{bmatrix} V_1 \\ V_2 \end{bmatrix} = \begin{bmatrix} 1 \\ 0 \end{bmatrix}$$

and the solution is $V_1 = 5/3$ V, $V_2 = 2$ V. If we need the currents through the three conductances, they are

$$I_{G1} = G_1 V_1 = \frac{5}{3} \text{ A}$$

$$I_{G2} = 2(V_1 - V_2) = 2\left(\frac{5}{3} - 2\right) = -\frac{2}{3} \text{ A}$$

$$I_{G3} = G_3 V_2 = 6 \text{ A}$$

The sign of I_{G2} indicates that the current actually flows from node 2 to node 1.

Example 1.

Write the nodal equations for the network in Fig. 4.1.3 and find the nodal voltages.
   The equations are

$$3V_1 - 2V_2 + 4(V_1 - V_2) = J$$
$$-2V_1 + 3V_2 = 0$$

In matrix form

$$\begin{bmatrix} 7 & -6 \\ -2 & 3 \end{bmatrix} \begin{bmatrix} V_1 \\ V_2 \end{bmatrix} = \begin{bmatrix} 1 \\ 0 \end{bmatrix}$$

FIGURE 4.1.3   Network with one VC.

The determinant $D = 21 - 12 = 9$,

$$N_1 = \begin{vmatrix} 1 & -6 \\ 0 & 3 \end{vmatrix} = 3$$

$$N_2 = \begin{vmatrix} 7 & 1 \\ -2 & 0 \end{vmatrix} = 2$$

and the solutions are $V_1 = 3/9$ and $V_2 = 2/9$.

Example 2.

The network in Fig. 4.1.4 has two independent and two voltage-controlled current sources. Find the nodal voltages.

The equations are

$$3V_1 - 2V_2 - g_2 V_2 = 1$$
$$-2V_1 + 5V_2 + g_1(V_1 - V_2) = 2$$

We have intentionally retained the symbols for the controlled sources. Inserting their values

$$\begin{bmatrix} 3 & -4 \\ 0 & 3 \end{bmatrix} \begin{bmatrix} V_1 \\ V_2 \end{bmatrix} = \begin{bmatrix} 1 \\ 2 \end{bmatrix}$$

The system determinant is $D = 9$, $N_1 = 11$ and $N_2 = 6$. The solutions are $V_1 = 11/9$ and $V_2 = 6/9$.

Example 3.

Do the same as for Fig. 4.1.2 but retain all elements as variables.

The equations are

$$(G_1 + G_2)V_1 - G_2 V_2 - g_2 V_2 = 1$$
$$-G_2 V_2 + (G_2 + G_3)V_2 + g_1(V_1 - V_2) = 2$$

FIGURE 4.1.4   Network with two VC.

or in matrix form

$$\begin{bmatrix} G_1 + G_2 & -g_2 - G_2 \\ g_1 - G_2 & G_1 + G_3 - g_1 \end{bmatrix} \begin{bmatrix} V_1 \\ V_2 \end{bmatrix} = \begin{bmatrix} J_1 \\ J_2 \end{bmatrix}$$

When we perform the necessary steps, we get

$$D = G_1 G_2 + G_1 G_3 + G_2 G_3 - G_1 g_1 - G_2 g_2 + g_1 g_2$$
$$N_1 = J_1 (G_2 + G_3 - g_1) + J_2 (g_2 + G_2)$$
$$N_2 = J_1 (G_2 - g_1) + J_2 (G_1 + G_2)$$

A network with three nodes is in Fig. 4.1.5. Using the same rules as above, we write for node 1

$$(G_1 + G_2 + G_6) V_1 - G_2 V_2 - G_6 V_3 - g_1 (V_2 - V_3) = J_1$$

for node 2

$$-G_2 V_1 + (G_2 + G_3 + G_4) V_2 - G_4 V_3 - g_2 V_1 + g_3 (V_1 - V_2) = 0$$

and for node 3

$$-G_6 V_1 - G_4 V_2 + (G_4 + G_5 + G_6) V_3 + g_1 (V_2 - V_3) + g_2 V_1 = J_2$$

The transconductances of the dependent sources must stay on the left side, the

FIGURE 4.1.5   Network with three VC.

independent sources are transferred on the right side of the equations. Collect the terms multiplied by the same voltage,

$$(G_1 + G_2 + G_6)V_1 - (G_2 + g_1)V_2 + (g_1 - G_6)V_3 = J_1$$
$$(g_3 - G_2 - g_2)V_1 + (G_2 + G_3 + G_4 - g_3)V_2 - G_4V_3 = 0$$
$$(g_2 - G_6)V_1 + (g_1 - G_4)V_2 + (G_4 + G_5 + G_6 - g_1)V_3 = J_2$$

In matrix form:

$$
\begin{bmatrix}
G_1 + G_2 + G_6 & -G_2 - g_1 & g_1 - G_6 \\
g_3 - G_2 - g_2 & G_2 + G_3 + G_4 - g_3 & -G_4 \\
g_2 - G_6 & g_1 - G_4 & G_4 + G_5 + G_6 - g_1
\end{bmatrix}
\begin{bmatrix}
V_1 \\ V_2 \\ V_3
\end{bmatrix}
=
\begin{bmatrix}
J_1 \\ 0 \\ J_2
\end{bmatrix}
$$

If we substitute the values and solve, we get

$$V_1 = \frac{156}{423} = 0.3688 \text{ V}$$

$$V_2 = \frac{112}{423} = 0.2648 \text{ V}$$

$$V_3 = \frac{129}{423} = 0.3050 \text{ V}$$

Example 4.

Write the nodal equations and the matrix equation for the network in Fig. 4.1.6. Use the numerical values and find the nodal voltages.

FIGURE 4.1.6   Network with two VC.

The equations are

$$6V_1 - 1V_2 - 2V_3 - g_1V_2 = J$$
$$-1V_1 + 4V_2 - 3V_3 + g_2V_1 = 0$$
$$-2V_1 - 3V_2 + 7V_3 = 0$$

Inserting the values for g_1 and g_2 we get

$$\begin{bmatrix} 6 & -3 & -2 \\ 1 & 4 & -3 \\ -2 & -3 & 7 \end{bmatrix} \begin{bmatrix} V_1 \\ V_2 \\ V_3 \end{bmatrix} = \begin{bmatrix} 1 \\ 0 \\ 0 \end{bmatrix}$$

The system determinant is $D = 107$ and the numerators are $N_1 = 19$, $N_2 = -1$ and $N_3 = 5$. The voltages are $V_1 = 19/107$, $V_2 = -1/107$ and $V_3 = 5/107$.

If you have access to an IBM-PC computer, use the program from Appendix A to confirm the above numerical solutions. Do so also for the other examples in this Chapter.

4.2. CURRENT-CONTROLLED VOLTAGE SOURCE

The current-controlled voltage source, CV, is the second of the four possible controlled sources. It is easy to incorporate into the mesh equations and is shown in Fig. 4.2.1. The input terminals on the left are connected by a short circuit and a current I flows through this connection from node i to node j. It is obvious that the voltages of the two input terminals are equal,

$$V_i - V_j = 0 \qquad\qquad (4.2.1)$$

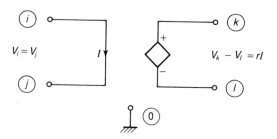

FIGURE 4.2.1   Current-controlled voltage source, CV.

and the currents flowing away from the terminals i and j are

$$I_i = I$$
$$I_j = -I$$

(4.2.2)

A voltage source is connected between the output terminals k and l, its voltage depending on the amount of current flowing through the short circuit of the input terminals. Because the voltages are measured with respect to ground, the equation describing the output properties of the CV is

$$V_k - V_l = rI$$

(4.2.3)

The constant r is called the *transresistance* and is measured in ohms. If no current flows through the controlling short circuit, the voltage difference of the controlled terminals is zero and the dependent voltage source behaves as a short circuit. We recall that an independent voltage source delivering zero voltage also behaves as a short circuit.

Consider the network in Fig. 4.2.2. Because all sources, dependent and independent, are voltages, we can use the mesh method. The current I in the first mesh controls the voltage of the CV on the right. In the first step we realize that $I = I_1$ and we write this into the figure. The KVL equation for the first mesh is

$$R_1 I_1 + R_2(I_1 - I_2) - E = 0$$

and for the second

$$R_2(I_2 - I_1) + R_3 I_2 - rI_1 = 0$$

The dependent source is handled exactly the same way as an independent source, with the only difference that it will stay on the left side. Rearranging the terms and transferring the independent voltage E on the right side we get

$$(R_1 + R_2)I_1 - R_2 I_2 = E$$
$$(-R_2 - r)I_1 + (R_2 + R_3)I_2 = 0$$

(4.2.4)

FIGURE 4.2.2   Network with one CV.

If we use numerical values for the elements, the equations become

$$3I_1 - 2I_2 = 1$$
$$-4I_1 + 5I_2 = 0$$

or in matrix form

$$\begin{bmatrix} 3 & -2 \\ -4 & 5 \end{bmatrix} \begin{bmatrix} I_1 \\ I_2 \end{bmatrix} = \begin{bmatrix} 1 \\ 0 \end{bmatrix}$$

Using matrix algebra we find $I_1 = 5/7$ and $I_2 = 4/7$.

Example 1.

For the network in Fig. 4.2.3a write the mesh equations, calculate the mesh currents and the output voltage.

The controlling current, indicated in the short circuit in the middle, is equal to $I = I_1 - I_2$ (because both I and I_1 are in the same direction while I_2 has opposite direction). This expression replaces the current I in Fig. 4.2.3b for which we write the mesh equations

$$3I_1 = E$$
$$3I_2 - 3I = 0$$

If we now substitute for I, we get the equations

$$3I_1 = E$$
$$-3I_1 + 6I_2 = 0$$

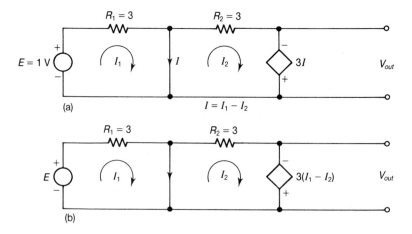

(a)

$I = I_1 - I_2$

(b)

FIGURE 4.2.3   Network with one CV.

or

$$\begin{bmatrix} 3 & 0 \\ -3 & 6 \end{bmatrix}\begin{bmatrix} I_1 \\ I_2 \end{bmatrix} = \begin{bmatrix} 1 \\ 0 \end{bmatrix}$$

The result is $D = 18$, $N_1 = 6$, $N_2 = 3$ and the currents are $I_1 = 6/18$ and $I_2 = 3/18$. The output voltage is taken across the dependent source and is equal to

$$V_{out} = -3(I_1 - I_2) = -\frac{1}{2}$$

Example 2.

In Fig. 4.2.4 calculate the voltage across the resistor R_2 using the mesh equations. We first note that $I = I_1$.
    The equations are

$$(R_1 + R_3)I_1 - R_3 I_2 = E$$
$$-R_3 I_1 + (R_2 + R_3)I_2 + 2I = 0$$

Inserting numerical values

$$\begin{bmatrix} 2 & -1 \\ 1 & 3 \end{bmatrix}\begin{bmatrix} I_1 \\ I_2 \end{bmatrix} = \begin{bmatrix} 1 \\ 0 \end{bmatrix}$$

The solutions are $D = 7$, $N_1 = -1$, $N_2 = -2$ and the voltage across R_2 is $V_{out} = -4/7$.

Example 3.

The network in Fig. 4.2.5 has three meshes. Calculate the indicated voltage V_{out}.
    In the first step we realize that the controlling current is equal to $I = I_2 - I_3$. In addition we note that all we need to calculate is the current I_3 and add the voltage across the resistor

FIGURE 4.2.4   Network with one CV.

FIGURE 4.2.5   Network with two sources and one CV.

R_4 to the voltage of the independent source E_2. The mesh equations are

$$I_1 + 2I = 2$$
$$-2I + 2I_2 - I_3 = 0$$
$$-I_2 + 2I_3 = -1$$

Inserting for the current I we get the system

$$\begin{bmatrix} 1 & 2 & -2 \\ 0 & 0 & 1 \\ 0 & -1 & 2 \end{bmatrix} \begin{bmatrix} I_1 \\ I_2 \\ I_3 \end{bmatrix} = \begin{bmatrix} 2 \\ 0 \\ -1 \end{bmatrix}$$

The third numerator is obtained as

$$N_3 = \begin{vmatrix} 1 & 2 & 2 \\ 0 & 0 & 0 \\ 0 & -1 & -1 \end{vmatrix}$$

Since the second and third column are identical, the determinant is zero and I_3 is zero. The output voltage will be equal to the voltage $E_2 = 1$ V.

Example 4.

Consider Fig. 4.2.6 and find the voltage V shown in the figure.

For the solution we first realize that the controlling current $I = -I_2$. The mesh equations are

$$3I_1 - 2I_2 + 3I = 2$$
$$-2I_1 + 5I_2 - 3I = -4$$

Substituting for I and inserting numerical values

$$\begin{bmatrix} 3 & -5 \\ -2 & 8 \end{bmatrix} \begin{bmatrix} I_1 \\ I_2 \end{bmatrix} = \begin{bmatrix} 2 \\ -4 \end{bmatrix}$$

FIGURE 4.2.6   Network with two sources and one CV.

The results are $D = 14$, $N_1 = -4$, $N_2 = -8$ and the currents are $I_1 = -4/14$ and $I_2 = -8/14$. The voltage V can be calculated in several ways; the simplest is to take the voltage of E_1 and subtract the voltage across the resistor R_1. The result is

$$V = E_1 - 1I_1 = 2 + \frac{4}{14} = \frac{32}{14} \text{ V}$$

We can obtain the same result by taking $V = E_2 + R_3 I_2$.

A more complicated network is in Fig. 4.2.7a. We will show that a small amount of preprocessing makes the problem as simple as the above ones. The network has two current-controlled voltage sources; the controlling currents are denoted by I_A and I_B, the controlled voltage sources by V_A and V_B. First we realize that $I_A = I_1 - I_2$ and $I_B = I_2$; we write these values to the controlled sources, as shown in Fig. 4.2.7b. Once this has been done, all the necessary information has been transferred to the dependent sources and we can write the mesh equations. For the first mesh

$$(R_1 + R_2)I_1 - R_2 I_2 - R_1 I_3 = E$$

For the second

$$-R_2 I_1 + (R_2 + R_3 + R_4)I_2 - R_3 I_3 + r_A(I_1 - I_2) = 0$$

and for the third

$$-R_1 I_1 - R_3 I_2 + (R_1 + R_3 + R_5)I_3 - r_A(I_1 - I_2) - r_B I_2 = 0$$

The terms multiplied by the same currents are collected:

$$(R_1 + R_2)I_1 - R_2 I_2 - R_1 I_3 = E$$
$$(-R_2 + r_A)I_1 + (R_2 + R_3 + R_4 - r_A)I_2 - R_3 I_3 = 0$$
$$(-R_1 - r_A)I_1 + (-R_3 + r_A - r_B)I_2 + (R_1 + R_3 + R_5)I_3 = 0$$

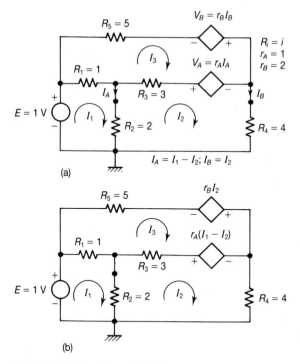

(a)

(b)

FIGURE 4.2.7   Network with two CV: (a) Original. (b) With information transferred to the dependent source.

and the equations are now in their final form. If we substitute numerical values ($R_i = i$, $r_A = 1$, $r_B = 2$, $E = 1$), we will get the system

$$3I_1 - 2I_2 - I_3 = 1$$
$$-I_1 + 8I_2 - 3I_3 = 0$$
$$-2I_1 - 4I_2 + 9I_3 = 0$$

In matrix form

$$\begin{bmatrix} 3 & -2 & -1 \\ -1 & 8 & -3 \\ -2 & -4 & 9 \end{bmatrix} \begin{bmatrix} I_1 \\ I_2 \\ I_3 \end{bmatrix} = \begin{bmatrix} 1 \\ 0 \\ 0 \end{bmatrix}$$

The solution is $I_1 = 60/130$, $I_2 = 15/130$ and $I_3 = 20/130$.

4.3. CURRENT-CONTROLLED CURRENT SOURCE

The current-controlled current source, CC, is shown in Fig. 4.3.1. The input terminals on the left are connected by a short circuit which senses the amount of current, I, flowing through it. The voltages at both input terminals are equal,

$$V_i - V_j = 0 \qquad (4.3.1)$$

The currents flowing away from the terminals i and j are

$$I_i = I$$
$$I_j = -I \qquad (4.3.2)$$

The output of the CC is formed by a current source connected between terminals k and l. The dependent source current is proportional to the current I of the input short circuit. The proportionality constant is usually denoted as α; it is dimensionless. The currents at the output terminals are

$$I_k = \alpha I$$
$$I_l = -\alpha I \qquad (4.3.3)$$

If no current flows through the controlling short circuit, the element does not deliver any current at its output, and behaves as the open circuit between the terminals k and l. Because the output of the CC is a current, it can be handled by the nodal formulation if we do some preprocessing steps.

Consider the network in Fig. 4.3.2a. The input terminals of the CC are indicated by the short circuit marked by the arrow and by I_A. The output of the CC is current αI_A. The controlling current I_A is equal to the current flowing through the conductance G_2. We can express this current as $I_A = G_2(V_1 - V_2)$ and write it into the network, see Fig. 4.3.2b. We are now ready to write the nodal equations. For

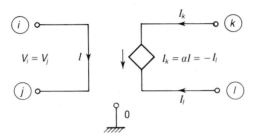

FIGURE 4.3.1   Current-controlled current source, CC.

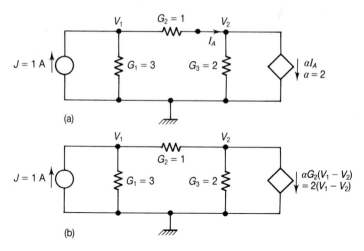

FIGURE 4.3.2   Network with one CC: (a) Original. (b) With information transferred to the dependent source.

node 1 the KCL is

$$G_1 V_1 + G_2(V_1 - V_2) - J = 0$$

and for node 2

$$G_2(V_2 - V_1) + G_3 V_2 + \alpha G_2(V_1 - V_2) = 0$$

The equations are rearranged

$$(G_1 + G_2)V_1 - G_2 V_2 = J$$
$$-G_2(1 - \alpha)V_1 + [G_3 + G_2(1 - \alpha)]V_2 = 0$$

and written in matrix form

$$\begin{bmatrix} (G_1 + G_2) & -G_2 \\ -G_2(1 - \alpha) & [G_3 + G_2(1 - \alpha)] \end{bmatrix} \begin{bmatrix} V_1 \\ V_2 \end{bmatrix} = \begin{bmatrix} J \\ 0 \end{bmatrix}$$

Using Cramer's rule we find the determinant of the matrix

$$D = G_1 G_2(1 - \alpha) + G_1 G_3 + G_2 G_3$$

At this moment, it is worth noting one important property of determinants. Intermediate steps lead to two terms $G_2^2(1 - \alpha)$, one with + and one with − sign; they cancel. We will remember that as long as we give every element a different subscript (or name), then it is *never* possible that the result would contain some element raised to power more than one. Should such a term remain in your result, you can be sure that you have made a mistake somewhere.

Using Cramer's rule we find the numerator

$$N_1 = J[G_3 + G_2(1 - \alpha)]$$

and the voltage V_1 is equal to

$$V_1 = \frac{N_1}{D} = \frac{J[G_3 + G_2(1 - \alpha)]}{G_1 G_2(1 - \alpha) + G_1 G_3 + G_2 G_3}$$

We find similarly

$$N_2 = J G_2(1 - \alpha)$$

and the voltage is

$$V_2 = \frac{N_2}{D} = \frac{J G_2(1 - \alpha)}{G_1 G_2(1 - \alpha) + G_1 G_3 + G_2 G_3}$$

Example 1.

Find the nodal voltages for the network in Fig. 4.3.3.

In this problem we first realize that the controlling current $I = G_3 V_2 = 4V_2$. We can now write the nodal equations

$$(G_1 + G_2)V_1 - G_2 V_2 - 12 V_2 = J_1$$
$$-G_2 V_1 + (G_2 + G_3)V_2 = J_2$$

Using numerical values

$$\begin{bmatrix} 3 & -13 \\ -1 & 5 \end{bmatrix} \begin{bmatrix} V_1 \\ V_2 \end{bmatrix} = \begin{bmatrix} 1 \\ 3 \end{bmatrix}$$

and the results are $D = 2$, $N_1 = 44$, $N_2 = 10$. The nodal voltages are $V_1 = 22$ and $V_2 = 5$.

FIGURE 4.3.3   Network with two sources and one *CC*.

FIGURE 4.3.4   Network with two sources and one *CC*.

Example 2.

Find the nodal voltages for the network in Fig. 4.3.4.

In the first step we realize that the controlling current $I = (V_1 - V_2)G_2$. The nodal equations are

$$(G_1 + G_2)V_1 - G_2 V_2 + \alpha I = J_1$$
$$-G_2 V_1 + (G_2 + G_3)V_2 - \alpha I = J_2$$

Inserting numerical values

$$\begin{bmatrix} 9 & -6 \\ -6 & 10 \end{bmatrix}\begin{bmatrix} V_1 \\ V_2 \end{bmatrix} = \begin{bmatrix} 2 \\ 3 \end{bmatrix}$$

The results are $D = 54$, $N_1 = 38$, $N_2 = 39$ and the voltages are $V_1 = 38/54$ and $V_2 = 39/54$.

The next network in Fig. 4.3.5a has three nodes and three current-controlled current sources. All conductances have, for simplicity, their values equal to the subscript. The controlling currents are denoted I_A, I_B, I_C and are marked in the figure by arrows. The dependent sources are marked as $1I_A$, $2I_B$ and $3I_C$. Looking at the figure we see that the controlling currents are

$$I_A = G_3 V_2 = 3V_2$$
$$I_B = G_4(V_2 - V_3) = 4V_2 - 4V_3$$
$$I_C = G_2(V_1 - V_2) = 2V_1 - 2V_2$$

The network is redrawn in Fig. 4.3.5b. In this figure the controlling currents are omitted because the information has been transferred to the controlled sources. All sources in the network are current sources and we use nodal formulation. For

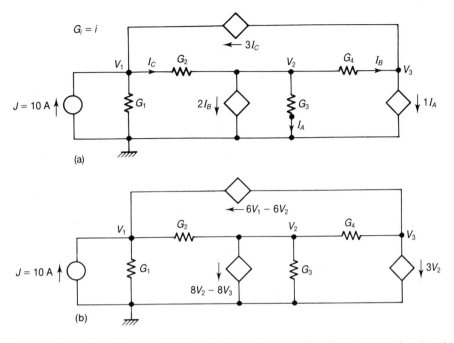

FIGURE 4.3.5   Network with three CC: (a) Original. (b) With information transferred to the dependent sources.

the first node:

$$3V_1 - 2V_2 - (6V_1 - 6V_2) = 10$$

For the second node:

$$-2V_1 + 9V_2 - 4V_3 + (8V_2 - 8V_3) = 0$$

and for the third node:

$$-4V_2 + 4V_3 + (3V_2) + (6V_1 - 6V_2) = 0$$

In matrix form

$$\begin{bmatrix} -3 & 4 & 0 \\ -2 & 17 & -12 \\ 6 & -7 & 4 \end{bmatrix} \begin{bmatrix} V_1 \\ V_2 \\ V_3 \end{bmatrix} = \begin{bmatrix} 10 \\ 0 \\ 0 \end{bmatrix}$$

The determinant of the system is $D = -208$, the numerators are $N_1 = -160$, $N_2 = -640$ and $N_3 = -880$. This makes the voltages $V_1 = N_1/D = 0.7692$ V, $V_2 = N_2/D = 3.0769$ V and $V_3 = N_3/D = 4.2308$ V.

Example 3.

Find nodal voltages for the network in Fig. 4.3.6a.

The network has one current-controlled current source and one voltage-controlled current source. First we realize that the current I is given by the expression $I = G_2(V_1 - V_2)$. Although it is not necessary to redraw the network, we will do so for clarification, as shown in Fig. 4.3.6b. As before, the controlling current is now omitted because the information has been transferred to the dependent source. The equations are

$$3V_1 - 2V_2 + 10V_1 - 10V_2 = 1$$
$$-2V_1 + 9V_2 - 4V_3 = 0$$
$$-4V_2 + 9V_3 - 10V_1 + 10V_2 + 3V_1 - 3V_2 = 0$$

Collecting terms

$$\begin{bmatrix} 13 & -12 & 0 \\ -2 & 9 & -4 \\ -7 & 3 & 9 \end{bmatrix} \begin{bmatrix} V_1 \\ V_2 \\ V_3 \end{bmatrix} = \begin{bmatrix} J \\ 0 \\ 0 \end{bmatrix}$$

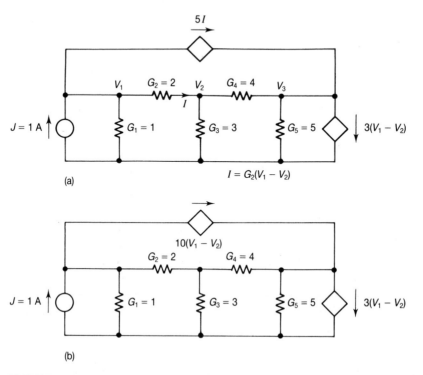

(a)

(b)

FIGURE 4.3.6   Network with one CC and one VC: (a) Original. (b) With information transferred to the dependent sources.

The solutions are $D = 657$, $N_1 = 93$, $N_2 = 46$, $N_3 = 57$ and the voltages are the appropriate ratios.

4.4. VOLTAGE-CONTROLLED VOLTAGE SOURCE

The voltage-controlled voltage source, VV, is shown in Fig. 4.4.1. Its input terminals sense the difference of the voltages at nodes i and j and the output is formed by a voltage source whose voltage is μ times the difference $V_i - V_j$. The amplification constant μ is dimensionless. The currents at the input terminals are zero,

$$I_i = I_j = 0 \qquad (4.4.1)$$

The output voltage, taken between nodes k and l, is

$$V_k - V_l = \mu(V_i - V_j) \qquad (4.4.2)$$

If the voltage difference $V_i - V_j = 0$, the output voltage is zero and the voltage source becomes a short circuit between terminals k and l. Because the VV is a special type of a voltage source, it can be handled by the mesh formulation if we do some preprocessing steps.

A small example, Fig. 4.4.2a, will clarify how we should proceed. The network has two meshes and we have already indicated the clockwise current directions by arrows. The voltage V_A across the resistor R_1 indicates the input terminals of the VV; this voltage, multiplied by μ, appears as a voltage source on the right of the figure. Since only the current I_1 flows through R_1, the controlling voltage is equal to $V_A = R_1 I_1$. We have redrawn the network in Fig. 4.4.2b and transferred this information to the dependent source. This preprocessing, written into the figure, helps us write the mesh equations directly by using the rules we learned in Chapter 2. For the first mesh we have

$$(R_1 + R_2)I_1 - R_2 I_2 = E$$

FIGURE 4.4.1   Voltage-controlled voltage source, VV.

(a)

(b)

FIGURE 4.4.2   Network with one VV: (a) Original. (b) With information transferred to the dependent source.

and for the second

$$-R_2I_1 + (R_2 + R_3)I_2 + \mu R_1 I_1 = 0$$

As always, we put the equations into matrix form

$$\begin{bmatrix} R_1 + R_2 & -R_2 \\ \mu R_1 - R_2 & R_2 + R_3 \end{bmatrix} \begin{bmatrix} I_1 \\ I_2 \end{bmatrix} = \begin{bmatrix} E \\ 0 \end{bmatrix}$$

The determinant is

$$D = R_1 R_2 (1 + \mu) + R_1 R_3 + R_2 R_3$$

and the currents are

$$I_1 = \frac{E(R_2 + R_3)}{R_1 R_2 (1 + \mu) + R_1 R_3 + R_2 R_3}$$

$$I_2 = \frac{E(R_2 - \mu R_1)}{R_1 R_2 (1 + \mu) + R_1 R_3 + R_2 R_3}$$

$$V = (I_1 - I_2)R_2$$

(a)

(b)

FIGURE 4.4.3   Network with one VV: (a) Original. (b) With information transferred to the dependent sources.

Example 1.

The network in Fig. 4.4.3a has one voltage-controlled voltage source with the controlling voltage taken across R_2. Find the output voltage by using the mesh method.

For the solution we realize that the controlling voltage $V = R_2(I_1 - I_2)$. To simplify the steps we transfer this information into a redrawn network Fig. 4.4.3b. The mesh equations are now

$$5I_1 - 3I_2 = 10$$
$$-3I_1 + 8I_2 + 9I_1 - 9I_2 = 0$$

or

$$\begin{bmatrix} 5 & -3 \\ 6 & -1 \end{bmatrix} \begin{bmatrix} I_1 \\ I_2 \end{bmatrix} = \begin{bmatrix} 10 \\ 0 \end{bmatrix}$$

For the output voltage we need I_2. The determinant is $D = 13$, $N_2 = -60$ and the output voltage is

$$V_{out} = 5 \cdot \left(-\frac{60}{13} \right) = -\frac{300}{13}$$

Example 2.

The network in Fig. 4.4.4a has two voltage-controlled voltage sources. Find the voltage at the output.

FIGURE 4.4.4   Network with two VV: (a) Original. (b) With information transferred to the dependent sources.

We first determine the expressions for the controlling voltages: $V_A = I_2 R_3 = 4I_2$, $V_B = R_1 I_1 = 2I_1$. The network is redrawn in Fig. 4.4.4b with the information transferred to the dependent sources. The mesh equations are

$$(R_1 + R_2)I_1 - R_2 I_2 + 12I_2 = E$$
$$-R_2 I_1 + (R_2 + R_3)I_2 + 4I_1 - 12I_2 = 0$$

Inserting numerical values

$$\begin{bmatrix} 5 & 9 \\ 1 & -5 \end{bmatrix} \begin{bmatrix} I_1 \\ I_2 \end{bmatrix} = \begin{bmatrix} 1 \\ 0 \end{bmatrix}$$

For the output voltage we need I_1. The determinant is $D = -34$, $N_1 = -5$ and the current is $I_1 = 5/34$. The output voltage is $V_{out} = 4I_1 = 20/34$.

A network with three meshes is in Fig. 4.4.5a. It has two voltage-controlled voltage sources and the controlling voltages are V_A and V_B. In order to simplify insertions, the resistors are given values equal to their subscripts. The amplification constants were chosen to be $\mu_A = 2$ and $\mu_B = 3$. The mesh currents are indicated in the figure.

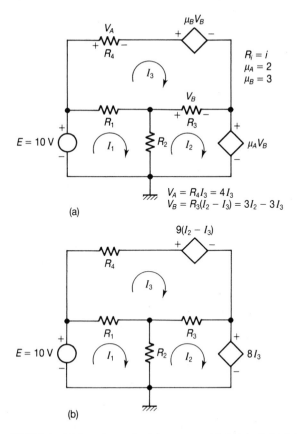

FIGURE 4.4.5   Network with two VV: (a) Original. (b) With information transferred to the dependent sources.

Similarly as in the above example, we first determine the controlling voltages in terms of the elements and the mesh currents. They are

$$V_A = R_4 I_3 = 4I_3$$
$$V_B = R_3(I_2 - I_3) = 3I_2 - 3I_3$$

These values are transferred into Fig. 4.4.5b. The figure is now prepared for the mesh equations:

$$3I_1 - 2I_2 - I_3 = 10$$
$$-2I_1 + 5I_2 - 3I_3 + (8I_3) = 0$$
$$-I_1 - 3I_2 + 8I_3 + (9I_2 - 9I_3) = 0$$

Terms in brackets indicate the dependent voltage sources. Rearranging the terms we get

$$
\begin{bmatrix}
3 & -2 & -1 \\
-2 & 5 & 5 \\
-1 & 6 & -1
\end{bmatrix}
\begin{bmatrix}
I_1 \\
I_2 \\
I_3
\end{bmatrix}
=
\begin{bmatrix}
10 \\
0 \\
0
\end{bmatrix}
$$

Cramer's rule provides $D = -84$, $N_1 = -350$, $N_2 = N_3 = -70$ and the currents are $I_1 = 4.16667$, $I_2 = I_3 = 0.83333$.

Example 3.

The network in Fig. 4.4.6a has three meshes, two independent voltage sources and two voltage-controlled voltage sources. Find the nodal voltage V. We first define the voltages V_A and V_B in terms of the mesh currents. They are

$$V_A = (I_2 - I_3)R_4 = 2I_2 - 2I_3$$
$$V_B = (I_1 - I_2)R_2 = 3I_1 - 3I_2$$

The values are transferred to the dependent sources, as given in Fig. 4.4.6b. The mesh

(a)

(b)

FIGURE 4.4.6   Network with two VV: (a) Original. (b) With information transferred to the dependent sources.

equations are

$$(R_1 + R_3)I_1 - R_3I_2 + 6(I_2 - I_3) = 1$$
$$-R_2I_1 + (R_2 + R_3 + R_4)I_2 - R_4I_3 - 6(I_2 - I_3) - 3(I_1 - I_2) = 0$$
$$-R_4I_2 + (R_4 + R_5)I_3 = -2$$

Inserting values:

$$\begin{bmatrix} 6 & 3 & -6 \\ -6 & 3 & 6 \\ 0 & -2 & 4 \end{bmatrix} \begin{bmatrix} I_1 \\ I_2 \\ I_3 \end{bmatrix} = \begin{bmatrix} 1 \\ 0 \\ -2 \end{bmatrix}$$

To obtain the indicated voltage we need I_1 and $V = E_1 - R_1I_1$. The determinant of the system matrix is $D = 144$, $N_1 = -48$ and $I_1 = -1/3$. The voltage

$$V = 1 - 3 \cdot \left(-\frac{1}{3}\right) = 2$$

4.5. SUMMARY OF DEPENDENT SOURCES

This chapter explained all possible cases of dependent sources; they are very important because they are used to model properties of many practical devices, like transistors. We now summarize our development, point out restrictions and indicate how the restrictions can be removed.

Our present knowledge shows that current sources (controlled or independent) are more suitable for nodal formulation, but only the voltage-controlled current source can be used without any additional steps. If we wish to use nodal formulation for a network with current-controlled current sources, we must first express the controlling currents in terms of nodal voltages.

Voltage sources (controlled or independent) are suitable for mesh formulation but only the current-controlled voltage sources can be used without modifications. If we wish to use mesh formulation for a network with a voltage-controlled voltage source, we must first express the controlling voltage in terms of mesh currents.

These restrictions are quite serious because we cannot yet analyze networks with a mixture of current and voltage sources. There are two ways out of this problem. The easier one, to be explained in the next chapter, applies some transformations and redrawings of the networks. A better one, but requiring more knowledge, uses a method which is neither nodal nor mesh. It is called the modified nodal formulation and is explained in Chapter 17.

A student having access to an IBM-PC computer should now start using the NETOPT program for the solution of linear networks. Description on how to use it is in Chapter 18.

PROBLEMS CHAPTER 4

All problems of this Chapter can be solved by the program NETOPT; for instruction see Chapter 18.

P.4.1 Use mesh equations to find the currents.

P.4.2 Use mesh equations to find the currents.

P.4.3 Use mesh equations to find the currents.

P.4.4 Use mesh equations to find the currents.

P.4.5 Use nodal formulation to get the nodal voltages.

P.4.6 Use nodal formulation to get the nodal voltages.

P.4.7 Use mesh formulation to get the currents.

P.4.8 Use mesh equations to get the currents.

P.4.9 Use nodal equations to get the nodal voltages.

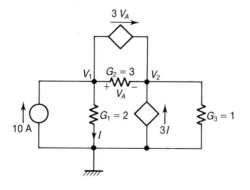

P.4.10 Use nodal equations to get the nodal voltages.

P.4.11 Use nodal equations to get the output voltage V_{out}.

P.4.12 Find the output voltage V_{out}.

P.4.13 Find the output voltage V_{out}.

P.4.14 Find the output voltage V_{out}.

P.4.15 Find the output voltage V_{out}.

P.4.16 Find the output voltage V_{out}.

P.4.17 Find the two nodal voltages.

P.4.18 Find the two nodal voltages.

P.4.19 Find the two nodal voltages.

P.4.20 Find the two nodal voltages.

P.4.21 Find the two nodal voltages.

P.4.22 Find the output voltage.

P.4.23 Find the output voltage.

P.4.27 Calculate the output voltage.

P.4.24 Calculate the output voltage.

P.4.28 Calculate the output voltage.

P.4.25 Calculate the output voltage.

P.4.29 Calculate the output voltage.

P.4.26 Calculate the output voltage.

P.4.30 Calculate the output voltage.

P.4.31 Calculate the output voltage.

P.4.32 Calculate the output voltage.

CHAPTER 5

Network Transformations

5.0. INTRODUCTION

We now know most of the elements which can be used in resistive networks. We also know two analysis methods (nodal and mesh) and how to write the network equations, but there are cases where our knowledge is not sufficient. We still do not know how to handle situations when we have a mixture of voltage and current sources.

This chapter explains several important concepts: transformation of sources, splitting of sources and transformation of networks by the Thévenin and Norton theorems. It also demonstrates the law of superposition, valid for any linear network, and the concept of input and output resistance.

Until now, we have treated the nodal and mesh methods as equally important; they are not. The last section of this chapter gives important recommendations which we follow in the rest of this book.

If you have access to an IBM-PC computer, use the program NETOPT to check all problems of this chapter (as well as the following ones). Instructions relating to the material covered so far are in Chapter 18, Sections 1, 2 and 3.

5.1. TRANSFORMATION OF SOURCES

All the previous examples that we dealt with had only one type of source. If they were current sources, we used nodal formulation. If they were voltage sources, we used mesh formulation. This situation does not occur in most practical cases and we must find a way to incorporate all kinds of sources (dependent or independent) into either of the two formulations.

One step which helps in this direction is the transformation of voltage sources into current sources and vice versa. In this section we will assume that each voltage source is connected in series with a resistor or each current source in parallel with the same resistor. The two cases are sketched in Fig. 5.1.1. On the left is a voltage source with the resistor R_S in series; the combination is connected to a load resistor R_L. On the right is a current source with the same resistor in parallel and with the same load resistor. We wish to find a transformation by which we can transform either of the two cases into the other one. In both cases, the same current must flow into the loading resistor, R_L, and the same voltage must appear across it.

Consider first the network on the left. It is a voltage divider and the voltage V is equal to

$$V = \frac{R_L}{R_S + R_L} E \qquad (5.1.1)$$

In the case of a current source, the two resistors are in parallel and we add their conductances to get the overall conductance,

$$G = \frac{1}{R_S} + \frac{1}{R_L} = \frac{R_S + R_L}{R_S R_L} \qquad (5.1.2)$$

Using nodal formulation for just one node, we get for the network on the right

$$\frac{R_S + R_L}{R_S R_L} V = J \qquad (5.1.3)$$

Because V must be equal in both situations, we substitute Eq. (5.1.1) into (5.1.3) and get $E = J R_S$. This looks very much like Ohm's law but it actually expresses transformation of sources. We indicate this fact by writing the subscript EQ into

FIGURE 5.1.1   Transformation of sources.

the following two equations

$$E_{EQ} = J_{EQ} R_S$$
$$J_{EQ} = E_{EQ} G_S$$

(5.1.4)

but we will not use the subscript later on. The transformation can be applied only in cases where a resistor is in series with the voltage source or in parallel with the current source.

We apply the transformation to the network in Fig. 5.1.2a. All resistor values are in ohms and for simplicity set equal to the subscript of the resistor. Figure 5.1.2b shows the transformation of the voltage source into a current source and we use the nodal formulation:

$$\begin{bmatrix} G_1 + G_2 + G_3 & -G_3 \\ -G_3 & G_3 + G_4 \end{bmatrix} \begin{bmatrix} V_1 \\ V_2 \end{bmatrix} = \begin{bmatrix} J_{EQ} \\ J \end{bmatrix}$$

Substituting numerical values

$$\begin{bmatrix} 11/6 & -1/3 \\ -1/3 & 7/12 \end{bmatrix} \begin{bmatrix} V_1 \\ V_2 \end{bmatrix} = \begin{bmatrix} 2 \\ 1 \end{bmatrix}$$

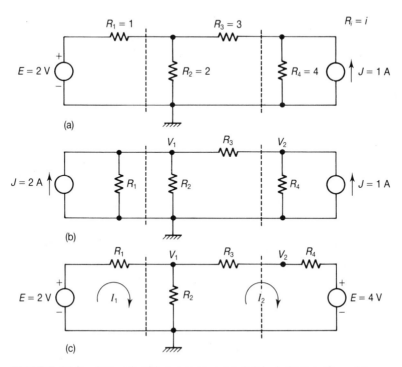

FIGURE 5.1.2   Network with two sources: (a) Original. (b) Transformed to current sources. (c) Transformed to voltage sources.

The determinant of the matrix is $D = 23/24$ and the voltages are $V_1 = 36/23 = 1.5652$ and $V_2 = 60/23 = 2.6087$.

We could also transform the current source on the right of Fig. 5.2.1a into a voltage source and get the network in Fig. 5.1.2c. Use of the mesh method leads to

$$\begin{bmatrix} 3 & -2 \\ -2 & 9 \end{bmatrix}\begin{bmatrix} I_1 \\ I_2 \end{bmatrix} = \begin{bmatrix} 2 \\ -4 \end{bmatrix}$$

The solution is $I_1 = 10/23$ and $I_2 = -8/23$. The negative sign indicates that the current I_2 actually flows the other way. We can confirm that the two results are equivalent. For instance, we can calculate the voltage V_1 either as $V_1 = E - R_1 I_1$ or as $V_1 = R_2(I_1 - I_2)$. In both cases we get $V_1 = 1.5652$.

Example 1.

A somewhat more complicated situation is in Fig. 5.1.3a where we first transform the voltage source into the current source, Fig. 5.1.3b, and then use the nodal formulation. The matrix equation is

$$\begin{bmatrix} G_1 + G_2 + G_3 & -G_3 + g_1 \\ -G_3 + g_2 & G_3 + G_4 - g_2 \end{bmatrix}\begin{bmatrix} V_1 \\ V_2 \end{bmatrix} = \begin{bmatrix} EG_1 \\ 0 \end{bmatrix}$$

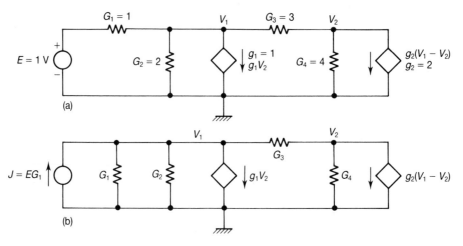

FIGURE 5.1.3   Network with dependent sources: (a) Original. (b) Transformed for nodal formulation.

Substituting numerical values

$$\begin{bmatrix} 6 & -2 \\ -1 & 5 \end{bmatrix}\begin{bmatrix} V_1 \\ V_2 \end{bmatrix} = \begin{bmatrix} 1 \\ 0 \end{bmatrix}.$$

The determinant is $D = 28$ and the voltages are $V_1 = 5/28$, $V_2 = 1/28$.

The transformations can be applied equally to dependent and independent sources. We will use the network in Fig. 5.1.4a to demonstrate transformation of a dependent current source into a dependent voltage source. Using (5.1.4) we redraw the network as shown in Fig. 5.1.4b. Note that these transformations did not modify the part of the network where we sense the controlling current of the CC and the controlling voltage of the VV. After this transformation we realize that the controlling current of the CC is $I = I_1 - I_2$ and that the controlling voltage of the VV is $V_1 = R_2(I_1 - I_2)$. We write these values into the network in Fig. 5.1.4c. All the necessary information has now been transferred into the last figure for which we write the mesh equations:

$$(R_1 + R_2)I_1 - R_2I_2 = E$$

$$(\mu R_2 - R_2 - \alpha R_4)I_1 + (R_2 - \mu R_2 + R_3 + R_4 + \alpha R_4)I_2 = 0$$

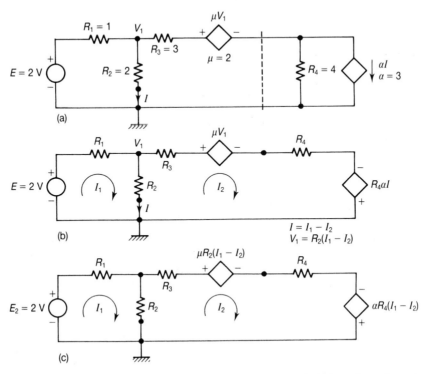

FIGURE 5.1.4   Network with dependent sources: (a) Original. (b) Transformed to voltage sources. (c) Final form for mesh formulation.

Collecting terms multiplied by the same currents

$$
\begin{bmatrix} (R_1 + R_2) & -R_2 \\ (-R_2 + \mu R_2 - \alpha R_4) & (R_2 - \mu R_2 + R_3 + R_4 + \alpha R_4) \end{bmatrix} \begin{bmatrix} I_1 \\ I_2 \end{bmatrix} = \begin{bmatrix} E \\ 0 \end{bmatrix}
$$

Substituting numerical values

$$
\begin{bmatrix} 3 & -2 \\ -10 & 17 \end{bmatrix} \begin{bmatrix} I_1 \\ I_2 \end{bmatrix} = \begin{bmatrix} 2 \\ 0 \end{bmatrix}
$$

The result is $I_1 = 34/31$ and $I_2 = 20/31$.

Example 2.

In this example, Fig. 5.1.5a, we transform the voltage sources into current sources. The VV on the right is transformed into a current source in parallel with G_5. The voltage source on the left is transformed into a current source in parallel with G_1. We note again that the transformations did not modify the part of the network where we sense the current of the VV and CC. We express the current I in terms of the voltage V_2, $I = G_4 V_2$ and write it instead of I. All these transformations are transferred into Fig. 5.1.5b for which we write

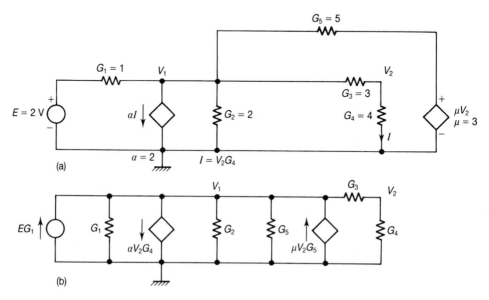

FIGURE 5.1.5   Network with voltage and current sources: (a) Original. (b) Transformed for nodal formulation.

the nodal equations

$$(G_1 + G_2 + G_3 + G_5)V_1 - G_3V_2 + \alpha V_2G_4 - \mu V_2 G_5 = EG_1$$
$$-G_3V_1 + (G_3 + G_4)V_2 = 0$$

or

$$\begin{bmatrix} (G_1 + G_2 + G_3 + G_5) & (-G_3 + \alpha G_4 - \mu G_5) \\ -G_3 & G_3 + G_4 \end{bmatrix}\begin{bmatrix} V_1 \\ V_2 \end{bmatrix} = \begin{bmatrix} EG_1 \\ 0 \end{bmatrix}$$

Using numerical values

$$\begin{bmatrix} 11 & -10 \\ -3 & 7 \end{bmatrix}\begin{bmatrix} V_1 \\ V_2 \end{bmatrix} = \begin{bmatrix} 2 \\ 0 \end{bmatrix}$$

The solution is $V_1 = 14/47$ and $V_2 = 6/47$.

Example 3.

Apply both transformations to the network in Fig. 5.1.6a. First we transform the voltage sources into current sources, as shown in Fig. 5.1.6b. Resistor values must be converted

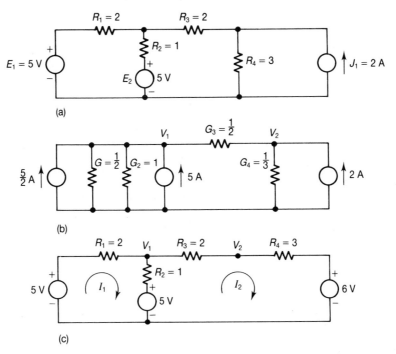

(a)

(b)

(c)

FIGURE 5.1.6 Network with mixed sources: (a) Original. (b) Transformed for nodal formulation. (c) Transformed for mesh formulation.

into conductances and the two nodal equations, in matrix form, are

$$\begin{bmatrix} 2 & -1/2 \\ -1/2 & 5/6 \end{bmatrix}\begin{bmatrix} V_1 \\ V_2 \end{bmatrix} = \begin{bmatrix} 15/2 \\ 2 \end{bmatrix}$$

The solutions are $D = 17/12$, $N_1 = 29/4$, $N_2 = 31/4$, $V_1 = 87/17$ and $V_2 = 93/17$. Figure 5.1.6c shows the transformation of the current source on the right into a voltage source. The mesh equations are

$$\begin{bmatrix} 3 & -1 \\ -1 & 6 \end{bmatrix}\begin{bmatrix} I_1 \\ I_2 \end{bmatrix} = \begin{bmatrix} 0 \\ -1 \end{bmatrix}$$

The solutions are $D = 17$, $N_1 = -1$, $N_2 = -3$ and the currents are $I_1 = -1/17, I_2 = -3/17$. To check correctness of both solutions we calculate the voltage V_1 by subtracting the voltage across the resistor R_1 from the voltage of the voltage source on the left: $V_1 = E_1 - R_1 I_1 = 87/17$. The voltage $V_2 = V_1 - R_3 I_2 = 93/17$.

We now summarize the rules when we can apply the transformation. We will make the rules general by using the word *element* instead of a *resistor*. As we will see later, the rules apply to capacitors and inductors as well.

1. The voltage source (dependent or independent) has one element in series with it.
2. The current source (dependent or independent) has one element in parallel with it.
3. Transformation of the part of the circuit where we sense the controlling voltage or current is not permitted without special care. This is explained in detail in Section 2.
4. Cases with more complicated connections are handled by source splitting, the subject of Section 3.

5.2. TRANSFORMATION OF CONTROLLING TERMINALS

Transformations of sources are always possible, but sometimes this may simultaneously involve transformation of a voltage or current which controls another source. To be specific, consider Fig. 5.2.1a where we wish to transform the voltage source into a current source. If we do so we lose all information about the current I which controls the current source.

A bit of preprocessing will solve the problem. All we have to do is determine the current I *before* the transformation. In the given network the controlling current is

$$I = G_1(E - V_1)$$

and we write this information to the dependent source, see Fig. 5.2.1b. Once this has been done we are free to transform the voltage source, as is done in

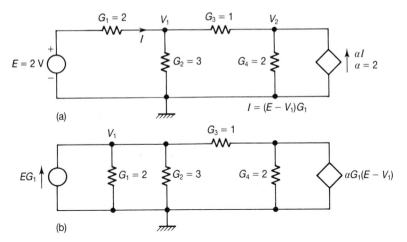

FIGURE 5.2.1   Transformation of controlling branch: (a) Given network. (b) Modified for nodal formulation.

Fig. 5.2.1b, and we can proceed with analysis:

$$(G_1 + G_2 + G_3)V_1 - V_2 G_3 = EG_1$$
$$-G_3 V_1 + (G_3 + G_4)V_2 - \alpha G_1(E - V_1) = 0$$

Transferring the independent source in the second equation to the right we arrive at

$$(G_1 + G_2 + G_3)V_1 - G_2 V_2 = EG_1$$
$$(\alpha G_2 - G_3)V_1 + (G_3 + G_4)V_2 = \alpha G_2 E$$

Inserting numerical values and writing in matrix form:

$$\begin{bmatrix} 6 & -1 \\ 3 & 3 \end{bmatrix} \begin{bmatrix} V_1 \\ V_2 \end{bmatrix} = \begin{bmatrix} 4 \\ 8 \end{bmatrix}$$

The solutions are $D = 21$, $N_1 = 20$, $N_2 = 36$, $V_1 = 20/21$ and $V_2 = 36/21$.

Another case where special attention must be paid is the network in Fig. 5.2.2a with a voltage-controlled current source. The controlling voltage is taken across R_1 and we wish to apply nodal formulation. The preprocessing first establishes that the controlling voltage is equal to

$$V = (E - V_1)G_1$$

The network is redrawn, the information transferred to the dependent current source. We are now free to transform the voltage source into a current source, see

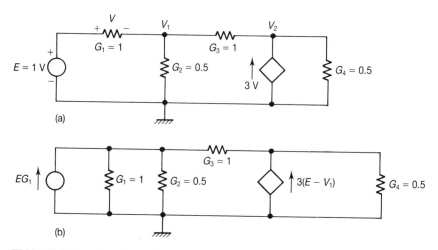

FIGURE 5.2.2   Transformation of controlling branch: (a) Given network. (b) Modified for nodal formulation.

Fig. 5.2.2b. The nodal equations are

$$2.5V_1 - V_2 = 1$$
$$-V_1 + 1.5V_3 - 3(E - V_1) = 0$$

As always, the independent voltage source goes to the right, the system becomes

$$\begin{bmatrix} 2.5 & -1 \\ 2 & 1.5 \end{bmatrix} \begin{bmatrix} V_1 \\ V_2 \end{bmatrix} = \begin{bmatrix} 1 \\ 3 \end{bmatrix}$$

and the solutions are $D = 5.75$, $N_1 = 4.5$ and $N_2 = 5.5$.

5.3. SPLITTING OF SOURCES

Source transformations help us to modify networks if the voltage source is connected in series with one resistor or if the current source is in parallel with one resistor. This may not always be the case. Splitting of sources is the last step we need in order to prepare any network for analysis by either the nodal or mesh method.

We start with the example shown in Fig. 5.3.1a. The voltage source is connected to two resistors and we cannot immediately apply the transformation into a current source. We know that an ideal voltage source can deliver any current and that we normally do not connect voltage sources in parallel, but in this situation we intentionally replace one voltage source by two voltage sources with *the same voltage*. Once we have two such sources, we can separate them and still have the

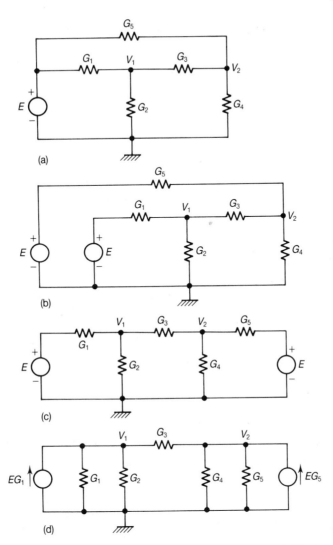

FIGURE 5.3.1   Example of voltage source splitting: (a) Original. (b) After splitting. (c) Redrawn. (d) Transformed for nodal formulation.

same voltages on the left side of G_1 and G_5. This separation is done in Fig. 5.3.1b and each conductance is in series with its own voltage source. The network is redrawn by taking G_5 to the right, as shown in Fig. 5.3.1c. We could apply mesh formulation to this network by converting conductances into resistances, but it would lead to a 3×3 system matrix. In the last step we transform the voltage sources into current sources and the network in Fig. 5.3.1d is prepared for nodal

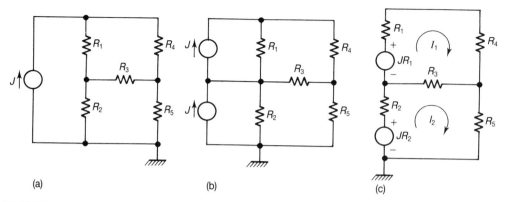

FIGURE 5.3.2   Example of current source splitting: (a) Original. (b) After splitting. (c) Transformed for mesh formulation.

formulation:

$$\begin{bmatrix} G_1 + G_2 + G_3 & -G_3 \\ -G_3 & G_3 + G_4 + G_5 \end{bmatrix} \begin{bmatrix} V_1 \\ V_2 \end{bmatrix} = \begin{bmatrix} EG_1 \\ EG_5 \end{bmatrix}$$

The splitting in the above network was done as an exercise; we could have used the mesh method directly.

A similar splitting can also be applied to current sources. An example is in Fig. 5.3.2a. The current source can be split into two current sources in series if *both deliver the same current*. This has been done in Fig. 5.3.2b where now each source is connected in parallel with its own resistor. We could have used nodal formulation from the beginning but it would lead to a 3×3 system matrix. If we transform the current sources into voltage sources, as shown in Fig. 5.3.2c, we get a network suitable for mesh formulation and the system matrix will be only 2×2:

$$\begin{bmatrix} R_1 + R_3 + R_4 & -R_3 \\ -R_3 & R_2 + R_3 + R_5 \end{bmatrix} \begin{bmatrix} I_1 \\ I_2 \end{bmatrix} = \begin{bmatrix} JR_1 \\ JR_2 \end{bmatrix}$$

More complicated situations can also occur. Figure 5.3.3a indicates splitting of one voltage source into two or three voltage sources, each in series with a resistor. Figure 5.3.3b shows splitting of one current source into two or three current sources, each in parallel with a resistor.

Figure 5.3.4a shows a network with two current sources and one voltage source in the middle. We wish to prepare this network for nodal formulation. The voltage source is split into two voltage sources in Fig. 5.3.4b and each is transformed into a current source as shown in Fig. 5.3.4c. The nodal equations are

$$\begin{bmatrix} G_1 + G_2 + G_5 & -G_5 \\ -G_5 & G_3 + G_4 + G_5 \end{bmatrix} \begin{bmatrix} V_1 \\ V_2 \end{bmatrix} = \begin{bmatrix} J_1 + EG_2 \\ J_2 + EG_3 \end{bmatrix}$$

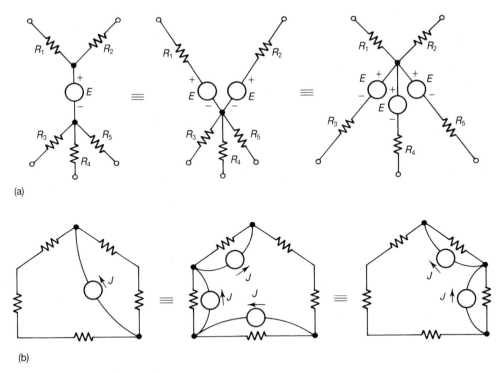

(a)

(b)

FIGURE 5.3.3   Splitting of sources: (a) Voltage sources. (b) Current sources.

Example 1.

Consider the network in Fig. 5.3.5 with one voltage source and one current source. Apply voltage splitting and current splitting to obtain the solution.

Figures 5.3.6a through c show the steps for voltage splitting and transformation to current sources. It leads to the network 5.3.6c and the nodal matrix equation is

$$\begin{bmatrix} 4 & 0 \\ 0 & 6 \end{bmatrix}\begin{bmatrix} V_1 \\ V_2 \end{bmatrix} = \begin{bmatrix} -1 \\ 4 \end{bmatrix}$$

with the solution $V_1 = -1/4$ and $V_2 = 2/3$.

Splitting of the current source and transformation into voltage sources is shown in Figs 5.3.7a and b. This leads to the system

$$\begin{bmatrix} 4/3 & -4/3 \\ -4/3 & 25/12 \end{bmatrix}\begin{bmatrix} I_1 \\ I_2 \end{bmatrix} = \begin{bmatrix} 5/3 \\ -7/6 \end{bmatrix}$$

The solutions are $D = 1$, $N_1 = 23/12$, $N_2 = 2/3$ and the currents are $I_1 = 23/12$ and $I_2 = 2/3$. As a check we can calculate the voltage $V_1 = E - R_1(I_1 - I_2) = -1/4$ and $V_2 = 1/2 + R_4 I_2 = 2/3$, the same result as above.

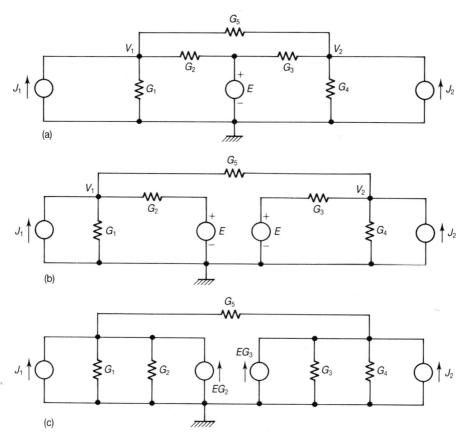

FIGURE 5.3.4 Example of voltage source splitting. (a) Original. (b) After splitting. (c) Transformed for nodal formulation.

FIGURE 5.3.5 Network with a voltage and a current source.

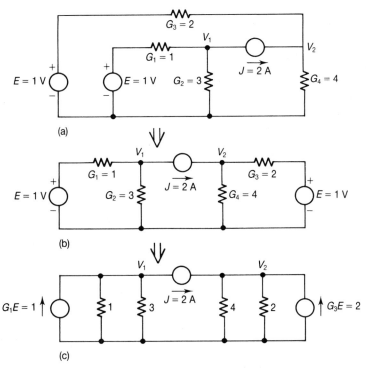

FIGURE 5.3.6   Network in Fig. 5.3.5:(a) After source splitting. (b) Redrawn. (c) Transformed for nodal formulation.

Figures 5.3.8a and b indicate the steps when we use splitting for both sources. The network becomes decoupled and each nodal voltage is obtained separately; the result is the same as above.

Example 2.

We wish to solve the network in Fig. 5.3.9a by the mesh method. This requires transformation of the current sources into voltage sources. The splitting of the current source in the middle of the network is shown in Fig. 5.3.9b. On the right, the current sources in parallel are combined into one and the source transformation is applied, as shown in Fig. 5.3.9c. If we use numerical values, Fig. 5.3.9d, we get

$$\begin{bmatrix} 3 & -2 & 0 \\ -2 & 14 & -7 \\ 0 & -7 & 13 \end{bmatrix} \begin{bmatrix} I_1 \\ I_2 \\ I_3 \end{bmatrix} = \begin{bmatrix} 1 \\ 14 \\ -6 \end{bmatrix}$$

and the solution is $I_1 = 413/347$, $I_2 = 446/347$ and $I_3 = 80/347$. We note that it would have

FIGURE 5.3.7   Network in Fig. 5.3.5: (a) After splitting. (b) Transformed for mesh formulation.

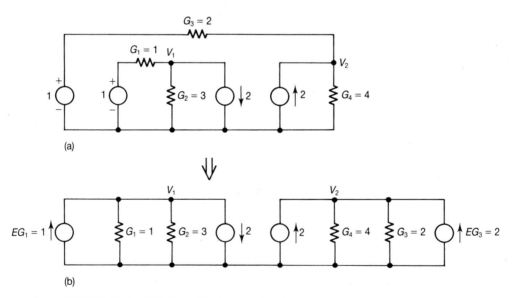

FIGURE 5.3.8   Splitting of both sources in Fig. 5.3.5: (a) After splitting. (b) Transformed for nodal formulation.

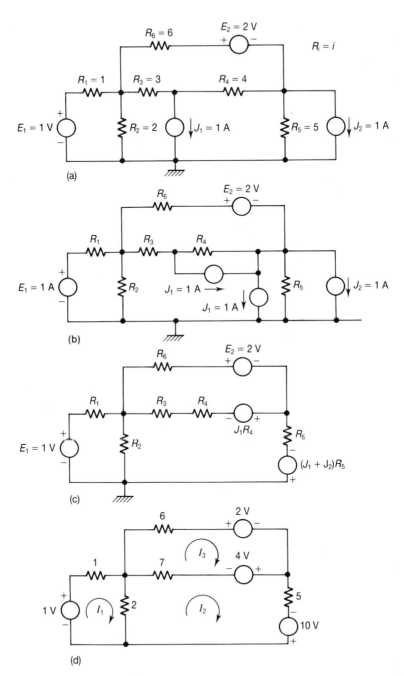

FIGURE 5.3.9   Network with several sources: (a) Original. (b) After splitting of the current source. (c) After transformation. (d) Final form for mesh formulation.

been easier to transform the voltage source into a current source, but we wanted to show the splitting of current sources.

5.4. SUPERPOSITION PRINCIPLE

Linear networks obey the important law of superposition: If the network has several independent sources, then we can calculate the contribution of each of them individually, as if the others were not present, and add the results. In order to do it correctly, we must properly remove the sources which we are not considering at that moment: replace *independent* voltage sources by a short circuit and remove *independent* current sources from the network. We have stressed the word *independent*, because we must never remove the dependent sources. They stay in the network without any change.

At present we know only resistive networks and dc signals (voltages delivered by batteries) but the law is completely general and applies for any signals, as long as the network is linear.

Consider the network in Fig. 5.4.1. It has two independent current sources and one VC. The matrix equation is

$$\begin{bmatrix} G_1 + G_2 & -G_2 \\ -G_2 - g & G_2 + G_3 + g \end{bmatrix} \begin{bmatrix} V_1 \\ V_2 \end{bmatrix} = \begin{bmatrix} J_1 + J_2 \\ -J_2 \end{bmatrix}$$

a. We solve first the case $J_2 = 0$:

$$\begin{bmatrix} 3 & -2 \\ -4 & 7 \end{bmatrix} \begin{bmatrix} V_1 \\ V_2 \end{bmatrix} = \begin{bmatrix} 1 \\ 0 \end{bmatrix}$$

The result is $V_1^a = 7/13$ and $V_2^a = 4/13$.

FIGURE 5.4.1   Demonstrating superposition principle.

b. In the next step we set $J_1 = 0$ and solve the system

$$\begin{bmatrix} 3 & -2 \\ -4 & 7 \end{bmatrix} \begin{bmatrix} V_1 \\ V_2 \end{bmatrix} = \begin{bmatrix} 2 \\ -2 \end{bmatrix}$$

with the result $V_1^b = 10/13$ and $V_2^b = 2/13$. Adding the results we get

$$V_1 = V_1^a + V_1^b = \frac{7}{13} + \frac{10}{13} = \frac{17}{13}$$

$$V_2 = V_2^a + V_2^b = \frac{4}{13} + \frac{2}{13} = \frac{6}{13}$$

To show that the addition indeed gives a correct result, we now leave both sources in place and solve

$$\begin{bmatrix} 3 & -2 \\ -4 & 7 \end{bmatrix} \begin{bmatrix} V_1 \\ V_2 \end{bmatrix} = \begin{bmatrix} 3 \\ -2 \end{bmatrix}$$

The two results are equal, as expected.

There are several important conclusions which we can draw from the above considerations and which are obvious only because we are using matrix algebra. The rules would be quite difficult to discover if we had used equation elimination.

1. Independent sources appear in the right-hand column of the matrix equation only.
2. Dependent sources appear in the system matrix.
3. Changing the nodes where the source is applied does not influence the matrix of the system.
4. The determinant of the system matrix (denominator of the solution) remains the same irrespective where we have the independent sources.

The fourth conclusion is perhaps the most important one, because it saves a lot of calculations.

Our next network in Fig. 5.4.2 is suitable for mesh formulation. It has one current-controlled voltage source. The controlling current is indicated in the figure as I and we see that it is equal to the circulating current I_2. This information is transferred into the figure. The matrix equation is

$$\begin{bmatrix} R_1 + R_2 & -R_2 + r \\ -R_2 & R_2 + R_3 - r \end{bmatrix} \begin{bmatrix} I_1 \\ I_2 \end{bmatrix} = \begin{bmatrix} E_1 \\ -E_2 \end{bmatrix}$$

Note again that the dependent source appears in the system matrix while the

FIGURE 5.4.2   Demonstrating superposition principle.

independent sources are in the right-hand side vector only. We can set either E_1 or E_2 equal to zero, solve and add the results; this is left to the student as an exercise. If we use both sources simultaneously, we get $I_1 = 3/9$ and $I_2 = -4/9$. We also note that setting $E_1 = 0$ is equivalent to short-circuiting this source; the same applies for E_2.

In a normal situation we solve the network with all its sources because this gives immediately the final result. We will see in the following two sections of this chapter that there may sometimes be special reasons why we should apply the sources one at a time. Even if we do so, we should still remember that the *system matrix does not change* and we must evaluate the system determinant only once.

5.5. INPUT AND OUTPUT RESISTANCE

More complicated transformations require the concept of an input resistance (conductance) at various points of the network. The concept gives us additional information about the network and will be needed in the next section. The word *input* relates to any two terminals of the network where we can connect an independent source. The expression *input resistance* describes the ratio of the voltage across the input terminals, divided by the current flowing into the network through the same terminals. Sometimes we also use the expression *output resistance*; it is nothing but an input resistance calculated at the terminals where we normally take the output of the network.

Any pair of terminals where we can connect an independent source can be considered as an input terminal pair. The first example, shown in Fig. 5.5.1, will deal with independent current sources. The network has three resistors, one voltage-controlled current source and three independent current sources, not yet connected to the network.

If we connect *only one of the sources* and calculate the voltage across that particular source, we obtain the input resistance, R_{in}, as the ratio of the voltage across the source to the current delivered by the source.

Let us attach the first current source, J_1. Since one terminal of this source is grounded, the voltage across it will be V_1 and the input resistance will be $R_{in, 1} = V_1/J_1$. If we disconnect the first source and attach instead the second source,

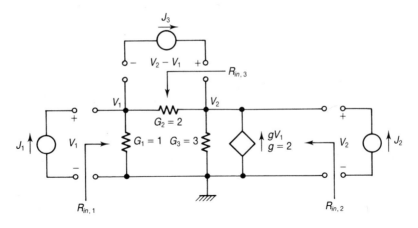

FIGURE 5.5.1   Input and output resistances, nodal formulation.

J_2, we can calculate V_2, the voltage across this source, and obtain the input resistance at this second pair of terminals as the ratio $R_{in,\,2} = V_2/J_2$. Finally, if we attach the third current source, J_3, the voltage across it will be $V_2 - V_1$ and the input resistance at this third pair of terminals will be $R_{in,\,3} = (V_2 - V_1)/J_3$. The important point is that we must *always remove all independent current sources except one* and calculate the voltage across the source which is left in the network. We can also proceed by preparing the equations with all independent sources simultaneously, but always set all except one equal to zero.

Let us now find the three input resistances. In all three cases, the network is the same and the dependent source, which is part of it, *must not be removed*. If all three independent sources are connected to the network, the system equation is

$$\begin{bmatrix} 3 & -2 \\ -4 & 5 \end{bmatrix} \begin{bmatrix} V_1 \\ V_2 \end{bmatrix} = \begin{bmatrix} J_1 - J_3 \\ J_2 + J_3 \end{bmatrix} \tag{5.5.1}$$

The determinant of the matrix is always the same, no matter which sources we remove from the right-hand side: $D = 7$. Setting $J_2 = J_3 = 0$ we find the voltage across the first source to be

$$V_1 = \frac{5J_1}{7}$$

The input resistance, looking into the network from the left, is

$$R_{in,\,1} = \frac{V_1}{J_1} = \frac{5}{7} \tag{5.5.2}$$

If we keep only J_2, we must solve

$$\begin{bmatrix} 3 & -2 \\ -4 & 5 \end{bmatrix}\begin{bmatrix} V_1 \\ V_2 \end{bmatrix} = \begin{bmatrix} 0 \\ J_2 \end{bmatrix} \qquad (5.5.3)$$

The solution is

$$V_2 = \frac{3J_2}{7}$$

and the input resistance at the terminals on the right is

$$R_{in,\,2} = \frac{3}{7} \qquad (5.5.4)$$

Keeping only the third source, J_3, we must solve

$$\begin{bmatrix} 3 & -2 \\ -4 & 5 \end{bmatrix}\begin{bmatrix} V_1 \\ V_2 \end{bmatrix} = \begin{bmatrix} -J_3 \\ J_3 \end{bmatrix} \qquad (5.5.5)$$

but now we need both $V_1 = 3J_3/7$ and $V_2 = 1J_3/7$. The input resistance at the terminals on top is

$$R_{in,\,3} = \frac{V_2 - V_1}{J_3} = \frac{2}{7} \qquad (5.5.6)$$

If the network has independent voltage sources, we can also calculate the input resistance (or conductance) by evaluating the amount of current flowing from the source into the network. We will use the network Fig. 5.5.2. It has one current-controlled voltage source. With both sources in place, the matrix equation is

$$\begin{bmatrix} R_1 + R_2 & -R_2 \\ -R_2 + r & R_2 + R_3 - r \end{bmatrix}\begin{bmatrix} I_1 \\ I_2 \end{bmatrix} = \begin{bmatrix} E_1 - E_2 \\ E_2 \end{bmatrix} \qquad (5.5.7)$$

The input resistance for the source E_1 is calculated by setting $E_2 = 0$ and by finding the mesh current I_1. Using numerical values

$$\begin{bmatrix} 3 & -2 \\ 0 & 3 \end{bmatrix}\begin{bmatrix} I_1 \\ I_2 \end{bmatrix} = \begin{bmatrix} E_1 \\ 0 \end{bmatrix} \qquad (5.5.8)$$

The determinant is $D = 9$ and the current I_1 is obtained by Cramer's rule

$$I_1 = \frac{3E_1}{9}$$

FIGURE 5.5.2   Input resistance for network with dependent sources.

from which we get the input conductance

$$G_{in,1} = \frac{I_1}{E_1} = \frac{3}{9} \tag{5.5.9}$$

The input resistance is, of course, the inverse.

In order to obtain the input conductance for the second source, we set $E_1 = 0$.

$$\begin{bmatrix} 3 & -2 \\ 0 & 3 \end{bmatrix}\begin{bmatrix} I_1 \\ I_2 \end{bmatrix} = \begin{bmatrix} -E_2 \\ E_2 \end{bmatrix} \tag{5.5.10}$$

Since the current flowing into the network from this source is equal to $I_2 - I_1$, we must solve for both these currents. The system determinant is already known and using Cramer's rule we get $I_1 = -E_2/9$ and $I_2 = 3E_2/9$. The current flowing into the network is

$$I_{in} = I_2 - I_1 = \frac{4E_2}{7}$$

and the input conductance is

$$G_{in} = \frac{I_{in}}{E_2} = \frac{I_2 - I_1}{E_2} = \frac{4}{7}$$

5.6. THÉVENIN AND NORTON THEOREMS

Transformation of sources explained in Section 1 of this chapter can be extended to more complicated situations. Consider Fig. 5.6.1 where we have one network with independent sources and one external network connected to it on the right. We wish to transform the left network into another one which will behave exactly

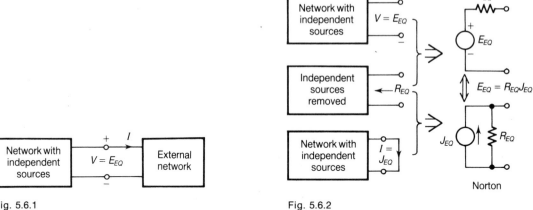

Fig. 5.6.1

Fig. 5.6.2

FIGURE 5.6.1   General network configuration.

FIGURE 5.6.2   Thévenin and Norton equivalents.

equivalently, no matter how we change the network on the right. This can be achieved only if the network on the left can deliver the same current, I, while maintaining the same voltage, V, across the indicated terminals.

Two transformations exist, called the Thévenin and the Norton transformations. Schematically, both are given in Fig. 5.6.2. The upper part replaces the network which has independent sources by one voltage source and one resistor in series. This is called the Thévenin transformation. The lower part replaces the same network by a combination of a current source in parallel with the same resistor. This is called the Norton transformation. The two transformed sources on the right are also coupled by the equation which we have already derived in Section 1 for the transformation of the voltage source into a current source.

To get the Thévenin transformation, we must perform two calculations, as we explain with the help of Fig. 5.6.1.

1. Disconnect the external network and calculate the voltage at the indicated terminals by taking into account all independent and dependent sources of the network on the left. The voltage we obtain, $V = E_{EQ}$, will be the value of the equivalent Thévenin voltage source.

2. Set all independent sources of the network on the left equal to zero (but keep all dependent sources) and attach one independent source at the terminals. If it is a voltage source, E, calculate the current which it is delivering, I. If it is a current source, J, calculate the voltage across it, V. The Thévenin equivalent

resistance is

$$R_{EQ} = \frac{E}{I}$$

if we used a voltage source, or

$$R_{EQ} = \frac{V}{J}$$

if we used a current source.

The Thévenin transformation is particularly easy to obtain if the network has a form suitable for nodal equations. We will see it on the example in Fig. 5.6.3a. The network has one current source, $J_1 = 2$ A, one voltage-controlled current source and three resistors given by their conductance values. Output of the network is on the right. We have also indicated in the figure another (dotted) current source which we will use to calculate the output resistance. Keeping both sources in place we write

$$\begin{bmatrix} G_1 + G_2 & -G_2 \\ -G_2 - g & G_2 + G_3 \end{bmatrix} \begin{bmatrix} V_1 \\ V_2 \end{bmatrix} = \begin{bmatrix} J_1 \\ J_2 \end{bmatrix}$$

In order to obtain the output voltage, we first remove J_2 by setting it equal to

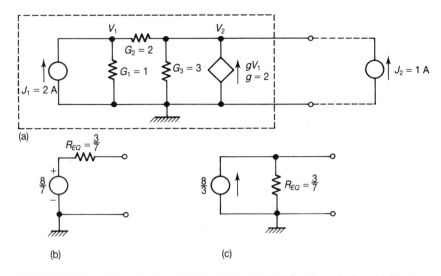

(a)

(b)   (c)

FIGURE 5.6.3   Network with a VC: (a) Original. (b) Thévenin equivalent. (c) Norton equivalent.

zero. This is equivalent to removing the source from the network and making an open circuit in its place, as required by the rule which gets us the equivalent voltage source. Inserting numerical values

$$\begin{bmatrix} 3 & -2 \\ -4 & 5 \end{bmatrix}\begin{bmatrix} V_1 \\ V_2 \end{bmatrix} = \begin{bmatrix} 2 \\ 0 \end{bmatrix}$$

The determinant of the matrix is $D = 7$ and the solution is $V_2 = E_{EQ} = 8/7$.

In the next step we remove the original independent source, J_1, by setting its value equal to zero, and connect the second source, $J_2 = 1$. This leads to

$$\begin{bmatrix} 3 & -2 \\ -4 & 5 \end{bmatrix}\begin{bmatrix} V_1 \\ V_2 \end{bmatrix} = \begin{bmatrix} 0 \\ J_2 \end{bmatrix}$$

The only difference from the previous case is in the right-hand side. We do not have to calculate the system determinant again since it is already known from the previous step. We find V_2 and the ratio $V_2/J_1 = R_{EQ} = 3/7$. The Thévenin equivalent of the network is in Fig. 5.6.3b. The Norton equivalent, if needed, is obtained by using Eq. (5.1.4); it is in Fig. 5.6.3c.

If the network is suitable for mesh equations, the Norton equivalent may be easier to obtain. Consider Fig. 5.6.4a and replace the output short-circuit with another independent voltage source, to be later used for calculation of the output conductance. Note in particular the positions of the + and − signs with respect

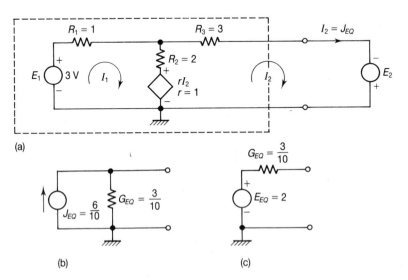

(a)

(b)   (c)

FIGURE 5.6.4   Network with a CV: (a) Original. (b) Norton equivalent. (c) Thévenin equivalent.

to the mesh current; its polarity must be such that it would deliver a current of the same direction into the network.

Keeping first both independent sources in the network, we write the mesh equations:

$$(R_1 + R_2)I_1 - R_2I_2 + rI_2 = E_1$$
$$-R_2I_1 + (R_2 + R_3)I_2 - rI_2 = E_2$$

and inserting numerical values

$$\begin{bmatrix} 3 & -1 \\ -2 & 4 \end{bmatrix} \begin{bmatrix} I_1 \\ I_2 \end{bmatrix} = \begin{bmatrix} E_1 \\ E_2 \end{bmatrix}$$

Set $E_2 = 0$ and calculate the current flowing through this source: $I_2 = J_{EQ}$. This requires the solution of

$$\begin{bmatrix} 3 & -1 \\ -2 & 4 \end{bmatrix} \begin{bmatrix} I_1 \\ I_2 \end{bmatrix} = \begin{bmatrix} 3 \\ 0 \end{bmatrix}$$

The determinant of the system is $D = 10$ and $J_{EQ} = 6/10$. In the next step we set $E_1 = 0$ and keep E_2. We must solve the system

$$\begin{bmatrix} 3 & -1 \\ -2 & 4 \end{bmatrix} \begin{bmatrix} I_1 \\ I_2 \end{bmatrix} = \begin{bmatrix} 0 \\ E_2 \end{bmatrix}$$

and obtain $I_2 = 3E_2/10$. The ratio $I_2/E_2 = 3/10$ is the output conductance. The Norton equivalent is in Fig. 5.6.4b and the Thévenin equivalent in Fig. 5.6.4c.

Example 1.

Find the Thévenin equivalent for the network in Fig. 5.6.5a.

We select nodal formulation and transform the voltage source into a current source. Simultaneously we must express on the original network the controlling current $I = G_1(E - V_1)$. The network is redrawn in b with another, auxiliary source on the right. The nodal equations with all sources kept in the equations as symbols are

$$7V_1 - 2V_2 = 2E_1 + J_1$$
$$-2V_1 + 2V_2 - 4E_1 + 4V_1 = J_2$$

and we must solve

$$\begin{bmatrix} 7 & -2 \\ 2 & 2 \end{bmatrix} \begin{bmatrix} V_1 \\ V_2 \end{bmatrix} = \begin{bmatrix} 2E_1 + J_1 \\ 4E_1 + J_2 \end{bmatrix}$$

FIGURE 5.6.5 Network with a *CC*: (a) Original. (b) With the necessary sources. (c) Thévenin equivalent. (d) Norton equivalent.

The determinant is $D = 18$.

    a. In the first step set $J_2 = 0$, insert for values for the other sources and solve for V_2. The result is $N_2 = 20$ and $V_2 = E_{EQ} = 20/18$.

    b. In the next step set $E_1 = 0$, $J_1 = 0$ and calculate again V_2. The result is now $N_2 = 7J_2$ and $V_2/J_2 = R_{EQ} = 7/18$.

The Thévenin and Norton equivalents are in Fig. 5.6.5c and d respectively.

    Let us summarize the steps: Both transformations require two solutions of the same network.

1. Calculate either the output voltage (Thévenin) or the output current (Norton) using all independent sources in the network.
2. Remove all independent sources, retain all dependent sources and calculate the resistance of the network from the output terminals.
3. If necessary, use Eq. (5.1.4) to get the desired form of the equivalent network.

    In practical applications, it is useful to do the steps by first preparing the system matrix with all sources, even the auxiliary one for calculation of the resistance from the output terminals. This can save us a lot of difficulties:

1. We avoid mistakes by removing dependent sources.
2. Removal of the various independent sources is done simply by setting their values equal to zero in the right hand side vector.
3. We avoid unnecessary steps by realizing that the system determinant is the same for both solutions.

Although the two transformations are very practical because they simplify the source network, the resulting sources are equivalent only as far as the external network is concerned. This should be very clearly understood. The transformations are *not* equivalent in terms of power delivered by the sources. If there is no external network, the Thévenin equivalent does not consume any power, because there is no current flowing from the source. The Norton equivalent *does* draw power from the source if it is disconnected from any external network.

5.7. OPERATIONAL AMPLIFIERS

There exists one more ideal and widely used element which we have not covered yet. It is the ideal operational amplifier, OPAMP. Its application will be covered fully in Chapter 12, but the student will probably need some understanding earlier in his studies. For this reason we give here a simple, yet very efficient method for the analysis of networks with operational amplifiers.

An ideal operational amplifier is a special voltage-controlled voltage source with two input terminals, i and j, as given in Fig. 5.7.1a. The voltages at these points are taken and their difference, $V_i - V_j$, is multiplied by the gain, A. Thus the output of the dependent voltage source, V_o, whose *one terminal is grounded* as shown, will be

$$V_o = A(V_i - V_j) \qquad (5.7.1)$$

The usual symbol for an operational amplifier, OPAMP, is in Fig. 5.7.1b. It is a common practice to mark the terminal i with a $+$ sign and terminal j with a $-$ sign, as shown, but these signs do not mean that the first voltage is positive and the second negative. They only reflect the fact that in Eq. (5.7.1) the first one is taken positively and the second one is subtracted. For the same reason it is also an accepted practice to write Eq. (4.5.1) in the form

$$V_o = A(V_+ - V_-) \qquad (5.7.2)$$

which we will be using in the following.

The definition of the operational amplifier is not yet finished. In order to have an *ideal* OPAMP we assume that the gain is *infinitely large*. Infinity is always troublesome; we cannot display it and we cannot use it in calculators or in

FIGURE 5.7.1   Operational amplifier: (a) Equivalent network. (b) Symbol.

computers. To get out of this difficulty we recall from mathematics how to handle a variable which may become infinitely large. We first divide the equation by this variable and only afterwards let it grow without bound. Anything finite, divided by such variable, will become zero. Applying these steps to Eq. (5.7.2) we first obtain

$$\frac{V_o}{A} = V_+ - V_- \tag{5.7.3}$$

and then apply the limiting step. In this case the expression on the left becomes zero and

$$0 = V_+ - V_-$$

or

$$V_+ = V_- \tag{5.7.4}$$

This is a very important conclusion. It tells us that the voltage at the + terminal will be *exactly* equal to the voltage at the − terminal and *no other equation is needed*. In other words, although the ideal OPAMP is in principle a voltage-controlled voltage source, we never use its equation.

There are some additional advantages: we can use nodal formulation and we need not even write any equation for the output node of the OPAMP. All the necessary information has been transferred to the statement that the input terminals are at the same potential. We will indicate that there is no need to write KCL at the output node by marking this node with a cross. We must still indicate the output voltage, V_{out}.

Several simple examples will clarify the procedure. If you are eager to learn more about networks with OPAMPS, turn to Chapter 12. Its initial sections will be relatively easy to follow; all you need is nodal formulation which we have already covered in detail. ·

Consider the network in Fig. 5.7.2a. The voltage source on the left is transformed into a current source by the Thévenin transformation, as shown in Fig. 5.7.2b. The output voltage, V_{out} is marked, but the node is crossed out to indicate that we will not write the KCL there. The + terminal is grounded and is at zero potential; the − terminal must be, according to Eq. (5.7.4), at the same zero potential. This voltage is written at the terminal. Proceeding our standard

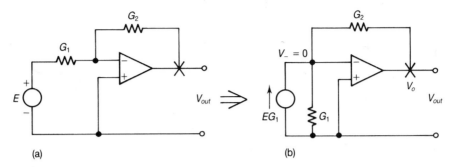

FIGURE 5.7.2   OPAMP inverting network: (a) Original. (b) After Thévenin transformation.

way we write the current balance for the only node which does not have the cross:

$$V_-(G_1 + G_2) - G_2 V_{out} = EG_1 \qquad (5.7.5)$$

However, we know that the voltage V_- is zero and any finite expression multiplied by zero becomes zero. Thus the first term is zero and we did not even have to write the term. We could have written directly

$$-G_2 V_{out} = EG_1$$

The solution is

$$V_{out} = -\frac{G_1}{G_2} E$$

Example 1.

Write the equations and find the output voltage for the network in Fig. 5.7.3.

The voltage at the + terminal is equal to E and according to Eq. (5.7.4) the same voltage must be at the other input terminal. This we express by writing *the same* voltage at the − terminal, as done in Fig. 5.7.3. The output terminal is crossed out, but has the voltage V_{out}. We need to write the balance of currents at the − terminal only:

$$E(G_1 + G_2) - G_2 V_{out} = 0$$

FIGURE 5.7.3   Noninverting OPAMP amplifier.

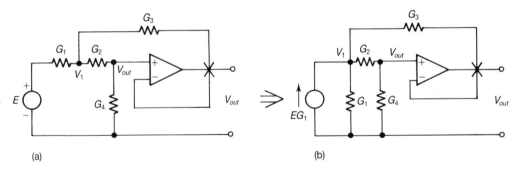

FIGURE 5.7.4   Network with OPAMP as an amplifier with unity gain: (a) Original. (b) Independent voltage source transformed into a current source.

The solution is

$$V_{out} = \left(1 + \frac{G_1}{G_2}\right) E$$

Example 2.

Write the nodal equations for the network in Fig. 5.7.4a.

We first transform the source on the left into a current source. This leads to Fig. 5.7.4b. The output voltage, due to the short circuit, appears at the − terminal, and due to the fact that the element is an ideal OPAMP, the same voltage will be at the + terminal, where we write the voltage V_{out}. We have two nodes which are not crossed out and where we must write the KCL:

$$(G_1 + G_2 + G_3)V_1 - G_2V_{out} - G_3V_{out} = EG_1$$
$$-G_2V_1 + (G_2 + G_4)V_{out} = 0$$

As usual, we put the equations into matrix form:

$$\begin{bmatrix} G_1 + G_2 + G_3 & -(G_2 + G_3) \\ -G_2 & G_2 + G_4 \end{bmatrix} \begin{bmatrix} V_1 \\ V_{out} \end{bmatrix} = \begin{bmatrix} EG_1 \\ 0 \end{bmatrix}$$

The determinant is $D = G_1(G_2 + G_4) + G_4(G_2 + G_3)$ and $N_{out} = EG_1G_2$. The output voltage is

$$V_{out} = \frac{G_1G_2}{G_1(G_2 + G_4) + G_4(G_2 + G_3)} E$$

Example 3.

Find the output voltage for the network in Fig. 5.7.5a.

Splitting the source on the left into two sources and transforming them into current sources leads to the network in Fig. 5.7.5b. V_{out} is marked, the node crossed out. The remaining two nodes *must be at the same potential*, which we express by writing the same

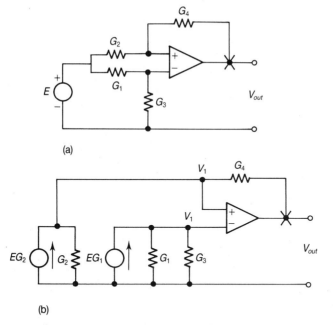

FIGURE 5.7.5   Simple network with OPAMP: (a) Original. (b) After Thévenin transformation.

voltage, V_1, at both. For these nodes we write separate nodal equations:

$$(G_2 + G_4)V_1 - G_4 V_{out} = EG_2$$

$$(G_1 + G_3)V_1 = EG_1$$

In matrix form

$$\begin{bmatrix} G_2 + G_4 & -G_4 \\ G_1 + G_3 & 0 \end{bmatrix} \begin{bmatrix} V_1 \\ V_{out} \end{bmatrix} = \begin{bmatrix} EG_2 \\ EG_1 \end{bmatrix}$$

The determinant is $D = G_4(G_1 + G_3)$, for the numerator we must find the determinant

$$N_{out} = \begin{vmatrix} G_2 + G_4 & EG_2 \\ G_1 + G_3 & EG_1 \end{vmatrix} = E(G_1 G_4 - G_2 G_3)$$

and the solution is

$$V_{out} = \frac{G_1 G_4 - G_2 G_3}{G_4(G_1 + G_3)} E$$

All networks with ideal OPAMPs, modeled as in Fig. 5.7.1, can be solved by this method. The only condition is that one end of the dependent voltage source, see Fig. 5.7.1b, must be grounded. The method can be extended to *all* voltage sources, dependent or independent, as long as one of their terminals is grounded. Detailed explanations are in Chapter 12, but here we will use the method for the problems solved above.

Fig. 5.7.6

Fig. 5.7.7

FIGURE 5.7.6   Analyze the network in Fig. 5.7.2a without Thévenin transformation.

FIGURE 5.7.7   Analyze the network in Fig. 5.7.7a without Thévenin transformation.

Example 4.

Analyze the network in Fig. 5.7.2a by omitting the Thévenin transformation.

The network is redrawn in Fig. 5.7.6, but now two nodes are crossed out, because they have voltage sources which have the other terminal grounded. The equation is written for only one node, omitting the term which would be multiplied by zero:

$$-G_2 V_{out} - G_1 E = 0$$

which leads to the same result as above.

Example 5.

Use the method for the network in Fig. 5.7.5.

The network is redrawn in Fig. 5.7.7. Crossed out are the two nodes with the voltage sources and the equations are written as the remaining nodes. They are the same as before,

$$(G_2 + G_4)V_1 - G_4 V_{out} = EG_2$$

$$(G_1 + G_3)V_1 = EG_1$$

Example 6.

Write the equations for the network in Fig. 5.7.8 by keeping all elements as variables. Put the equations into matrix form. Afterwards insert numerical values and solve for V_{out}. Let the input voltage be $E = 1$ V.

The crosses are at two nodes and voltages are written at all nodes. We write the equations in the sequence of numbers in circles:

$$(G_1 + G_2 + G_3)V_1 - G_2 V_2 - G_3 V_{out} = EG_1$$

$$-G_2 V_1 + (G_2 + G_4)V_2 = 0$$

$$(G_5 + G_6 + G_7)V_2 - G_7 V_{out} = EG_5$$

FIGURE 5.7.8   Analyze without Thévenin transformation.

In matrix form:

$$
\begin{bmatrix}
G_1 + G_2 + G_3 & -G_2 & -G_3 \\
-G_2 & G_2 + G_4 & 0 \\
0 & G_5 + G_6 + G_7 & -G_7
\end{bmatrix}
\begin{bmatrix}
V_1 \\
V_2 \\
V_{out}
\end{bmatrix}
=
\begin{bmatrix}
EG_1 \\
0 \\
EG_5
\end{bmatrix}
$$

Inserting numerical values:

$$
\begin{bmatrix}
4 & -1 & -2 \\
-1 & 2 & 0 \\
0 & 6 & -1
\end{bmatrix}
\begin{bmatrix}
V_1 \\
V_2 \\
V_{out}
\end{bmatrix}
=
\begin{bmatrix}
E \\
0 \\
2E
\end{bmatrix}
$$

This solution is $V_1 = 6/5 = 1.2$ V, $V_2 = 3/5 = 0.6$ V, $V_{out} = 8/5 = 1.6$ V.

5.8. RECOMMENDATIONS

The various methods we discussed so far give us the possibility of analyzing any network by either the nodal or the mesh formulation. In some cases this involves quite a bit of redrawing and preprocessing before we are able to write the equations. The purpose of these steps is to reduce the number of equations to a minimum because the solution process is the most time consuming.

In our explanations we treated both the nodal and mesh formulations equally, but they are not equally important. As long as the networks are small, like our examples, it does not matter much which one we select. If we go to larger networks, then the nodal method is much more practical. Its results are voltages which are usually easier to measure than currents. Even more importantly, the nodal method is general while the mesh method applies to planar networks only, that is to networks which can be drawn on paper without any element crossing some other element. It is sometimes quite difficult to establish that a large network is planar.

We will remember the following rules of thumb for our studies and for all hand analyses:

1. If all sources in a planar network are voltage sources then it is probably easiest to use the mesh formulation.
2. If the sources are mixed, it is generally better to transform the voltage sources into current sources and apply nodal formulation. In such cases use conductances, G, to handle resistors.

The preprocessing which we must go through and the redrawing of figures which is sometimes necessary are reasonable when we are learning the tricks of the trade. In addition, the transformations sometimes reduce the size of the system, a point of utmost importance in hand calculations. Unfortunately, the transformations are also sources of possible errors and programs for analysis never use them. Another method, called modified nodal, is preferred; it will be explained in Chapter 17.

PROBLEMS CHAPTER 5

Solve the problems by hand but check results by running NETOPT. Refer to Sections 18.2 and 18.3 for information. Remember that symbolic option of NETOPT automatically assumes that the source has unit value (1 V for a voltage source and 1 A for the current source).

P.5.1 Transform the voltage source into a current source and calculate the nodal voltage V_1.

P.5.2 Transform the voltage source into a current source and calculate the nodal voltage V_1.

P.5.3 Transform the voltage into a current source, calculate V_1 and I.

P.5.4 Add the two voltages on the left, transform all voltage sources into current sources and obtain V_1.

P.5.5 Transform the voltage sources into current sources and calculate V_A.

P.5.6 Transform the current source into a voltage source and calculate the voltages V_1 and V_2.

P.5.7 Transform the current sources into voltage sources and calculate I_A.

P.5.8 Transform the current sources into voltage sources and calculate the current I.

P.5.9 Transform the current sources into voltage sources and calculate V_1 and V_2.

P.5.10 Transform the current sources into voltage sources. Calculate the voltage across G_2.

P.5.11 Use voltage splitting of the voltage source on the left and calculate the voltage across R_3.

P.5.12 Use source splitting and then transform the voltage sources into current sources. Apply nodal formulation to get the voltage across R_3.

P.5.13 Use voltage source splitting, transform the sources into current sources and calculate the voltage across R_2.

P.5.14 Apply current source splitting to J_2 and transform the current sources into voltage sources and use mesh formulation. Calculate the three nodal voltages. Note that the network is directly suitable for nodal formulation, but we wish to practice something else.

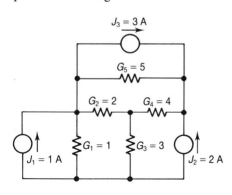

P.5.15 Calculate the Thévenin and Norton equivalents. The auxiliary source is indicated by dotted connection. Write the equations with both sources simultaneously but in the solution always use only one of them.

P.5.16 Do the same as in P.5.15.

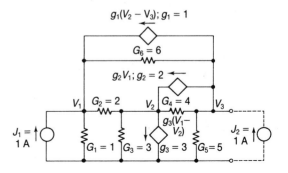

P.5.17 Do the same as in P.5.15.

P.5.18 Do the same as in P.5.15.

P.5.19 Do the same as in P.5.15.

P.5.20 Do the same as in P.5.15.

P.5.21 Express the current I in terms of E, V_1 and G_E, transform the voltage source into a current source and use nodal formulation to calculate the nodal voltages V_1 and V_2. Keep α as a variable.

P.5.22 Use voltage splitting to obtain the voltage V. Also find the Thévenin equivalent for the indicated output.

P.5.23 Find the Thévenin and Norton equivalent.

P.5.24 Find the output voltage by first transforming the voltage source into a current source.

P.5.25 Do the same as in P.5.24.

P.5.26 Find the output voltage by transforming first the voltage source into a current source and applying nodal formulation.

P.5.27 Do the same as in P.5.26.

P.5.28 Find the Thévenin and Norton equivalents.

P.5.29 Find the Thévenin and Norton equivalents.

P.5.30 Use voltage splitting and nodal formulation to calculate the voltages V_1 and V_2. Keep G_3 as a variable.

P.5.31 Solve P.5.1 by keeping all resistors as variables.

P.5.32 Solve P.5.2 through P.5.5 by keeping the constants of the dependent sources as variables.

P.5.33 In P.5.7 keep the conductances and the α of the CC as variables. Check your derivations by running the symbolic option in NETOPT.

P.5.34 In P.5.13 keep the constant r of the CV as a variable and take the output across the dependent source. Select such value of r that the output voltage will be 10 V.

P.5.35 Transform the voltage source and use nodal formulation to get the nodal voltages V_1 and V_2.

P.5.36 Transform the voltage source, use nodal formulation and obtain V_{out}.

P.5.37 Use voltage source splitting and transform the voltage sources into current sources. Apply nodal formulation. All resistors have unit values. Obtain the nodal voltages V_2 and V_3.

P.5.38 Use nodal formulation after transforming the voltage source. Obtain V_{out}.

P.5.41 Apply voltage splitting to find the indicated nodal voltages. Also use current splitting to get the result.

P.5.39 Use nodal formulation to get the output voltage.

P.5.42 Use nodal formulation to solve for the output. As a matter of interest, the network in the dashed box is one of the possible models of a bipolar transistor.

P.5.40 Obtain the output voltage.

P.5.43 Transform the dependent voltage source into a current source and solve for V_3 using nodal formulation.

P.5.44 Use transformations to get a network suitable for nodal formulation and find V_{out}. Next transform the network to make it suitable for mesh formulation and find the same output.

P.5.45 First transform the dependent voltage source into a current source and calculate the voltages V_1 and V_2. Find I and obtain the output. Also transform the independent current source into a voltage source, use mesh formulation and get the same.

P.5.46 Transform the dependent current source into a voltage source and use mesh formulation to get the output voltage. Next transform the independent voltage source into a current source, use mesh formulation and obtain the same output voltage.

P.5.47 Analyze the network and find an expression for V_{out} in terms of all conductances as variables. How does G_5 influence the result? Explain.

P.5.48 Find V_{out} in terms of G and J.

P.5.49 Two conductances are given numerically and two are kept as variables. Find the expression for the output voltage and select the two conductances such that $V_{out} = 2E$.

CHAPTER 6

Capacitors and Inductors

6.0. INTRODUCTION

This chapter introduces two elements of fundamental importance: capacitors and inductors. The relationships of their currents and voltages are more complicated than those for the resistors and are expressed by differentiation and integration. We also define two new variables, the charge of the capacitor and the flux of the inductor. For the sake of completeness we give the basic relationships for nonlinear elements but all our discussions will use linear elements. On the simplest cases we show how one can set up differential equations for networks containing capacitors, inductors and resistors, but we will not go into details. While it is possible to set up such differential equations for very simple networks, it is quite difficult to do so for large networks, especially if they have dependent sources. We will later develop a unified method for networks of arbitrary sizes and with arbitrary components; it will be based on the theory of the Laplace transformations.

6.1. CAPACITORS

A capacitor is an element composed of two conducting materials, separated by an insulator. We will first say a few words about its use and how it can be manufactured.

Technological advances constantly change the way capacitors are produced, but the student can make a reasonable capacitor quite easily by using the simplest materials available to him. The insulating material can be paper. If it is waxed, even better. The conducting material can be tin foil from a chocolate bar. Take a narrow strip of paper, place somewhat narrower tin foil strips on each side and put another strip of paper on top. Connect a wire to each of the foils and roll together making sure that the two wires or foils do not touch. The result is a useful capacitor, similar to capacitors produced commercially for many years. If we connect the wires to the terminals of a car battery, a certain amount of electrons will be transferred to the capacitor. If we disconnect it from the battery, the electrons will stay and we will be able to measure a voltage on the two leads. If the insulating material is a good one, the electrons will stay for a long time; thus the capacitor can store energy. For an initial understanding of its usefulness let us say that an ideal capacitor stops dc current but lets changing currents pass through, making it easier for signals which are fast. The property is somewhat similar to the resistance, but a resistance which is small for fast signals and infinite for dc.

If electrons are stored in the capacitor, we say that there is charge on it. The charge is denoted by q or $q(t)$ (if we wish to indicate dependence on time) and is measured in coulombs, C. The symbol for a capacitor is in Fig. 6.1.1a.

In a general case, the charge is a function of the voltage across the capacitor,

$$q(t) = f[v(t)] \tag{6.1.1}$$

and we can plot q as function of v, as shown in Fig. 6.1.2. The current through the capacitor is the derivative of the charge with respect to time,

$$i(t) = \frac{dq(t)}{dt} \tag{6.1.2}$$

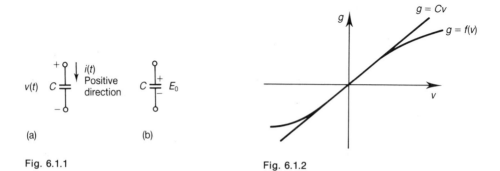

(a)   (b)

Fig. 6.1.1   Fig. 6.1.2

FIGURE 6.1.1   Symbol for a capacitor: (a) Voltage and current orientation. (b) Initial voltage E_0.

FIGURE 6.1.2 q–v characteristic of capacitors.

We can also express the charge as function of the current,

$$q(t) = \int_{-\infty}^{t} i(\tau)\, d\tau \qquad (6.1.3)$$

but it is inconvenient to work with an integral having its lower bound at $-\infty$. We prefer to split Eq. (6.1.3) into two integrals

$$q(t) = \int_{-\infty}^{0} i(\tau)\, d\tau + \int_{0}^{t} i(\tau)\, d\tau, \qquad (6.1.4)$$

denote the first integral as initial charge at time $t = 0$ and use the symbol q_0 or $q(0)$,

$$q(t) = q_0 + \int_{0}^{t} i(\tau)\, d\tau \qquad (6.1.5)$$

These expressions are valid for any capacitor.

A special case is the linear capacitor for which the charge is directly proportional to the voltage,

$$q(t) = Cv(t) \qquad (6.1.6)$$

and the proportionality constant is called the *capacitance*, C. It is measured in farads, F. The linear relationship of the voltage and charge is indicated in Fig. 6.1.2 by the straight line.

The current through a linear capacitor is given by

$$i(t) = C\frac{dv(t)}{dt} \qquad (6.1.7)$$

Positive direction of current is related to the voltage across the capacitor by the same rules as before: If the current flows from $+$ to $-$, then it is taken as positive. These conventions are also illustrated in Fig. 6.1.1a.

If we invert Eq. (6.1.7), we get an expression for the voltage on a linear capacitor

$$v(t) = \frac{1}{C}\int_{-\infty}^{t} i(\tau)\, d\tau \qquad (6.1.8)$$

Similarly as above, we split this integral into two,

$$v(t) = \frac{1}{C}\int_{-\infty}^{0} i(\tau)\, d\tau + \frac{1}{C}\int_{0}^{t} i(\tau)\, d\tau$$

denote the first integral by E_0 and call it the *initial* voltage on the capacitor,

$$v(t) = E_0 + \frac{1}{C} \int_0^t i(\tau)\, d\tau \tag{6.1.9}$$

We are using the letter E because, as we will see later, this initial voltage acts, in some sense, like an independent source. The way we will indicate the initial voltage in networks is shown in Fig. 6.1.1b.

The power relationship of the voltage across the capacitor and the current flowing through it is given by

$$p(t) = v(t)i(t) \tag{6.1.10}$$

The energy stored in a capacitor is given by the integral of Eq. (6.1.10),

$$w_C(t) = \int_{-\infty}^t p(\tau)\, d\tau = C \int_{-\infty}^t v(\tau)\frac{dv(\tau)}{d\tau}\, d\tau = C \int_{-\infty}^t v(\tau)\, dv(\tau) = \frac{1}{2} Cv^2(t) \tag{6.1.11}$$

6.2. CONNECTIONS OF CAPACITORS

Capacitors can be connected in parallel or in series and we now derive the rules to obtain the resulting capacitor values.

Consider first the parallel connection shown in Fig. 6.2.1. It is some kind of a current divider because all capacitors have the same voltage but different currents can flow through them. Let the current source change with time which we indicate by writing $j(t)$. The current flowing through each capacitor is given by Eq. (6.1.7):

$$i_1(t) = C_1 \frac{dv(t)}{dt}$$

$$i_2(t) = C_2 \frac{dv(t)}{dt} \tag{6.2.1}$$

$$i_3(t) = C_3 \frac{dv(t)}{dt}$$

FIGURE 6.2.1   Parallel connection of capacitors.

The respective currents add and give the overall current flowing into the combination,

$$j(t) = i_1(t) + i_2(t) + i_3(t)$$

$$= (C_1 + C_2 + C_3)\frac{dv(t)}{dt} \tag{6.2.2}$$

We can denote

$$C_{parallel} = C_1 + C_2 + C_3 \tag{6.2.3}$$

and obtain the total capacitance for a parallel connection. It is equal to the sum of the individual capacitances. The formula can be extended to any number of capacitors connected in parallel.

Series connection of capacitors is shown in Fig. 6.2.2. The combination resembles a voltage divider because the same current flows through all elements but individual voltages can differ. For each capacitor we can write

$$v_1(t) = \frac{1}{C_1} \int_0^t i(\tau)\, d\tau$$

$$v_2(t) = \frac{1}{C_2} \int_0^t i(\tau)\, d\tau \tag{6.2.4}$$

$$v_3(t) = \frac{1}{C_3} \int_0^t i(\tau)\, d\tau$$

The overall voltage is equal to

$$e(t) = v_1(t) + v_2(t) + v_3(t) \tag{6.2.5}$$

FIGURE 6.2.2   Series connection of capacitors.

Inserting from Eq. (6.2.4) we get

$$e(t) = \left(\frac{1}{C_1} + \frac{1}{C_2} + \frac{1}{C_3} \right) \int_0^t i(\tau)\, d\tau$$

and the overall capacitance for series connection is

$$\frac{1}{C_{series}} = \frac{1}{C_1} + \frac{1}{C_2} + \frac{1}{C_3} \qquad (6.2.6)$$

The formula is clearly valid for any number of capacitors connected in series.

If we compare these networks with similar resistive networks we realize that the capacitor is in some sense similar to the conductance, because the values add for parallel connections. For series connections we must add inverted values of capacitances and then take the reciprocal of this sum to get the equivalent capacitance.

6.3. INDUCTORS

An inductor is usually manufactured as a coil of insulated wire, often with an iron core or with a material which has properties similar to iron. The symbol for an inductor is in Fig. 6.3.1a.

Similarly as in the case of the capacitor, we give first an informal and very simplified explanation of its application. If we wish to have a good inductor, we must use a wire which has low resistance, for instance silver. Suppose that we manage to produce an inductor with a wire that has no resistance. Then the inductor will have properties which are opposite to those of a capacitor. It lets dc current pass through, but presents increasingly larger "resistance" to rapidly changing signals around the inductor. If a current flows through the inductor, it creates around it magnetic field which stores energy and which can have influence on other inductors placed nearby. This property is used in the construction of transformers but we postpone such details. The word we normally use is *flux*, we denote it by Φ and measure it in units called *webers*, Wb.

(a)   (b)

FIGURE 6.3.1   Symbol for an inductor: (a) Voltage and current orientations. (b) Initial current J_0.

In a general case, the flux is a function of the current flowing through the inductor,

$$\Phi(t) = f[i(t)] \tag{6.3.1}$$

We can plot this relationship, as shown in Fig. 6.3.2. The voltage across the inductor is defined as the derivative of the flux with respect to time,

$$v(t) = \frac{d\Phi(t)}{dt} \tag{6.3.2}$$

We can also invert this relationship and get

$$\Phi(t) = \int_{-\infty}^{t} v(\tau)\, d\tau \tag{6.3.3}$$

It is not convenient to work with the infinite bounds. We usually split this integral into two

$$\Phi(t) = \int_{-\infty}^{0} v(\tau)\, d\tau + \int_{0}^{t} v(\tau)\, d\tau \tag{6.3.4}$$

and denote the first integral as an initial flux, $\Phi(0)$ or Φ_0,

$$\Phi(t) = \Phi_0 + \int_{0}^{t} v(\tau)\, d\tau \tag{6.3.5}$$

These expressions are valid for any inductor.

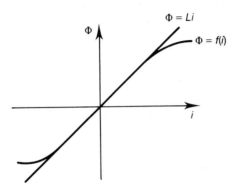

FIGURE 6.3.2 Φ–i characteristic of inductors.

A special case is the linear inductor. In this case the flux is directly proportional to the current,

$$\Phi(t) = Li(t) \tag{6.3.6}$$

and the proportionality constant is called the *inductance, L*. It is measured in henries, H. This relationship is represented by the straight line in Fig. 6.3.2.

The voltage across a linear inductor is given by

$$v(t) = L\frac{di(t)}{dt} \tag{6.3.7}$$

and is related to the positive direction of the current by the same rules as always: the current is positive if it flows from $+$ to $-$. The convention is also illustrated in Fig. 6.3.1a.

If we invert Eq. (6.3.7), we get an expression for the current through a linear inductor

$$i(t) = \frac{1}{L}\int_{-\infty}^{t} v(\tau)\, d\tau \tag{6.3.8}$$

Similarly as above, we split this integral into two integrals,

$$i(t) = \frac{1}{L}\int_{-\infty}^{0} v(\tau)\, d\tau + \frac{1}{L}\int_{0}^{t} v(\tau)\, d\tau$$

denote the first integral by J_0 and call it the *initial* current through the inductor,

$$i(t) = J_0 + \frac{1}{L}\int_{0}^{t} v(\tau)\, d\tau \tag{6.3.9}$$

We are using the letter J because this initial current acts, in some way, similarly as an independent current source. We will mark the initial current through the inductor as shown in Fig. 6.3.1b.

The power relationship of the voltage across the inductor and the current flowing through it is

$$p(t) = v(t)i(t) \tag{6.3.10}$$

The inductor is an energy storing device and the energy stored in it is given by the integral of Eq. (6.3.10),

$$w_L(t) = \int_{-\infty}^{t} p(\tau)\, d\tau = L\int_{-\infty}^{t} i(\tau)\frac{di(\tau)}{dt}\, d\tau = L\int_{-\infty}^{t} i(\tau)\, di(\tau) = \frac{1}{2}Li^2(t) \tag{6.3.11}$$

The reader may have noticed that the material of this section is remarkably similar to the material of the first section of this chapter. This is no coincidence: the capacitor and the inductor are indeed very similar in the sense that all we do is exchange voltages for currents and vice versa.

6.4. CONNECTIONS OF INDUCTORS

Inductors can be connected in series or in parallel and we will derive the rules to obtain the resulting inductance values.

We first consider the series connection, shown in Fig. 6.4.1. It resembles the voltage divider: the same current flows through the individual elements but the voltages across them differ. The sum of individual voltages is equal to the voltage of the source, $e(t)$. Using Eq. (6.3.7) we write the voltage across each inductor

$$v_1(t) = L_1 \frac{di(t)}{dt}$$

$$v_2(t) = L_2 \frac{di(t)}{dt} \qquad (6.4.1)$$

$$v_3(t) = L_3 \frac{di(t)}{dt}$$

The respective voltages add and give the overall voltage across the combination

$$e(t) = v_1(t) + v_2(t) + v_3(t) \qquad (6.4.2)$$

Inserting from Eq. (6.4.1)

$$e(t) = (L_1 + L_2 + L_3) \frac{di(t)}{dt}$$

FIGURE 6.4.1   Series connection of inductors.

We can denote

$$L_{series} = L_1 + L_2 + L_3 \tag{6.4.3}$$

and obtain the total inductance for the series connection. It is equal to the sum of the individual inductances. The formula can be extended to any number of inductors connected in series.

Parallel connection of inductors is shown in Fig. 6.4.2. The combination resembles a current divider where the voltage is common to all elements but the currents through the individual elements differ. For each inductor we can write

$$i_1(t) = \frac{1}{L_1} \int_0^t v(\tau) \, d\tau$$

$$i_2(t) = \frac{1}{L_2} \int_0^t v(\tau) \, d\tau \tag{6.4.4}$$

$$i_3(t) = \frac{1}{L_3} \int_0^t v(\tau) \, d\tau$$

The total current is equal to

$$j(t) = i_1(t) + i_2(t) + i_3(t) \tag{6.4.5}$$

Inserting from Eq. (6.4.4) we get

$$j(t) = \left(\frac{1}{L_1} + \frac{1}{L_2} + \frac{1}{L_3} \right) \int_0^t v(\tau) \, d\tau$$

and the overall inductance for parallel connection is

$$\frac{1}{L_{parallel}} = \frac{1}{L_1} + \frac{1}{L_2} + \frac{1}{L_3} \tag{6.4.6}$$

The formula is clearly valid for any number of inductors connected in parallel.

FIGURE 6.4.2   Parallel connection of inductors.

If we compare the connections of inductors with similar connections of resistors, we see that inductors are in some sense similar to resistors, because their inductances are added in series connections. In parallel connections we must add the inverted values of individual inductances. We thus again observe the dual nature of capacitors and inductors.

6.5. FIRST-ORDER NETWORKS

Responses of networks to suddenly changing signals have been traditionally found by solving the appropriate differential equations. In introductory texts, the explanations were always restricted to very primitive cases, the first- and second-order networks. This is no coincidence; if the networks are larger and have dependent sources, it is quite difficult to set up the appropriate differential equations. Moreover, it is not possible to unify this *time domain* method with other types of analyses. A more general and much more versatile way is to use the Laplace transformation and we strongly recommend its use. In an attempt to unify the theory, we structured this text so that emphasis is given to the Laplace transformation.

However, since tradition is also of importance, and some instructors may prefer to go the traditional way, we devote the rest of this chapter to the solutions of simple networks by means of the differential equations. We assume that the theory of linear differential equations has already been covered elsewhere.

Until now we always assumed that the source connected to a network is a *dc* source, i.e. a source of constant voltage or current, for instance from a battery. This also tacitly assumed that the connection happened a long time ago. Very often we also wish to know what is going on from the moment we connect the source.

Let us now see what takes place if we connect a source of constant voltage of E volts to a capacitor. The capacitor will charge immediately to the voltage $v = E$. This immediate charge means that the current must be some very short impulse. Afterwards, there will be no current, because

$$i_C(t) = C \frac{dv(t)}{dt} = 0$$

and the derivative of a constant is zero. Similarly, if we attach a source of constant current to an inductor, an impulse of voltage will occur across the inductor but afterwards the voltage will be zero because

$$v_L(t) = L \frac{di(t)}{dt} = 0$$

and $i(t)$ is constant. If the network is complicated, we must set up the appropriate equations; this we will demonstrate on several simple cases.

Consider first the network in Fig. 6.5.1; it is easy to see that the same current flows through the resistor and through the inductor. The network looks like a

Fig. 6.5.1

Fig. 6.5.2

FIGURE 6.5.1   Series R–L network.

FIGURE 6.5.2   Parallel R–C network.

voltage divider and the sum of voltages across the individual elements must be equal to the voltage of the source. The voltages across the elements are given by the equations

$$v_L = L \frac{di}{dt}$$

$$v_R = Ri$$

(6.5.1)

and, because of KVL,

$$v_L + v_R = e(t)$$

(6.5.2)

Inserting from Eq. (6.5.1) we get

$$L \frac{di}{dt} + Ri = e(t)$$

or, after dividing by L,

$$\frac{di}{dt} + \frac{R}{L} i = \frac{e(t)}{L}$$

(6.5.3)

This is a first-order differential equation, prepared for the solution.

Another network is shown in Fig. 6.5.2, a parallel connection of C, R and $j(t)$. In this network all elements have the same voltage and the sum of currents through C and G must be equal to the current delivered by the source. For the individual elements we have

$$i_C = C \frac{dv}{dt}$$

$$i_R = \frac{v}{R}$$

(6.5.4)

and we also have

$$i_C + i_R = j(t) \qquad (6.5.5)$$

Inserting from (6.5.4)

$$C\frac{dv}{dt} + \frac{v}{R} = j(t)$$

and dividing by C

$$\frac{dv}{dt} + \frac{1}{RC}v = \frac{j(t)}{C} \qquad (6.5.6)$$

This differential equation is of the same type as Eq. (6.5.3); we will use it in the following.

Step 1. Start with the *homogeneous* equation, which is our original differential equation, Eq. (6.5.6), with its right-hand side replaced by zero.

$$\frac{dv}{dt} + \frac{1}{RC}v = 0 \qquad (6.5.7)$$

Assume that the possible solution has the form

$$v_H(t) = k_1 e^{st} \qquad (6.5.8)$$

The derivative is

$$\frac{dv_H(t)}{dt} = k_1 s e^{st} \qquad (6.5.9)$$

and we insert both Eq. (6.5.8) and Eq. (6.5.9) into Eq. (6.5.7):

$$k_1 s e^{st} + \frac{1}{RC} k_1 e^{st} = 0$$

This can be rewritten as

$$k_1 e^{st}\left(s + \frac{1}{RC}\right) = 0$$

which is true only if

$$s = -\frac{1}{RC}$$

The product RC was given the name *time constant*

$$\tau = RC \qquad (6.5.10)$$

Substitution into Eq. (6.5.8) gives the homogeneous solution

$$v_H(t) = k_1 e^{-t/\tau} \tag{6.5.11}$$

Step 2. For any given input signal we must find, in addition, the *particular* solution. This involves a more general form than the signal actually has. For instance, if the signal is $j(t) = J$, a constant, we assume the particular solution in the form

$$v_P(t) = A \tag{6.5.12}$$

where A is also a constant, but not necessarily equal to J. The derivative of Eq. (6.5.12) is

$$\frac{dv_P(t)}{dt} = 0 \tag{6.5.13}$$

Inserting both these expressions into Eq. (6.5.6) we obtain

$$0 + \frac{1}{RC} A = \frac{J}{C}$$

or

$$A = RJ \tag{6.5.14}$$

Step 3. The complete solution is the sum of the two solutions

$$v(t) = v_H(t) + v_P(t) = k_1 e^{-t/\tau} + RJ \tag{6.5.15}$$

Step 4. The unknown constant, k_1, is determined from the initial voltage across the capacitor, E_0. At $t = 0$ the voltage $v(0) = E_0$. Inserting into Eq. (6.5.15)

$$E_0 = k_1 + RJ$$

or

$$k_1 = E_0 - RJ$$

The complete solution is

$$v(t) = E_0 e^{-t/\tau} - RJe^{-t/\tau} + RJ \qquad t \geq 0 \tag{6.5.16}$$

The current which we suddenly changed from zero to the value J appears in the solution in the constant RJ. It forms the so called *steady state* component. The other two components are called *transients* because they change with time and eventually become zero.

We will consider separately three possible situations

a. No signal, $J = 0$, but nonzero initial voltage, and $E_0 \neq 0$.
   Equation (6.5.16) reduces to

$$v(t) = E_0 e^{-t/\tau} \qquad t \geq 0$$

The plot of this function for $E_0 = 1$ is in Fig. 6.5.3. It is also quite instructive to give a few of its values: for $t = \tau$ the response is $0.3678E_0$. For $t = 5\tau$ the function has the value $0.00674E_0$, only approximately 0.7% of its maximum value. We usually say that for all practical purposes the function is zero for $t > 5\tau$.

b. Zero initial voltage, $E_0 = 0$, nonzero dc signal.
   Equation (6.5.16) simplifies to

$$v(t) = RJ(1 - e^{-t/\tau}) \qquad t \geq 0$$

This curve is plotted in Fig. 6.5.4 for $RJ = 1$. It grows from zero and at

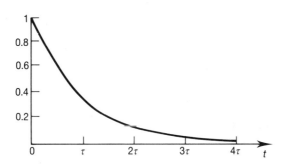

FIGURE 6.5.3   Function $f(t) = e^{-t/\tau}, t \geq 0$.

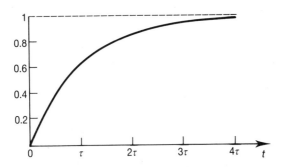

FIGURE 6.5.4   Function $f(t) = 1 - e^{-t/\tau}, t \geq 0$.

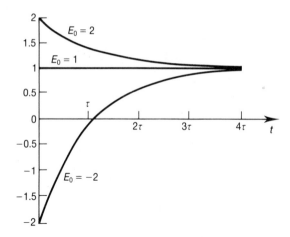

FIGURE 6.5.5   Functions $f(t) = E_0 e^{-t/\tau} + 1 - e^{-t/\tau}$, $t \geq 0$ for indicated initial values.

$t = \tau$ reaches approximately 63% of its final value. For most practical purposes it reaches its final value for times $t > 5\tau$.

c. Both the source and initial voltage exist.

This case is the full solution in Eq. (6.5.16). The curve will start at $v(0) = E_0$ and will reach the steady state value RJ. Figure 6.5.5 sketches three situations for $RJ = 1$: $E_0 = -2$, $E_0 = 1$ and $E_0 = 2$.

Example 1.

Find the response of the network in Fig. 6.5.2 for the input signal $j(t) = \cos 2t$. To simplify the steps let $R = 1\ \Omega$ and $C = 1$ F.

The differential equation has now the form

$$\frac{dv}{dt} + v = \cos 2t \tag{6.5.17}$$

The homogeneous solution is already known,

$$v_H(t) = k_1 e^{-t}$$

For the particular solution we assume a more general term than is the input signal,

$$v_P(t) = A \cos 2t + B \sin 2t$$

Its derivative is

$$\frac{dv_P(t)}{dt} = -2A \sin 2t + 2B \cos 2t$$

Inserting into Eq. (6.5.17) leads to

$$-2A \sin 2t + 2B \cos 2t + A \cos 2t + B \sin 2t = \cos 2t$$

Comparing terms on both sides of the equation we conclude that

$$-2A + B = 0$$

$$A + 2B = 1$$

with the solution $A = 1/5$, $B = 2/5$. The complete solution is

$$v(t) = v_H(t) + v_P(t) = k_1 e^{-t} + \frac{1}{5} \cos 2t + \frac{2}{5} \sin 2t \qquad (6.5.18)$$

Suppose that the initial voltage is E_0 at $t = 0$. Substituting into Eq. (6.5.18) gets us

$$E_0 = k_1 + \frac{1}{5}$$

or

$$k_1 = E_0 - \frac{1}{5}$$

The final result is

$$v(t) = E_0 e^{-t} - \frac{1}{5} e^{-t} + \frac{1}{5} \cos 2t + \frac{2}{5} \sin 2t \qquad t \geq 0$$

In this expression we have again the transient terms, those with e^{-t}, and steady state terms, those with $\cos 2t$ and $\sin 2t$.

It is instructive to consider at this moment how the capacitor and inductor behave at the time of switching, and at time which is very far from switching, theoretically infinitely far. A suddenly connected dc voltage will be described by the graph in Fig. 6.5.6a. If the voltage is equal to 1 V, then such a signal is called the *unit step*, $u(t)$. The derivative of this signal is zero everywhere except at time $t = 0$. If we look at the break from the left, the angle will change to $+90$ degrees and the tangent will become infinitely large. Immediately afterwards, in fact at the very same instant, the signal changes again, back to zero, and the derivative is again infinitely large, but with negative sign. Altogether, there will be an infinitely large impulse of zero duration; this impulse, Fig. 6.5.6b, was given the name *Dirac impulse*, $\delta(t)$, and we will deal with it extensively later. At this time all we have to realize is that the current through the capacitor, defined by the derivative $i = C \, dv/dt$, will be very large; the capacitor will behave as a short circuit. After long time, the signal will become a true dc signal and its derivative will be zero; the capacitor will behave as an open circuit. One more situation needs to be considered, the case when the capacitor has an initial voltage. At this time, and *only* at this time, it behaves as a voltage source. All three situations are collected

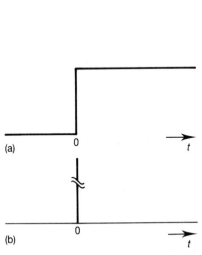

(a)   0 t

(b)   0 t

Fig. 6.5.6

Fig. 6.5.7

FIGURE 6.5.6   Simple signals: (a) Unit step, $u(t)$, and (b) Dirac impulse, $\delta(t)$.

FIGURE 6.5.7   Behavior of capacitors and inductors at $t = 0$ and $t = \infty$.

in Fig. 6.5.7 for both the capacitor and inductor. We suggest that you follow the above argument for the inductor to convince yourself about the validity of this figure.

We will now analyze the network in Fig. 6.5.8. The voltage across the capacitor is

$$v_C(t) = \frac{1}{C} \int i \, dt$$

and the sum of voltages around the loop is equal to

$$\frac{1}{C} \int i \, dt + iR = e(t)$$

In order to obtain a differential equation, we differentiate to get

$$\frac{di}{dt} + \frac{1}{RC} i = \frac{1}{R} \frac{de(t)}{dt} \tag{6.5.19}$$

Fig. 6.5.8

Fig. 6.5.9

FIGURE 6.5.8   Series $R-C$ network.

FIGURE 6.5.9   Parallel $R-L$ network.

We will return to this equation later, but now we derive a similar equation for the network in Fig. 6.5.9. Since

$$i_L = \frac{1}{L} \int v \, dt$$

the sum of currents will be

$$\frac{1}{L} \int v \, dt + \frac{1}{R} v = j(t)$$

This equation is differentiated and simplified to

$$\frac{dv}{dt} + \frac{R}{L} v = R \frac{dj(t)}{dt} \qquad (6.5.20)$$

We can see that Eq. (6.5.19) and Eq. (6.5.20) have the same form and we need to consider only one; we select Eq. (6.5.19).

The homogeneous equation is solved the same way as before. Denoting $\tau = RC$,

$$i_H(t) = k_1 e^{-t/\tau}$$

The right-hand side of Eq. (6.5.19) is the derivative of the input signal. If it is a suddenly connected dc voltage, then its derivative is zero and

$$i_P(t) = 0$$

The complete solution is

$$i(t) = k_1 e^{-t/\tau} \qquad (6.5.21)$$

For the initial voltage on the capacitor we make use of Fig. 6.5.7: at $t = 0$ the initially charged capacitor behaves as a voltage source and the initial current will be

$$i(0) = \frac{E - E_0}{R}$$

Substituting this initial condition into Eq. (6.5.21) leads to $(E - E_0)/R = k_1$. The solution is

$$i(t) = \frac{E - E_0}{R} e^{-t/\tau}$$

We suggest that you plot this response for $E = 1$ and for E_0 equal to $+2$ V and -2 V. You can use NETOPT to display the responses. We also suggest that you follow the above steps for the network in Fig. 6.5.9.

6.6. SECOND-ORDER NETWORKS

The name second-order networks applies to situations in which the solution of the homogeneous equation leads to a polynomial of second order. The networks may have one capacitor, one inductor and one or more resistors. However, second-order networks are also those having two capacitors (or two inductors), with one or more resistors. In this section we will indicate how we can set up and solve the differential equations. We will consider the series connection of R, L, C, as shown in Fig. 6.6.1, with initial voltage, E_0, on the capacitor and initial current, J_0, through the inductor.

The same current flows through all three elements and the voltages across them add. Thus

$$v_R + v_L + v_C = e(t)$$

or

$$Ri + \frac{1}{C} \int i \, dt + L \frac{di}{dt} = e(t) \tag{6.6.1}$$

FIGURE 6.6.1   Series RLC network with initial conditions.

Differentiating Eq. (6.6.1) we arrive at a second-order linear differential equation

$$\frac{d^2i}{dt^2} + \frac{R}{L}\frac{di}{dt} + \frac{1}{LC} = \frac{1}{L}\frac{de(t)}{dt} \tag{6.6.2}$$

The homogeneous equation is

$$\frac{d^2i}{dt^2} + \frac{R}{L}\frac{di}{dt} + \frac{1}{LC} = 0 \tag{6.6.3}$$

and we assume that its solution has the form

$$i = e^{st}$$

The derivatives are

$$\frac{di}{dt} = se^{st}$$

$$\frac{d^2i}{dt^2} = s^2e^{st}$$

Inserting these terms into Eq. (6.6.3) we get

$$e^{st}\left(s^2 + \frac{R}{L}s + \frac{1}{LC}\right) = 0$$

which can be satisfied only if the polynomial in the bracket is equal to zero:

$$s^2 + \frac{R}{L}s + \frac{1}{LC} = 0$$

This equation has two roots:

$$s_{1,2} = -\frac{R}{2L} \pm \left(\frac{R^2}{4L^2} - \frac{1}{LC}\right)^{1/2} \tag{6.6.4}$$

Depending on the values in this expression, we can have three situations:

a. If the expression under the square root, called the *discriminant*, is larger than zero, then we have two real roots, as given by Eq. (6.6.4).
b. If the discriminant is zero, we have one real double root.

$$s_{1,2} = -\frac{R}{2L} \tag{6.6.5}$$

c. If the discriminant is smaller than zero, we have two complex conjugate
roots

$$s_{1,2} = -\frac{R}{2L} \pm j\left(\frac{1}{LC} - \frac{R^2}{4L^2}\right)^{1/2} \tag{6.6.6}$$

In case (a) the homogeneous solution is

$$i_H(t) = k_1 e^{s_1 t} + k_2 e^{s_2 t} \tag{6.6.7}$$

In case (b) the homogeneous solution is

$$i_H(t) = t e^{st} \tag{6.6.8}$$

and in case (c) the homogeneous solution is

$$i_H(t) = k_1 e^{(\alpha + j\beta)t} + k_2 e^{(\alpha + j\beta)t} \tag{6.6.9}$$

where α and β are the real and imaginary parts in Eq. (6.6.6).

Particular solution depends, as always, on the input signal. If it is a suddenly
switched dc voltage, then we need its first derivative, as requested by Eq. (6.6.2).
For a dc source it is zero and there is no particular solution. The final result is

$$i(t) = i_H(t) \tag{6.6.10}$$

We must still determine the so-far unknown constants in the homogeneous
equations. The initial current is equal to the initial current of the inductor,

$$i(0) = J_0 \tag{6.6.11}$$

To find the derivative of $i(0)$ at $t = 0$ we go back to the original Eq. (6.6.1). At
$t = 0$ we have

$$RJ_0 + E_0 + L\frac{di(t)}{dt}\bigg|_{t=0} = E$$

and from here

$$\frac{di(t)}{dt}\bigg|_{t=0} = \frac{E - E_0 - RJ_0}{L} \tag{6.6.12}$$

For the cases (a) or (c) we have found

$$i(t) = k_1 e^{s_1 t} + k_2 e^{s_2 t} \tag{6.6.13}$$

and differentiating

$$\frac{di}{dt} = s_1 k_1 e^{s_1 t} + s_2 k_2 e^{s_2 t} \tag{6.6.14}$$

Setting $t = 0$ in Eq. (6.6.13) and Eq. (6.6.14) and using the initial conditions Eq. (6.6.11) and Eq. (6.6.12), we arrive at a set of two linear equations in two unknowns:

$$0 = k_1 + k_2$$

$$\frac{E - E_0 - RJ_0}{L} = s_1 k_1 + s_2 k_2$$

Their solution provides the necessary k_1 and k_2.

The case (b) leads to the solution

$$i(t) = k_1 t e^{st}$$

which we differentiate

$$\frac{di}{dt} = k_1 e^{st} + k_1 t s e^{st}$$

In this second equation we set $t = 0$ and use for the left-hand side Eq. (6.6.12), with the result

$$k_1 = \frac{E - E_0 - RJ_0}{L}$$

Inserting the constants into Eq. (6.6.7), Eq. (6.6.8) or Eq. (6.6.9) will give the solution of the problem. We suggest that you run NETOPT to see the responses. Use various initial conditions and various values R and for simplicity set $L = 1$ H, $C = 1$ F.

PROBLEMS CHAPTER 6

P.6.1 If the capacitor has the value $C = 3$ F and the voltage across it changes with the function $v_C(t) = 3 \sin 5t$, what is the function of the current $i_C(t)$ flowing through it?

P.6.3 The sketch shows the function of $v_C(t)$ for a capacitor having the value $C = 2$ F. Find the function of current, $i_C(t)$, flowing through it.

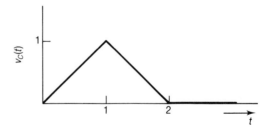

P.6.2 If the capacitor has the value $C = 3$ F and the current through it changes with the function $i_C(t) = 3 \sin 5t$, what is the function of the voltage $v_C(t)$ across it?

P.6.4 Do the same as in P.6.3 for this $v_C(t)$.

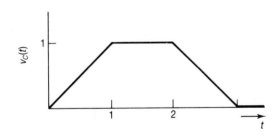

P.6.5a–c. Find equivalent capacitances of the networks.

(a)

(b)

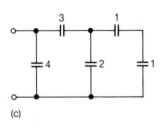

(c)

P.6.6 Find equivalent capacitance of the combination.

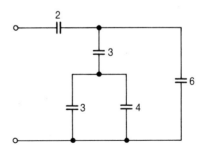

P.6.7 If the inductor has the value $L = 2$ H and $i_L(t) = 3 \sin 4t$, what is the function $v_L(t)$?

P.6.8 If the inductor has the value $L = 2$ H and $v_L(t) = 3 \sin 4t$, what is the current $i_L(t)$?

P.6.9 The sketch shows the function of $i_L(t)$ for an inductor having the value $L = 2$ H. Find the function of voltage, $v_L(t)$, appearing across it.

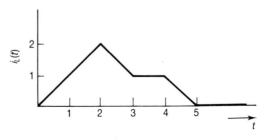

P.6.10 Do the same as in P.6.9 for this $i_L(t)$.

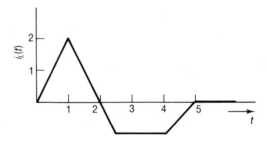

P.6.11a–c. Find equivalent inductances of the networks.

(a)

(b)

(c)

P.6.12 Find equivalent inductance for the combination.

If you wish to practice solutions of networks by differential equations, turn to the problems in Chapter 10. Their time domain solutions are given there.

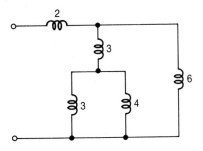

CHAPTER 7

Networks with Capacitors and Inductors

7.0. INTRODUCTION

This chapter covers the first steps in the analysis of linear networks in the Laplace transform. The concepts of impedance and admittance are defined and it is shown that analysis of networks which have resistors, dependent sources, inductors and capacitors, proceeds along exactly the same lines as were covered in the previous chapters. The only difference is that instead of only R or G we use other expressions. Examples show that in the Laplace transform the solutions of networks are expressed as ratios of two polynomials in the variable s. The only purpose of this chapter is to introduce to the student the necessary mechanics to analyze arbitrary networks. Detailed explanations and the theories are covered in subsequent chapters.

The symbolic option of the program NETOPT, explained in Section 18.5, can confirm all derivations done by hand.

7.1. IMPEDANCES AND ADMITTANCES

Laplace transform is a method used for solving linear differential equations and all linear networks can be solved by it. Its main feature is that it replaces differentiation by a multiplication with a special variable, s. It also replaces integration by the division by s and thus transforms transcendental operations into algebraic operations. For instance, Eq. (6.1.7), repeated here for convenience

$$i = C \frac{dv}{dt} \tag{7.1.1}$$

is replaced by

$$I = sCV \tag{7.1.2}$$

and Eq. (6.3.1)

$$v = L \frac{di}{dt} \tag{7.1.3}$$

is replaced by

$$V = sLI \tag{7.1.4}$$

If we take the integral expressions, then Eq. (6.1.8), repeated here

$$v = \frac{1}{C} \int_0^t i(\tau)\, d\tau \tag{7.1.5}$$

is replaced by

$$V = \frac{1}{sC} I \tag{7.1.6}$$

and Eq. (6.3.8), also repeated,

$$i = \frac{1}{L} \int_0^t v(\tau)\, d\tau \tag{7.1.7}$$

is replaced by

$$I = \frac{1}{sL} V \tag{7.1.8}$$

The reader may have noticed that the integral forms Eq. (7.1.5) and Eq. (7.1.7) were slightly modified and start from zero. In all the above expressions we have neglected the possibility that an initial charge can be stored on the capacitor or that an initial flux may be created around the inductor. This we do intentionally, to make the initial explanations as simple as possible. We will return to the full possibilities later.

At this moment we do not say what s exactly means and we will see later that it can be given different meanings, depending on whether we are interested in time responses (already briefly introduced in Chapter 6) or in frequency domain responses (to be explained later). Also, until now, the source was always a dc source; this may not be so in the future. From now on, E or J will be *symbols* for the two independent sources, essentially indicating that such terms should go on the right side of equations. Their exact meaning and/or values will differ from application to application and different substitutions will be needed.

Looking at the above equations, we realize that they all have the form of Ohm's law. In fact, we can collect the terms and write

$$Y = \begin{cases} G \\ sC \\ \dfrac{1}{sL} \end{cases} \tag{7.1.9}$$

The symbol Y is called the *admittance* and applies to all three forms. We can also collect the other set and write

$$Z = \begin{cases} R \\ \dfrac{1}{sC} \\ sL \end{cases} \tag{7.1.10}$$

The symbol Z applies for all three terms and is called the *impedance*. The set Eq. (7.1.9) is used whenever we write the nodal equations, the set Eq. (7.1.10) whenever we write the mesh equations. What is most important, all the analysis steps we covered in previous chapters are directly applicable to networks with capacitors and inductors. All we do is substitute the proper expression for the element.

7.2. NETWORKS WITH CAPACITORS

If the network contains only resistors and capacitors, then it is natural to use nodal formulation and transform all sources (dependent or independent) to current sources. As before, the resistors are expressed in their conductance equivalents.

The first network which we will analyze is shown in Fig. 7.2.1. The currents are $I_G = GV$, $I_C = sCV$ and the nodal equation is

$$(G + sC)V = J$$

FIGURE 7.2.1   GC network with a current source.

FIGURE 7.2.2   RC network with a voltage source.

The unknown voltage V is

$$V = \frac{1}{G + sC}\, J$$

We stress again that, depending on application, s as well as J may be later replaced by something else.

Suppose that instead of the current source we have a voltage source and the network in Fig. 7.2.2. In this case it is easy to use the mesh method and write

$$\left(R + \frac{1}{sC}\right)I = E$$

The current is

$$I = \frac{sC}{sCR + 1}\, E$$

and the output voltage is given by

$$V_{out} = Z_C I = \frac{1}{sCR + 1}\, E$$

The network could also be looked upon as a voltage divider. Using impedances we have

$$V_{out} = \frac{Z_C}{Z_R + Z_C}\, E = \frac{1}{sCR + 1}\, E$$

Another network is in Fig. 7.2.3a, with a voltage source, two capacitors and two resistors. The voltage source is first transformed into a current source as shown in Fig. 7.2.3b. The two nodal equations are

$$(G_1 + sC_1 + G_2)V_1 - G_2V_2 = EG_1$$
$$-G_2V_1 + (G_2 + sC_2)V_2 = 0$$

FIGURE 7.2.3   Network with two capacitors: (a) Original. (b) Transformed for nodal formulation.

The equations can be put into matrix form

$$
\begin{bmatrix} (G_1 + G_2 + sC_1) & -G_2 \\ -G_2 & G_2 + sC_2 \end{bmatrix} \begin{bmatrix} V_1 \\ V_2 \end{bmatrix} = \begin{bmatrix} EG_1 \\ 0 \end{bmatrix}
$$

Inserting numerical values

$$
\begin{bmatrix} 2s + 7 & -4 \\ -4 & 3s+4 \end{bmatrix} \begin{bmatrix} V_1 \\ V_2 \end{bmatrix} = \begin{bmatrix} 3E \\ 0 \end{bmatrix}
$$

Solution proceeds exactly the same way as we have been doing before, only the steps are now somewhat more complicated because we must keep the s variable as a letter. The determinant of the system matrix is

$$
D = \begin{vmatrix} 2s + 7 & -4 \\ -4 & 3s+4 \end{vmatrix} = (2s + 7)(3s + 4) - 16 = 6s^2 + 29s + 12
$$

The numerator for V_1 is

$$
N_1 = \begin{bmatrix} 3E & -4 \\ 0 & 3s+4 \end{bmatrix} = (9s + 12)E
$$

and the numerator for V_2 is

$$
N_2 = \begin{bmatrix} 2s + 7 & 3E \\ -4 & 0 \end{bmatrix} = 12E
$$

The solution in the Laplace domain is

$$V_1 = \frac{9s + 12}{6s^2 + 29s + 12} E$$

$$V_2 = \frac{12}{6s^2 + 29s + 12} E$$

Example 1.

The network in Fig. 7.2.4 has one capacitor and a voltage-controlled current source. Find the two nodal voltages.

The nodal equations are

$$(G_1 + sC)V_1 - sCV_2 = J$$

$$-sCV_1 + (G_2 + sC)V_2 - g(V_1 - V_2) = 0$$

In matrix form

$$\begin{bmatrix} G_1 + sC & -sC \\ -sC - g & G_2 + sC + g \end{bmatrix} \begin{bmatrix} V_1 \\ V_2 \end{bmatrix} = \begin{bmatrix} J \\ 0 \end{bmatrix}$$

Inserting numerical values

$$\begin{bmatrix} 2s + 2 & -2s \\ -2s - 3 & 2s + 6 \end{bmatrix} \begin{bmatrix} V_1 \\ V_2 \end{bmatrix} = \begin{bmatrix} J \\ 0 \end{bmatrix}$$

The denominator is $D = 10s + 12$, the numerators are $N_1 = (2s + 6)J$ and $N_2 = (2s + 3)J$. The voltages are

$$V_1 = \frac{2s + 6}{10s + 12} J$$

$$V_2 = \frac{2s + 3}{10s + 12}$$

FIGURE 7.2.4   Network with a VC.

Example 2.

The network in Fig. 7.2.5a has a current-controlled current source and we wish to obtain the voltages V_1 and V_2. Following the steps explained in Section 5.2 we first express the current I by using the original figure: $I = sC_1(E - V_1)$. As can be seen, nothing has changed as far as the analysis steps are concerned. Once this current has been determined and written to the dependent source, we can apply the source transformations, again exactly the same way as before. The transformed network is in Fig. 7.2.5b where we have written the controlling current to the controlled current source. The nodal equations are

$$(G_1 + sC_1 + sC_2)V_1 - sC_2V_2 = EsC_1$$
$$-sC_2V_1 + (sC_2 + sC_3 + G_2)V_2 - \alpha sC_1(E - V_1) = 0$$

Inserting numerical values and transferring the terms with E to the right-hand side we get the system

$$\begin{bmatrix} 4s + 3 & -2s \\ 2s & 3s + 4 \end{bmatrix} \begin{bmatrix} V_1 \\ V_2 \end{bmatrix} = \begin{bmatrix} 2sE \\ 4sE \end{bmatrix}$$

The solutions are: $D = 16s^2 + 25s + 12$, $N_1 = (14s^2 + 8s)E$ and $N_2(12s^2 + 12s)E$. The nodal voltages are

$$V_1 = \frac{14s^2 + 8s}{16s^2 + 25s + 12} E$$

$$V_2 = \frac{12s^2 + 12s}{16s^2 + 25s + 12} E$$

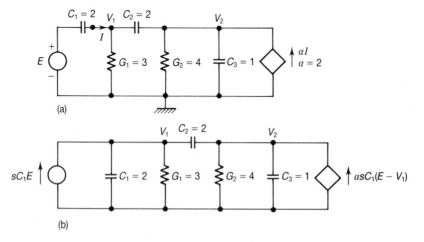

FIGURE 7.2.5   Network with a CC. (a) Original. (b) Transformed for nodal formulation.

Example 3.

The network in Fig. 7.2.6a has only voltage sources and thus can be immediately analyzed by the mesh method. It can also be transformed to make it suitable for nodal formulation. Use both methods to obtain the output voltage.

In the mesh method we first find that the controlling current is $I = I_1 - I_2$. The mesh equations are

$$\left(\frac{1}{2s} + 3\right)I_1 - 3I_2 = E$$

$$-3I_1 + \left(3 + \frac{1}{s}\right)I_2 + 2(I_1 - I_2) = 0$$

Collecting terms multiplied by the same current we get

$$\begin{bmatrix} \dfrac{6s + 1}{2s} & -3 \\ -1 & \dfrac{s + 1}{s} \end{bmatrix} \begin{bmatrix} I_1 \\ I_2 \end{bmatrix} = \begin{bmatrix} E \\ 0 \end{bmatrix}$$

The determinant is

$$D = \frac{6s^2 + 7s + 1}{2s^2} - 3 = \frac{7s + 1}{2s^2}$$

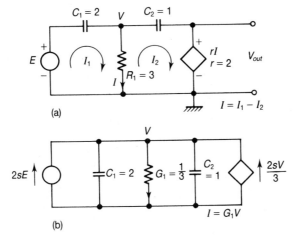

(a)

(b)

FIGURE 7.2.6   Network with a *CC*: (a) Original. (b) Transformed for nodal formulation.

The numerators are $N_1 = [(s + 1)/s]E$ and $N_2 = E$, the currents are

$$I_1 = \frac{2s^2 + 2s}{7s + 1} E$$

$$I_2 = \frac{2s^2}{7s + 1}$$

The output voltage is $V_{out} = rI = 2(I_1 - I_2)$ and the result is

$$V_{out} = \frac{4s}{7s + 1} E$$

The problem simplifies considerably if we transform the voltage sources into current sources and use nodal formulation. The network is redrawn in Fig. 7.2.6b, the controlling current is now $I = GV = (1/3)V$. Transferring this information to the dependent source we have to solve just one equation for the unknown V,

$$\left(3s + \frac{1}{3}\right)V - \frac{2s}{3}V = 2sE$$

with the result

$$V = \frac{6s}{7s + 1} E$$

This voltage is not yet the desired output voltage, it is the voltage at the node in the middle of Fig. 7.2.6a. However, since the output voltage is $V_{out} = rI = 2I$ and the current was determined as $I = GV$, all we have to do is collect the partial results and get

$$V_{out} = rGV = \frac{4s}{7s + 1} E$$

This is the same result as above.

7.3. NETWORKS WITH INDUCTORS

If the network contains only resistors, inductors and voltage sources, then it is natural to use the mesh formulation. In this case the resistors are expressed in terms of their resistances.

The first network we consider is in Fig. 7.3.1 and the mesh equation is

$$(R + sL)I = E$$

with the solution

$$I = \frac{E}{sL + R}$$

Fig. 7.3.1   Fig. 7.3.2

FIGURE 7.3.1   RL network with a voltage source.

FIGURE 7.3.2   GL network with a current source.

The voltage across the inductor is

$$V_L = sLI = \frac{sL}{sL + R} E$$

We could have solved this network also as a voltage divider by using

$$V_L = \frac{Z_L}{Z_R + Z_L} E$$

to get the result immediately.

A parallel combination of a current source and the same elements is in Fig. 7.3.2. In this case we apply nodal formulation:

$$\left(G + \frac{1}{sL}\right) V = J$$

and the solution is

$$V = \frac{sL}{sLG + 1} J$$

A network with two meshes is in Fig. 7.3.3 and here it also pays to use mesh analysis. The equations are

$$(R_1 + sL_1)I_1 - sL_1 I_2 = E$$

$$-sL_1 I_1 + (sL_1 + sL_2 + R_2)I_2 = 0$$

FIGURE 7.3.3   Network with two inductors.

Inserting numerical values

$$\begin{bmatrix} 5+s & -s \\ -s & 4s+10 \end{bmatrix}\begin{bmatrix} I_1 \\ I_2 \end{bmatrix} = \begin{bmatrix} E \\ 0 \end{bmatrix}$$

The determinant is

$$D = (5+s)(4s+10) - s^2 = 3s^2 + 30s + 50$$

and $N_2 = sE$. The current of the second mesh is

$$I_2 = \frac{s}{3s^2 + 30s + 50} E$$

and the output voltage is

$$V_{out} = sL_2 I_2 = \frac{3s^2}{3s^2 + 30s + 50} E$$

Additional examples show more complicated situations.

Example 1.

Find the output voltage of the network in Fig. 7.3.4.
    The network is suitable for mesh equations. The current controlling the CC is $I = I_2$.

FIGURE 7.3.4   Network with a CC.

The mesh equations are:

$$(3 + 2s)I_1 - 2sI_2 + 2I = E$$

$$-2sI_1 - 2I + (7s + 6)I_2 = 0$$

Inserting for I and collecting terms multiplied by the same variable

$$\begin{bmatrix} 3 + 2s & 2 - 2s \\ -2s & 7s + 4 \end{bmatrix} \begin{bmatrix} I_1 \\ I_2 \end{bmatrix} = \begin{bmatrix} E \\ 0 \end{bmatrix}$$

The determinant is

$$D = (3 + 2s)(7s + 4) - (-2s)(2 - 2s) = 10s^2 + 33s + 12$$

The numerator $N_2 = 2sE$ and the current of the second mesh is

$$I_2 = \frac{2s}{10s^2 + 33s + 12} E$$

The output voltage is

$$V_{out} = sL_2I_2 = \frac{10s^2}{10s^2 + 33s + 12} E$$

Example 2.

The network in Fig. 7.3.5a is suitable for nodal formulation and so we use it. Next we apply transformations of the sources and use mesh formulation.

The nodal equations are

$$\left(\frac{1}{3s} + \frac{1}{4}\right) V_1 - \frac{1}{4} V_2 = J$$

$$-\frac{1}{4} V_1 + \left(\frac{1}{4} + \frac{1}{2s}\right) V_2 - 2(V_1 - V_2) = 0$$

Collecting terms we get

$$\begin{bmatrix} \dfrac{3s + 4}{12s} & \dfrac{-1}{4} \\[3mm] \dfrac{-9}{4} & \dfrac{9s + 2}{4s} \end{bmatrix} \begin{bmatrix} V_1 \\ V_2 \end{bmatrix} = \begin{bmatrix} J \\ 0 \end{bmatrix}$$

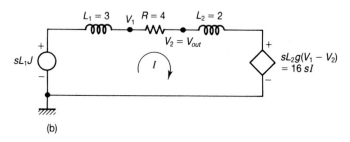

FIGURE 7.3.5   Network with a VC: (a) Original. (b) Transformed for mesh formulation.

The determinant is

$$D = \frac{3s + 4}{12s} \frac{9s + 2}{4s} - \frac{9}{16} = \frac{21s + 4}{24s^2}$$

The numerator $N_2 = (9/4)J$ and the output voltage is

$$V_{out} = \frac{54s^2}{21s + 4} J$$

Transformation of the sources reduces the network to only one mesh, shown in Fig. 7.3.5b. The controlling voltage is now $V_1 - V_2 = RI = 4I$ and the expression for the controlled source is written into the figure. The single mesh equation is

$$(5s + 4)I + 16sI = 3sJ$$

and the solution is

$$I = \frac{3s}{21s + 4} J$$

The output voltage, V_2, is marked in the figure and is obtained as

$$V_{out} = 16sI + sL_2I = 18sI = \frac{54s^2}{21s + 4} J$$

This is the same result as above.

Example 3.

The network in Fig. 7.3.6a has one voltage-controlled voltage source and one current-controlled voltage source. Find the output voltage using the mesh method.

In the first step we realize that the controlling variables are $I = I_2$ and $V_A = R_1(I_1 - I_2) = 3(I_1 - I_2)$. The values are transferred into Fig. 7.3.6b and the mesh equations are

$$(2s + 3)I_1 - 3I_2 + 4I_2 = E$$

$$-3I_1 + (7 + s)I_2 - 4I_2 - 9I_1 + 9I_2 = 0$$

Collecting terms

$$\begin{bmatrix} 2s + 3 & 1 \\ -12 & 12 + s \end{bmatrix} \begin{bmatrix} I_1 \\ I_2 \end{bmatrix} = \begin{bmatrix} E \\ 0 \end{bmatrix}$$

The determinant is

$$D = (2s + 3)(12 + s) + 12 = 2s^2 + 27s + 48$$

The numerators are $N_1 = (s + 12)E$ and $N_2 = 12E$. The currents are

$$I_1 = \frac{s + 12}{2s^2 + 27s + 48} E$$

$$I_2 = \frac{12}{2s^2 + 27s + 48} E$$

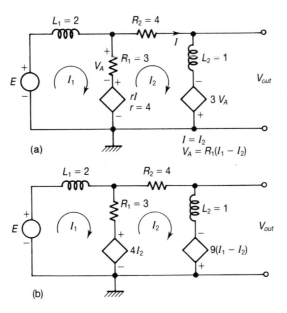

(a)

$$I = I_2$$
$$V_A = R_1(I_1 - I_2)$$

(b)

FIGURE 7.3.6   Network with dependent sources: (a) Original. (b) Transformed for mesh formulation.

The output can be calculated, for instance, as the voltage of the independent voltage source on the left, less the voltage across L_1, less the voltage across R_2. Thus $V_{out} = E - sL_1I_1 - R_2I_2$. The result is

$$V_{out} = \frac{3s}{2s^2 + 27s + 48} R$$

7.4. NETWORKS WITH CAPACITORS AND INDUCTORS

If the network contains both inductors and capacitors, we cannot avoid the unpleasant fractions, at least not with the methods we have learned so far. Theoretically, it does not matter which method we select, provided the network is planar. In general, it is advantageous to favor the nodal method.

The first network we will consider is a parallel combination of R, L and C, as shown in Fig. 7.4.1. This network has a special name, it is called the *parallel tuned circuit*. We will analyze it by the nodal method and by using G for the resistor:

$$\left(G + sC + \frac{1}{sL} \right) V = J$$

and the voltage is

$$V = \frac{sL}{s^2LC + sLG + 1} J$$

If we are interested in the currents through the elements, we obtain them by using $I_G = GV$, $I_C = sCV$ and $I_L = (1/sL)V$.

Another frequently used network is in Fig. 7.4.2, a series combination of the same elements. The network is called the *series tuned circuit* and mesh formulation

Fig. 7.4.1

Fig. 7.4.2

FIGURE 7.4.1   Parallel tuned circuit.

FIGURE 7.4.2   Series tuned circuit.

is clearly advantageous:

$$\left(R + sL + \frac{1}{sC}\right)I = E$$

and the current is

$$I = \frac{sC}{s^2LC + sRC + 1}E$$

If we wish to get the voltages across the elements, then they are obtained by $V_R = RI$, $V_L = sLI$, and $V_C = (1/sC)I$.

Example 1.

Analyze the network in Fig. 7.4.3a and find the output voltage.

It is advantageous to use nodal formulation, but first we must determine on the original network the controlling voltage. It is $V_A = E - V_1$. This is transferred into the modified network in Fig. 7.4.3b and we can proceed with the equations:

$$\left(2 + 3s + \frac{1}{s}\right)V_1 - \frac{1}{s}V_2 = 3sE$$

$$-\frac{1}{s}V_1 + \frac{1}{s}V_2 - 2E + 2V_1 = 0$$

(a)

(b)

FIGURE 7.4.3   GLC network with a VC: (a) Original. (b) Transformed for nodal formulation.

Transferring the term with E to the right we arrive at

$$
\begin{bmatrix}
\dfrac{3s^2 + 2s + 1}{s} & -\dfrac{1}{s} \\[3mm]
\dfrac{2s - 1}{s} & \dfrac{1}{s}
\end{bmatrix}
\begin{bmatrix}
V_1 \\ V_2
\end{bmatrix}
=
\begin{bmatrix}
3sE \\ 2E
\end{bmatrix}
$$

The determinant is

$$
D = \frac{3s^2 + 2s + 1}{s^2} + \frac{2s - 1}{s^2} = \frac{3s + 4}{s}
$$

For the numerator we must evaluate

$$
N_2 =
\begin{vmatrix}
\dfrac{3s^2 + 2s + 1}{s} & 3sE \\[3mm]
\dfrac{2s - 1}{s} & 2E
\end{vmatrix}
= \frac{7s + 2}{s} E
$$

and the output voltage is

$$
V_2 = \frac{7s + 2}{3s + 4} E
$$

Example 2.

Analyze the network in Fig. 7.4.4a by the nodal method.

In the first step we find the controlling current of the CC: it is $I = (E - V_1)/sL$. Writing

(a)

(b)

FIGURE 7.4.4   RLC network with a CC: (a) Original. (b) Transformed for nodal formulation.

this value to the dependent source and transforming the independent voltage source into a current source we arrive at the network Fig. 7.4.4b for which we can write

$$\left(\frac{1}{3s} + 1 + 2s\right)V_1 - 2sV_2 = \frac{E}{3s}$$

$$-2sV_1 + (2s + 2)V_2 - \frac{2}{3s}(E - V_1) = 0$$

Transferring the term with E to the right and collecting terms

$$\begin{bmatrix} \dfrac{6s^2 + 3s + 1}{3s} & -2s \\[2ex] \dfrac{2 - 6s^2}{3s} & 2s + 2 \end{bmatrix} \begin{bmatrix} V_1 \\ V_2 \end{bmatrix} = \begin{bmatrix} \dfrac{E}{3s} \\[2ex] \dfrac{2E}{3s} \end{bmatrix}$$

The determinant is

$$D = \frac{18s^2 + 12s + 2}{3s}$$

For the numerator we must find the determinant

$$N_2 = \begin{vmatrix} \dfrac{6s^2 + 3s + 1}{3s} & \dfrac{E}{3s} \\[2ex] \dfrac{2 - 6s^2}{3s} & \dfrac{2E}{3s} \end{vmatrix} = \frac{6s + 2}{3s} E$$

The result is

$$V_2 = \frac{6s + 2}{18s^2 + 12s + 2} E$$

Example 3.

For the network in Fig. 7.4.5 we wish to derive the output voltage as a function of the capacitor C. This means that we must keep the letter C in all equations and solutions. The network is in a form suitable for nodal formulation and the equations are:

$$\left(sC + 1 + \frac{1}{2s}\right)V_1 - \frac{1}{2s}V_2 = J$$

$$-\frac{1}{2s}V_1 + \left(\frac{1}{2s} + 2\right)V_2 - gV_1 = 0$$

FIGURE 7.4.5   RLC network with a VC.

In matrix form this leads to

$$
\begin{bmatrix}
\dfrac{2s^2C + 2s + 1}{2s} & -\dfrac{1}{2s} \\[3mm]
-\dfrac{1 + 2s}{2s} & \dfrac{4s + 1}{2s}
\end{bmatrix}
\begin{bmatrix} V_1 \\ V_2 \end{bmatrix}
=
\begin{bmatrix} J \\ 0 \end{bmatrix}
$$

The determinant is

$$
D = \frac{2s^2C + 2s + 1}{2s}\frac{4s + 1}{s} - \frac{1}{2s}\frac{1 + 2s}{2s} = \frac{8Cs^2 + s(8 + 2C) + 4}{4s}
$$

For the numerator we find $N_2 = [(2s + 1)/2s]\,J$ and the ratio gives the output voltage

$$
V_2 = \frac{2s + 1}{4Cs^2 + s(4 + C) + 2}\,J
$$

Example 4.

Find the output voltage for the network in Fig. 7.4.6 by keeping the inductance L as a variable element. Use the given numerical values for all other elements.

FIGURE 7.4.6   RLC network.

The network is suitable for nodal formulation and we write directly

$$\begin{bmatrix} 1 + s + \dfrac{1}{sL} & -\dfrac{1}{sL} \\[3mm] -\dfrac{1}{sL} & 1 + s + \dfrac{1}{sL} \end{bmatrix} \begin{bmatrix} V_1 \\ V_2 \end{bmatrix} = \begin{bmatrix} J \\ 0 \end{bmatrix}$$

The determinant, after a few intermediate steps, is

$$D = \frac{s^3 L + 2s^2 L + s(L + 2) + 2}{sL}$$

and the numerator for the output is $N_2 = J/sL$. The output voltage is

$$V_2 = \frac{1}{s^3 L + 2s_2 L + s(L + 2) + 2} J$$

All examples we solved lead to the ratio of two polynomials. Some had the highest power of s equal to 1, and such networks would be called *first order* networks. If the highest power of s of any of the polynomials is 2, we speak about a *second order* network and so on. Important is that analysis of all networks composed of the elements we introduced in Chapter 1 will *always* lead to the ratio of two polynomials in s, and all coefficients of these polynomials will be real.

PROBLEMS CHAPTER 7

Problems in this chapter lead to rational functions in the variable s. Such functions can be obtained by the symbolic option in the program NETOPT and you can confirm your results, see Chapter 18, Section 5. A few important limitations should be stressed. The symbolic option automatically assumes that the source has unit value. It also considers *only one* independent source, the first one given in the input file. Should you use the option for the Thévenin and Norton examples, then you must change the input files to reflect the two steps.

(a)

(b)

P.7.1a–d. Use nodal analysis for all problems; in c first transform the source into a current source. For each capacitor substitute $Y = sC$, for conductances use $Y = G$. Find V_{out}.

(c)

(d)

P.7.2 Dependent sources are handled exactly the same way as in resistive networks. Find the voltages V_1 and V_2 by using $Y = sC$ for the capacitors.

P.7.3 Find the output voltage $V_2 = V_{out}$.

P.7.4 Find the output voltage.

P.7.5 Mesh analysis is applied the same way as before, with $Z = 1/sC$ for capacitors and $Z = R$ for resistors. Find the output voltage.

P.7.6 Use nodal formulation and find V_{out}.

P.7.7 Thévenin or Norton equivalents are calculated the same way as before, only now the independent sources may become functions of s and the equivalent resistance or conductance is changed into an equivalent impedance or admittance. Find the Thévenin and Norton equivalents for this circuit.

P.7.8 Find the Thévenin and Norton equivalents.

P.7.9 Find the Thévenin and Norton equivalents.

P.7.10 Express the controlling current in terms of the input voltage and V_1, substitute for I in the dependent source and then transform the voltage source into a current source. Use nodal formulation to obtain $V_{out} = V_2$.

P.7.11 In nodal formulation use $Y = 1/sL$ for inductors and $Y = G$. Find V_1 and V_2.

P.7.12 In mesh formulation use $Z = sL$ for inductors and $Z = R$ for resistors. Find the output voltage V_{out}.

P.7.13 Calculate the voltage V_{out} using mesh formulation.

P.7.14 Find the voltage V. This can be done in three ways: conveniently by finding first I_1 and then calculating $V = 1 - sLI_1$. It can also be done by calculating $I = I_2$ and then $V = sLI$. Finally, less conveniently, it can be obtained by calculating both I_1 and I_2 and then $V = 3I_2 + 5(I_1 - I_2)$. In all cases you must get the same result. Perform all three steps.

P.7.15 Find the output voltage.

P.7.16 Find the output voltage.

P.7.17 Calculate V_1 and V_2.

P.7.18 Find both nodal voltages.

P.7.19 Find the Thévenin and Norton equivalents.

P.7.20 Calculate V_{out} using mesh formulation. Note that you must first express the current I in terms of the source voltage and V_1. Next you must substitute this current into the dependent source and only then can you transform the voltage source into a current source. The current source will be a function of s.

P.7.21 Calculate V_{out} by first transforming the voltage source into a current source. Note that you must use $Y = sC$ for the capacitor, $Y = 1/sL$ for the inductor and $Y = G$ for resistances.

P.7.22 Find the output voltage by using nodal formulation.

P.7.23 Find the output voltage by using nodal formulation.

P.7.24 Find the voltage V. This is a similar network as in P.7.14, only now we have an inductor and a capacitor. Find the voltage V the same three ways as in P.7.14.

P.7.25 Find the output voltage V_{out} using nodal formulation.

P.7.27 Calculate the Thévenin and Norton equivalents.

P.7.26 Find the output voltage V_{out} using nodal formulation.

P.7.28 Find the output voltage.

CHAPTER 8

Frequency Domain

8.0. INTRODUCTION

One of the typical signals used in electrical engineering is the sinusoidal signal. The electric power distributed in North America has sinusoidal form and the current changes its direction periodically 60 times per second. Europe uses alternating current distribution as well, but the frequency is 50 Hz. Unless there is power failure, the system is connected all the time. If you turn on the light or connect an electric motor, you disturb the system a little but in general we can say that the system is not changing and is, as we say, in a steady state. Occasionally, there may be power failure and then other processes take place; we call them transient.

This chapter will study methods for the analysis of systems with sinusoidal signals in steady state. For instructions how to use the program NETOPT for frequency domain solutions see Section 18.4.

8.1. SINUSOIDAL SIGNALS AND PHASORS

A sinusoidal signal belongs to a class of *periodic* functions which have the property that

$$v(t + nT) = v(t) \qquad n = \ldots, -2, -1, 0, 1, 2, \ldots \tag{8.1.1}$$

Here T is called the *period* of the signal, measured in seconds, s. The signal repeats itself after the period and we cannot discover which value of n applies to the instant we consider. The simplest of all periodic functions are the cosine and sine, defined by the equations

$$v(t) = V \cos (\omega t + \phi)$$
$$v(t) = V \sin (\omega t + \phi)$$

(8.1.2)

The letter v is used for voltages, i is used for currents. In Eq. (8.1.2) V is called the *amplitude*, t is time in seconds, s, ω is angular frequency measured in radians per second, rad/s, and ϕ is an angle. The angular frequency, ω, is related to the period, T, by the expression

$$\omega = \frac{2\pi}{T}$$

(8.1.3)

and the inverted value of the period

$$f = \frac{1}{T}$$

(8.1.4)

is called the *frequency* and is measured in hertz, Hz. The angle, ϕ, can be measured either in degrees or in radians, related by the correspondence of 2π and $360°$. Thus π corresponds to $180°$, $\pi/2$ corresponds to $90°$ and so on. In calculations we must always know which of these two units is being considered.

Figure 8.1.1a shows a circle with two vectors in it. The vectors rotate counterclockwise with constant and equal angular speed, ω, but we now stop the

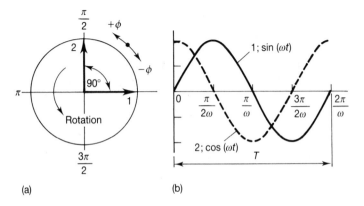

(a)   (b)

FIGURE 8.1.1   (a) Rotating vectors. (b) Equivalent sinusoids.

motion. The relative position of the two vectors does not change with rotation and both vectors stay 90° apart all the time. We can stop the rotation at any instant and get a picture similar to the one in Fig. 8.1.1a, only rotated by some angle.

Consider first the vector at position 1 and assume that you are a viewer positioned to the right of the line which is an extension of the vector. For the instant shown you will not see the vector, just one point. If the vector starts rotating, you will see a growing line until the vector reaches position 2, when you will see the longest line. Afterwards the line will start decreasing, will become zero, will start growing on the other side, reach its maximum value and start decreasing again. If we also keep track of the times needed to reach the particular projected point and draw these times on the horizontal axis, we will get a sinusoid, shown in Fig. 8.1.1b. A similar consideration for vector 2 will result in a cosine, as shown. The figure also shows that adding an angle ϕ results in a counterclockwise move by the given amount. Subtraction of ϕ moves the vector clockwise.

Rotate vector 1 counterclockwise by adding to it the angle $\phi = 90°$. It will reach the position 2 and will become $v(t) = V \sin (\omega t + 90°)$. However, as we have already stated above, this vector represents the cosine, $v(t) = V \cos \omega t$. It must follow that

$$\sin (\omega t + 90°) = \cos \omega t$$

which is a well known formula. We can consider similarly the vector 2 as the original one, shift it clockwise by $\phi = -90°$ and thus place it in the position of the original vector 1. This leads to another well known formula

$$\cos (\omega t - 90°) = \sin \omega t$$

Figure 8.1.2a shows three vectors. Their corresponding sinusoids are in b. Vector 1 is $v_1(t) = V \sin \omega t$, vector 2 is $v_2(t) = V \sin (\omega t + 45°)$, and vector 3 is $v_3(t) = -V \cos \omega t$. If we shift vector 1 backwards (clockwise) by 90°, we will get $V \sin (\omega t - 90°)$, but this is equal to $v_3(t)$. Thus another formula is obtained

$$\sin (\omega t - 90°) = -\cos \omega t$$

Similarly, if we start with vector 3 and shift it by $+90°$, it will become $v_1(t)$ and the resulting formula is

$$-\cos (\omega t + 90°) = \sin \omega t$$

The examples show that sines and cosines are actually a single function looked upon from different reference points. A summary of various trigonometric equations is in Table 8.1.1.

The words *leading* or *lagging* are often used when speaking about the various sinusoids or their equivalent vectors. Place yourself in Fig. 8.1.1a on the circle at

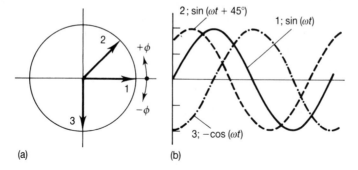

FIGURE 8.1.2   (a) Rotating vectors. (b) Equivalent sinusoids.

the point marked by π and start the counterclockwise rotation of both vectors. Vector 2 will reach your place first, and we would say that vector 2 is leading vector 1 by 90°. By the same token we could say that vector 1 is lagging vector 2 by 90°. These statements are correct but they depend on the point of observation. If you place your observation point between the two vectors, to the position marked by $\pi/4$, and start the counterclockwise rotation, then vector 1 will reach you first, and will thus be the leading vector. It will lead vector 2 by 270°.

TABLE 8.1.1   Trigonometric equivalences

$\sin (A + B) = \sin A \cos B + \cos A \sin B$	$\sin A \sin B = \dfrac{1}{2} [\cos (A - B) - \cos (A + B)]$
$\sin (A - B) = \sin A \cos B - \cos A \sin B$	$\cos A \cos B = \dfrac{1}{2} [\cos (A - B) + \cos (A + B)]$
$\cos (A + B) = \cos A \cos B - \sin A \sin B$	$\sin A \cos B = \dfrac{1}{2} [\sin (A - B) + \sin (A + B)]$
$\cos (A - B) = \cos A \cos B + \sin A \sin B$	$\sin A + \sin B = 2 \sin \left(\dfrac{A + B}{2}\right) \cos \left(\dfrac{A - B}{2}\right)$
$\sin (-A) = -\sin A$	$\sin A - \sin B = 2 \cos \left(\dfrac{A + B}{2}\right) \sin \left(\dfrac{A - B}{2}\right)$
$\cos (-A) = \cos A$	$\cos A + \cos B = 2 \cos \left(\dfrac{A + B}{2}\right) \cos \left(\dfrac{A - B}{2}\right)$
$\sin (A + 90°) = \cos A$	$\cos A - \cos B = -2 \sin \left(\dfrac{A + B}{2}\right) \sin \left(\dfrac{A - B}{2}\right)$
$\cos (A - 90°) = \sin A$	

The correspondence of variously shifted sine waves and their vectors has one big advantage: instead of drawing shifted sinusoids, which is time consuming and confusing, we can consider vectors like in Fig. 8.1.1a or 8.1.2a. Such vectors are also given a special name, the *phasors*. What we still need is a method which can describe the vector sizes and orientations mathematically. This is done by introducing the *plane of complex numbers*.

Consider a plane in which we mark the origin and two perpendicular axes. Usually the variable x is assigned to the horizontal axis and variable y to the vertical axis. This is not suitable for our situation because we must also take into account the fact that a vector which has rotated by 360° has reached its original position and not a new one. The problem is resolved by using complex numbers. In the plane of complex numbers, the so called *s-plane*, distances measured on the horizontal axis are marked by real numbers. Distances measured on the vertical axis are multiplied by the imaginary number

$$j = \sqrt{-1} \tag{8.1.5}$$

The notation differs from mathematics where we always use the letter i. Since in electrical engineering i usually denotes the current, it is accepted practice to use j for the imaginary unit. The properties of Eq. (8.1.5) are such that

$$
\begin{aligned}
j^2 &= -1 \\
j^3 &= -j = -\sqrt{-1} \\
j^4 &= 1
\end{aligned}
\tag{8.1.6}
$$

and the values will start repeating with further multiplications by j. Every multiplication by j represents a rotation by 90° counterclockwise. In Fig. 8.1.3 the vector v_1 is pointing to the right from the origin. If we multiply this vector by $j = \sqrt{-1}$, then in the complex plane it will rotate by 90° counterclockwise and

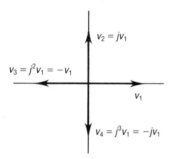

FIGURE 8.1.3   Rotations by 90°.

will point upwards as a vector $v_2 = jv_1$. If we multiply by another j, we will get $v_3 = j^2v_1 = -v_1$. Multiplication by another j will get us to $v_4 = j^3v_1 = -jv_1$. Still another multiplication by j will return us to the original position of the vector v_1.

All other vectors in the plane will have real and imaginary parts. Figure 8.1.4 shows four vectors, all starting at the origin and all having real and imaginary parts. Since all vectors always start at the same origin, we can alternatively talk about points in the complex plane, by tacitly understanding that these complex points represent the endpoints of the vectors. Thus the points in Fig. 8.1.4 are $s_1 = 3 + j2$, $s_2 = -1 + j3$ and so on. We thus arrive at the concept of the general complex number

$$s = a + jb \tag{8.1.7}$$

The real and imaginary parts are given special notation

$$a = \operatorname{Re}(s)$$
$$b = \operatorname{Im}(s) \tag{8.1.8}$$

Here $\operatorname{Im}(s)$ itself is a real number representing the distance in vertical direction. Thus a complex number can also be expressed by

$$s = \operatorname{Re}(s) + j \operatorname{Im}(s) \tag{8.1.9}$$

If the imaginary part of the complex number changes its sign, then we say that this new number is *complex conjugate* to the previous one and we mark it by a bar,

$$\bar{s} = \operatorname{Re}(s) - j \operatorname{Im}(s) \tag{8.1.10}$$

Complex numbers can be added, subtracted, multiplied or divided. Addition and subtraction are performed on the real and imaginary parts independently.

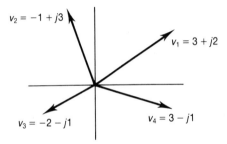

FIGURE 8.1.4 General vectors in complex plane.

For instance, if we have

$$s_1 = a_1 + jb_1$$

and

$$s_2 = a_2 + jb_2$$

then their sum is

$$s = s_1 + s_2 = (a_1 + a_2) + j(b_1 + b_2) \qquad (8.1.11)$$

and their difference is

$$s = s_1 - s_2 = (a_1 - a_2) + j(b_1 - b_2) \qquad (8.1.12)$$

For the vectors s and $\bar{s}$ we thus get

$$s + \bar{s} = a + jb + a - jb = 2a = 2 \, \text{Re} \, (s)$$
$$s - \bar{s} = a + jb - a + jb = j2b = j2 \, \text{Im} \, (s) \qquad (8.1.13)$$

Complex numbers can also be expressed by their distance from the origin, A, and by the angle, ϕ, shown in Fig. 8.1.5. The two descriptions are equivalent; the first one is in *Cartesian* or *rectangular* coordinates, the second in *polar* coordinates. It follows from the figure that

$$A = (a^2 + b^2)^{1/2}$$
$$\tan \phi = \frac{b}{a} \qquad (8.1.14)$$

or

$$\phi = \arctan \left(\frac{b}{a} \right) \qquad (8.1.15)$$

The distance, A, is also called the absolute value and we use the notation

$$A = |s|$$

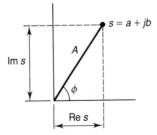

FIGURE 8.1.5   Rectangular and polar coordinates.

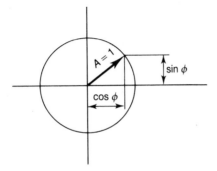

FIGURE 8.1.6   Addition and subtraction of vectors: (a) Addition. (b) Subtraction.

Example 1.

Figure 8.1.6a shows two vectors, $s_1 = 1 + j2$ and $s_2 = 3 + j1$. Their sum is $s = s_1 + s_2 = 4 + 3j$. Note the way this addition is accomplished by drawing the dotted lines which are parallel to the two vectors. Figure 8.1.6b shows two vectors $s_1 = 3 + 2j$ and $s_2 = 1 + 2j$. Their difference is $s = s_1 - s_2 = 2 + j0$. Note how this can be achieved graphically by first copying the vector to be subtracted in opposite direction and then using this new vector and the parallel lines with the first vector like in the addition.

Another set of useful formulas can be found by considering the unit circle in Fig. 8.1.7. Since the radius is equal to one, then the real part is $a = \cos \phi$ and the imaginary part is $b = \sin \phi$. The point on the unit circle is thus $s = \cos \phi + j \sin \phi$. If we increase the amplitude to A, we can reach any point in the complex plane and describe it by

$$s = A(\cos \phi + j \sin \phi) \tag{8.1.16}$$

FIGURE 8.1.7   Unit circle relation to sine and cosine.

There exists still another, more general expression for the term in the bracket:

$$e^{j\phi} = \cos \phi + j \sin \phi \qquad (8.1.17)$$

Using this formula we can express an arbitrary point in the complex plane by

$$s = Ae^{j\phi} \qquad (8.1.18)$$

and its complex conjugate by

$$\bar{s} = A(\cos \phi - j \sin \phi) = Ae^{-j\phi} \qquad (8.1.19)$$

Summarizing the development, we have three equivalent expressions for a point in the complex plane s:

$$\begin{aligned} s &= a + jb \\ &= Ae^{j\phi} \\ &= A(\cos \phi + j \sin \phi) \end{aligned} \qquad (8.1.20)$$

Addition and subtraction are always performed in Cartesian coordinates, the way we showed in Eq. (8.1.11) and Eq. (8.1.12). For multiplication and division we have two choices. In Cartesian coordinates the product is

$$s_1 \cdot s_2 = (a_1 + jb_1)(a_2 + jb_2) = (a_1 a_2 - b_1 b_2) + j(a_1 b_2 + a_2 b_1) \qquad (8.1.21)$$

In polar coordinates,

$$s_1 \cdot s_2 = A_1 e^{j\phi_1} A_2 e^{j\phi_2} = A_1 A_2 e^{j(\phi_1 + \phi_2)} \qquad (8.1.22)$$

Division in Cartesian coordinates is expressed by

$$\begin{aligned} \frac{s_1}{s_2} &= \frac{a_1 + jb_1}{a_2 + jb_2} = \frac{(a_1 + jb_1)(a_2 - jb_2)}{(a_2 + jb_2)(a_2 - jb_2)} \\ &= \frac{a_1 a_2 + b_1 b_2 + j(b_1 a_2 - a_1 b_2)}{a_2^2 + b_2^2} \end{aligned} \qquad (8.1.23)$$

and in polar coordinates by

$$\frac{s_1}{s_2} = \frac{A_1 e^{j\phi_1}}{A_2 e^{j\phi_2}} = \frac{A_1}{A_2} e^{j(\phi_1 - \phi_2)} \qquad (8.1.24)$$

Since today even modestly priced calculators have some kind of conversion between the Cartesian and polar coordinates, it does not matter much which of the two definitions is used. If a calculator is not available, then it is easier to work with the Cartesian formulas.

Example 2.

Find the product and ratio of the complex numbers $s_1 = 3 + j4$ and $s_2 = 2 - j3$.

a. The product in Cartesian coordinates is

$$s = s_1 + s_2 = (3 + 4j)(2 - 3j) = 6 + 8j - 9j + 12 = 18 - j1$$

In polar coordinates $s_1 = 5e^{j53.13}$, $s_2 = 3.60555e^{-j56.3099}$ and the result is $s = 18.027e^{-j3.18}$. The angle is given in degrees with decimal fractions.

b. The ratio in Cartesian coordinates is

$$s = \frac{s_1}{s_2} = \frac{3 + 4j}{2 - 3j} = \frac{(3 + 4j)(2 + 3j)}{(2 - 3j)(2 + 3j)}$$

$$= \frac{6 + 9j + 8j - 12}{4 + 9} = \frac{-6 + 17j}{13} = -0.4615 + j1.3077$$

In polar coordinates the result is the ratio of the two polar representations, $s = 1.387e^{j109.44}$, the angle given again in degrees with decimal fractions. Note that it does not matter whether we write $3j$ or $j3$.

Example 3.

Convert the following vectors in rectangular coordinates to polar coordinates. The angles are given in degrees with decimal fractions.

$$2 + j3 = 3.6055e^{j56.31}$$

$$5 - 2j = 5.385e^{-j21.801}$$

$$-3 - 4j = 5e^{-j126.87}$$

$$-2 + 1j = 2.236e^{j153.43}$$

Convert the following vectors in polar form into rectangular form. The angles are given in degrees.

$$5e^{j30} = 4.33 + j2.5$$

$$4e^{-j25} = 3.625 - j1.690$$

$$2e^{-j135} = -1.414 - j1.414$$

$$4e^{j160} = -3.759 + j1.368$$

The exponential expressions are also valid for functions of time,

$$Ae^{j(\omega t + \phi)} = A[\cos(\omega t + \phi) + j\sin(\omega t + \phi)] \qquad (8.1.25)$$

from which we also have

$$A\cos(\omega t + \phi) = \text{Re } Ae^{j(\omega t + \phi)}$$
$$A\sin(\omega t + \phi) = \text{Im } Ae^{j(\omega t + \phi)} \qquad (8.1.26)$$

The formulas enable us to use vectors when dealing with sinusoids. The left part of Eq. (8.1.25) can be rewritten in the form

$$Ae^{j(\omega t + \phi)} = [Ae^{j\phi}]e^{j\omega t} \qquad (8.1.27)$$

where the term in the square brackets represents a vector with amplitude A, rotated by the angle ϕ. Permanent rotation in counterclockwise direction is expressed by the term $e^{j\omega t}$. To describe a vector we often use bold letters and thus

$$\mathbf{A} = Ae^{j\phi} \qquad (8.1.28)$$

The name *phasor* is used for Eq. (8.1.28).

Suppose that we are asked to perform the addition

$$v(t) = 4\cos(\omega t - 20°) + 3\cos(\omega t + 30°) \qquad (8.1.29)$$

Applying Eq. (8.1.26) and Eq. (8.1.27), we express the functions in exponential form

$$v(t) = \text{Re } \{[4e^{-j20} + 3e^{j30}]e^{j\omega t}\} \qquad (8.1.30)$$

All we have to do is consider the terms in the square brackets, $4e^{-j20} + 3e^{j30}$. Direct addition is not possible but we can convert to rectangular form, perform the addition and then convert back to polar form:

$$4e^{-j20} + 3e^{j30} = (3.7588 - j1.36808) + (2.598 + j1.5) = 6.3568 + j0.13192$$
$$= 6.3582e^{j1.18885}$$

Using this result we go back to Eq. (8.1.27), replace the term in the square brackets

$$v(t) = \text{Re } \{[6.3582e^{j1.18885}]e^{j\omega t}\}$$

and return to Eq. (8.1.26) and to the cosine function

$$v(t) = 6.3582\cos(\omega t + 1.18885°)$$

Example 4.

Find the equivalent cosine for

$$v(t) = 5 \cos(\omega t + 30°) - 4 \sin(\omega t - 20°)$$

The second function must be first converted to a cosine:

$$v(t) = 5 \cos(\omega t + 30°) - 4 \cos(\omega t - 20° - 90°) = 5 \cos(\omega t + 30°) - 4 \cos(\omega t - 110°)$$

$$= \text{Re}\{[5e^{j30} - 4^{-j110}]e^{j\omega t}\}$$

We now deal with the part in the square brackets

$$5e^{j30} - 4e^{-j110} = (4.33 + j2.5) - (-1.368 - j3.7588) = 5.698 + j6.2588$$

$$= 8.464e^{j47.684}$$

Returning to cosine form we have the result

$$v(t) = 8.464 \cos(\omega t + 47.684°)$$

Example 5.

Find the sum of

$$v(t) = 2 \cos(\omega t + 30°) - 3 \cos(\omega t - 50°) + 4 \cos(\omega t + 130°)$$

The vectors to be added are

$$2e^{j30} - 3e^{-j50} + 4e^{j130} = (1.7321 + j) - (1.9284 - j2.2981) + (-2.5712 + j3.0642)$$

$$= -2.7675 + j6.3623 = 6.9381e^{j113.508}$$

and

$$v(t) = 6.9381 \cos(\omega t + 113.508°)$$

The above developments were done by using cosine functions, but the same applies to sine functions, with the only difference that the symbol Re is replaced by the symbol Im. Important is that we must select one type and use it throughout. Operations can be summarized in the following steps:

1. All functions on which we operate *must* have the same ω.
2. Convert all functions to the form $v(t) = V \cos(\omega t + \phi)$.
3. Find the vector representation $\mathbf{V} = Ve^{j\phi}$ and, if necessary, convert to rectangular form $\mathbf{V} = a + jb$.
4. Perform the necessary operations and convert the result, $\mathbf{R} = c + jd$, into polar form, $\mathbf{R} = R^{j\psi}$.
5. The resulting time function is $R \cos(\omega t + \psi)$.

The above points remain valid if the word cosine is replaced everywhere by the word sine. The operations are fundamental to frequency domain analysis of networks.

8.2. FREQUENCY DOMAIN ANALYSIS

In frequency domain analysis we are interested in the reaction of a linear network to a permanently acting cosine signal. By permanently acting we mean that any switching or changes of the signal have happened long ago, theoretically an infinitely long time ago. To proceed with the theory we introduce a few definitions based on the results of the previous section. Let the independent voltage source be represented by the signal and its corresponding phasor

$$e(t) = E \cos (\omega t + \phi)$$
$$\mathbf{E} = Ee^{j\phi}$$

$$(8.2.1)$$

For the independent current source we have

$$j(t) = J \cos (\omega t + \phi)$$
$$\mathbf{J} = Je^{j\phi}$$

$$(8.2.2)$$

Network voltages and currents are similarly defined

$$v(t) = V \cos (\omega t + \phi)$$
$$\mathbf{V} = Ve^{j\phi}$$

$$(8.2.3)$$

and

$$i(t) = I \cos (\omega t + \phi)$$
$$\mathbf{I} = Ie^{j\phi}$$

$$(8.2.4)$$

The bold letters formally denote that we are dealing with vectors.

In Chapter 7 we introduced the impedances and admittances for capacitors and inductors. We also stated that every independent source is represented by a letter E or J, to be replaced by appropriate substitution. In frequency domain analysis we do the following:

1. For any given angular frequency we replace s by $j\omega$.
2. Every independent source is replaced by its phasor.
3. The resulting voltages and currents are also phasors.
4. Return to the functions of time by using the above equivalences.

We will start our frequency domain studies by performing the operations on impedances and admittances. The resistor voltage and current are coupled by Ohm's law,

$$V_R = RI_R \tag{8.2.5}$$

Let the independent variable, in this case the current, be the function

$$i(t) = I \cos \omega t \tag{8.2.6}$$

The phasor of the current is

$$\mathbf{I} = I e^{j0} = I \tag{8.2.7}$$

The voltage phasor, $\mathbf{V}$, will be R times the current phasor and will point in the same direction as the current phasor. It is an accepted practice to draw the vector of the independent variable horizontally, pointing to the right. Thus for a resistor both the current and voltage phasors will point in the same direction, as shown in Fig. 8.2.1, where the phasors are drawn as two parallel lines for clarity. We say that the voltage across the resistor and the current through it are *in phase*. If we return to the original time function, we will have

$$v_R(t) = RI \cos \omega t \tag{8.2.8}$$

For an inductor we defined in Eq. (7.1.4)

$$V_L = sLI_L \tag{8.2.9}$$

After the substitution $s = j\omega$,

$$V_L = j\omega L I_L \tag{8.2.10}$$

Let I_L represent a vector pointing horizontally to the right. The voltage across

FIGURE 8.2.1   Voltage phasors for given current.

the inductor will be a phasor whose length will be ωL times the length of the current phasor, but will point upwards, because of the multiplication by j. In agreement with our discussions in Section 1 we will say that the voltage across the inductor *leads* the current through the inductor by 90°. Returning to functions of time, this means that

$$v_L(t) = \omega L \cos (\omega t + 90°) \tag{8.2.11}$$

The phasor $\mathbf{V}_L$ is plotted in Fig. 8.2.1.

Consider next the capacitor by using Eq. (7.1.6)

$$V_C = \frac{1}{j\omega C} I_C = -j \frac{1}{\omega C} I_C \tag{8.2.12}$$

We conclude that the voltage phasor of the capacitor will be $1/\omega C$ times the length of the current phasor but will point down and will *lag* the current by 90°. In time domain, the voltage across the capacitor will be

$$v_C(t) = \frac{1}{\omega C} \cos (\omega t - 90°) \tag{8.2.13}$$

The phasor of the voltage, $\mathbf{V}_C$, is plotted in Fig. 8.2.1.

Work with admittances is similar. For the resistor we have the other form of Ohm's law

$$I_R = GV_R \tag{8.2.14}$$

The independent variable is now the voltage, which we draw as a horizontal vector pointing to the right. The current will be G times the length of the voltage phasor and will point in the same direction, as shown in Fig. 8.2.2. For the capacitor we have, after the substitution,

$$I_C = j\omega C V_C \tag{8.2.15}$$

FIGURE 8.2.2   Current phasors for given voltage.

which means that the current phasor is ωC times the length of the voltage phasor and points upwards, due to the multiplication by j. We say that the capacitive current leads the voltage by 90°. The phasor is shown in Fig. 8.2.2. If we consider the inductor, then

$$I_L = \frac{1}{j\omega L} V_L = -j \frac{1}{\omega L} V_L \tag{8.2.16}$$

The current phasor will point downwards, as shown in Fig. 8.2.2, and its length will be $1/\omega L$ times the length of the voltage phasor.

If the network is more complicated, we use the same network analysis methods as in previous chapters. Let us demonstrate it on the network in Fig. 8.2.3. The source is $j(t) = 1 \cos \omega t$, its phasor is $\mathbf{J} = 1e^{j0} = 1$. We set up the equations exactly the same way as we have been doing until now by using admittances. We will keep the elements as variables and the symbol J for the source. Eventually, the phasor will replace the symbol. The matrix equation is

$$\begin{bmatrix} sC_1 + G & -G \\ -G & sC_2 + G \end{bmatrix} \begin{bmatrix} V_1 \\ V_2 \end{bmatrix} = \begin{bmatrix} J \\ 0 \end{bmatrix}$$

If we are asking for the output voltage, $V_2(s) = V_2$, and keep the variables throughout our solutions, then the determinant is

$$D = s^2 C_1 C_2 + sG(C_1 + C_2)$$

the numerator is

$$N_2 = GJ$$

and the output voltage is

$$V_2(s) = \frac{G}{s^2 C_1 C_2 + sG(C_1 + C_2)} J$$

FIGURE 8.2.3   Network with two capacitors.

Inserting element values,

$$V_2(s) = \frac{2}{3s^2 + 8s} J$$

Substituting $s = j\omega$ and $\mathbf{J} = J = 1$, as derived above,

$$V_2(j\omega) = \frac{2}{-3\omega^2 + 8j\omega}$$

This expression gives us the possibility to evaluate the response for various ω. If we select, for instance, $s = j\omega = j1$, then

$$V_2(j1) = \frac{2}{-3 + 8j} = 0.2341e^{-j110.556}$$

This indicates that for an input signal $j(t) = 1 \cos(1t)$ we get the output voltage $v_2(t) = 0.2341 \cos(1t - 110.556°)$.

We can equally well substitute $s = j1$ and $\mathbf{J} = J = 1$ at the very beginning of our work and get the system

$$\begin{bmatrix} 2+j & -2 \\ -2 & 2+3j \end{bmatrix} \begin{bmatrix} V_1 \\ V_2 \end{bmatrix} = \begin{bmatrix} 1 \\ 0 \end{bmatrix}$$

obtain the determinant

$$D = (2+j)(2+3j) - 4 = -3 + 8j$$

calculate the numerator

$$N_1 = \begin{vmatrix} 2+j & 1 \\ -2 & 0 \end{vmatrix} = 2$$

and get

$$V_2(j1) = \frac{2}{-3 + 8j}$$

which is the same result as above. As we can see, it is only a matter of convenience when we substitute $s = j\omega$ and the phasor value for the independent source.

Example 1.

The signal applied to the network in Fig. 8.2.4 is $e(t) = 3 \cos (2t + 30°)$. Thus the phasor for the source is $\mathbf{E} = 3e^{j30} = 2.598 + j1.5$ and $s = j2$. Calculate I_2 and V_{out} in two ways:

a. Insert element values but keep s as a variable and use the symbol E for the source. Obtain the result and then substitute for s and E.
b. Insert all numerical values at the beginning.

a. The mesh equations are

$$(1 + 2s)I_1 - 2sI_2 = E$$

$$-2sI_1 + (2 + 4s)I_2 = 0$$

In matrix form

$$\begin{bmatrix} 1 + 2s & -2s \\ -2s & 2 + 4s \end{bmatrix} \begin{bmatrix} I_1 \\ I_2 \end{bmatrix} = \begin{bmatrix} E \\ 0 \end{bmatrix}$$

The denominator is $D = 4s^2 + 8s + 2$, the numerator is $N_2 = 2sE$, the current is

$$I_2 = \frac{2sE}{4s^2 + 8s + 2}$$

and the output voltage is

$$V_{out}(s) = sL_2 I_2 = \frac{4s^2 E}{4s^2 + 8s + 2}$$

Now we can substitute $s = j2$ and $E = 2.598 + j1.5$. The output voltage will be

$$V_{out}(j2) = \frac{-16(2.598 + j1.5)}{-14 + j16} = 0.438 + j2.2148$$

Conversion to polar form provides $V_{out}(j2) = 2.2577e^{j78.8148}$ and the output voltage in time domain is $v_{out}(t) = 2.2577 \cos (2t + 78.8148)$.

FIGURE 8.2.4   Network with two inductors.

b. Inserting all numerical values immediately leads to

$$\begin{bmatrix} 1+j4 & -j4 \\ -j4 & 2+8j \end{bmatrix} \begin{bmatrix} I_1 \\ I_2 \end{bmatrix} = \begin{bmatrix} 2.598 + j1.5 \\ 0 \end{bmatrix}$$

The denominator is $D = -14 + j16$, the numerator is $N_2 = j4E = -6 + j10.392$. The current $I_2 = 0.5537 - j0.1095$ and the output voltage is $j4I_2 = 0.438 + j2.2148$, the same result as above.

We may occasionally encounter a problem in which the network has several sources, *all* with the same ω but with different amplitudes and phases. In such case we can proceed in several ways, as demonstrated on the network of Fig. 8.2.5. The first signal and its phasor are

$$j_1(t) = 2 \cos (2t + 45°)$$

$$\mathbf{J}_1 = 2e^{j45°} = 2(\cos 45° + j \sin 45°) = \sqrt{2} + j\sqrt{2}$$

The second signal and its phasor are

$$j_2(t) = 3 \cos (2t + 90°)$$

$$\mathbf{J}_2 = 3e^{j90°} = 3(\cos 90° + j \sin 90°) = 3(0 + j1) = 3j$$

In both cases $\omega = 2$. Using nodal formulation, the system equation is

$$\begin{bmatrix} 1+s & -s & 0 \\ -s & 2+3s & -2s \\ 0 & -2s & 3+2s \end{bmatrix} \begin{bmatrix} \mathbf{V}_1 \\ \mathbf{V}_2 \\ \mathbf{V}_3 \end{bmatrix} = \begin{bmatrix} J_1 \\ J_2 \\ 0 \end{bmatrix}$$

If we do the calculations by hand, then we can

FIGURE 8.2.5   Network with two current sources.

1. Keep the variables s, J_1 and J_2 and solve. Substitute into the result $s = j2$ and the above numerical values for J_1 and J_2.
2. Substitute $s = j2$ into the matrix, use the form $\sqrt{2} + j\sqrt{2}$ and $j3$ for the sources and do all calculations in complex algebra.

We will show both cases. In the first case we evaluate the determinant

$$D(s) = 12s^2 + 19s + 6$$

and by using Cramer's rule find

$$N(s) = 2s^2 J_1 + (2s + 2s^2)J_2$$

The output voltage is

$$V_3(s) = \frac{2s^2 J_1 + (2s + 2s^2)J_2}{12s^2 + 19s + 6}$$

If we now substitute the phasors and set $s = j2$, we obtain

$$V_3(j2) = \frac{-8(\sqrt{2} + j\sqrt{2}) + (4j - 8)3j}{-42 + j38} = \frac{-23.3137 - j35.3137}{-42 + j38} = 0.7471e^{j98.705°}$$

The steady state output is $v_3(t) = 0.7471 \cos (2t + 98.705°)$. If we start with the substitutions immediately, then we must solve

$$\begin{bmatrix} 1 + j2 & -j2 & 0 \\ -j2 & 2 + j6 & -j4 \\ 0 & -4j & 3 + j4 \end{bmatrix} \begin{bmatrix} V_1 \\ V_2 \\ V_3 \end{bmatrix} = \begin{bmatrix} \sqrt{2} + j\sqrt{2} \\ j3 \\ 0 \end{bmatrix}$$

The result is the same but operations with complex numbers are often more difficult.

We should mention that there exists still another possibility: invoking the law of superposition. In such case we find the determinant of the system and then solve twice, first by using J_1 only and next time using J_2 only. We will get two independent solutions and their sum will be equal to the complete steady-state frequency domain response. Try this possibility yourself.

Example 2.

The network in Fig. 8.2.6 has two independent sources and one current-controlled current source. Find the two nodal voltages. The phasors are $J_1 = 2$, $J_2 = 3$ and the controlling current, expressed in terms of the nodal voltage, is $I = G_1 V_1 = 0.5 V_1$. The controlled current is $\alpha I = 2I = 2 \times 0.5 V_1 = V_1$. The two nodal equations, with J denoting the symbol for a current source, are

$$(s + 0.5)V_1 + \alpha I = J_1$$

$$(0.2 + 2s)V_2 - \alpha I = J_2$$

Substituting for $\alpha I = V_1$ we get

$$\begin{bmatrix} s + 1.5 & 0 \\ -1 & 2s + 0.2 \end{bmatrix}\begin{bmatrix} V_1 \\ V_2 \end{bmatrix} = \begin{bmatrix} J_1 \\ J_2 \end{bmatrix}$$

The determinant is $D = 2s^2 + 3.2s + 0.3$,

$$N_1 = \begin{vmatrix} J_1 & 0 \\ J_2 & 2s + 0.2 \end{vmatrix} = (2s + 0.2)J_1$$

$$N_2 = \begin{vmatrix} s + 1.5 & J_1 \\ -1 & J_2 \end{vmatrix} = (s + 1.5)J_2 + J_1$$

Substituting numerical values for the phasors and $s = j2$ we obtain

$$V_1 = \frac{(0.4 + j8)2}{-7.7 + j6.4} = 0.48 - j0.64 = 0.8e^{-j53.13°}$$

$$V_2 = \frac{6.5 + j6}{-7.7 + j6.4} = -0.11621 - j0.8758 = 0.8835e^{-j97.558°}$$

$j_1(t) = 2 \cos 2t$ $j_2(t) = 3 \cos 2t$

FIGURE 8.2.6   Network with two current sources and a CC.

The resulting time functions are

$$v_1(t) = 0.8 \cos (2t - 53.13°)$$

$$v_2(t) = 0.98835 \cos (2t - 97.558°)$$

In the past, phasors were used extensively because electrical engineering dealt mostly with power distribution and thus with only one frequency. In present applications we deal usually with ranges of frequencies and this requires repeated evaluations for various $s = j\omega$.

Let us summarize the developments of this section.

1. Frequency domain analysis can proceed only if all independent sources have the same frequency.
2. All independent sources must be given as cosines or transformed into cosines.
3. Express all sources in terms of their phasors.
4. Replace every s by $j\omega$.
5. Use phasors for the independent sources.
6. Obtain the desired output as a voltage, V, or current, I. These results are phasors.
7. Return to the original cosine functions using Eq. (8.2.3) and Eq. (8.2.4).

The above steps remain valid if we replace everywhere the word cosine by sine.

8.3. RC AND RL NETWORKS

Frequency domain analysis may ask for the solution of one of the two typical cases:

1. What is the steady-state response of the network to a given signal or several given signals, all with the same frequency but with different amplitudes and phases.
2. What is the steady-state response of the network to one input signal, always with the same amplitude and phase, but with various frequencies.

The first case was covered in Section 2, the second will be the subject of this section. The signal is assumed to be the function $\cos \omega t$, the phasor is either $\mathbf{E} = E = 1$ or $\mathbf{J} = J = 1$ and s is replaced by $j\omega$ with a sequence of values of ω. For each frequency, the result of the analysis is a complex number, $R = a + jb$, as we have seen in Section 2. For each such number we can calculate the absolute value and the phase. Two plots can be prepared, both having on the horizontal axis either the frequency f in Hz, or ω in rad/s. If we plot on the vertical axis the absolute value, $|R|$, then we get the *amplitude response* of the network. If we plot the phase, then the curve is called the *phase response* of the network. In many cases the absolute value is not a convenient unit and we may decide to plot on the vertical axis the amplitude response in *decibels*, dB, which is $20 \log_{10} |R|$.

We will introduce the concept of frequency domain response by taking four simple RC and RL voltage dividers. The first network is in Fig. 8.3.1. Its output voltage is

$$V_{out}(s) = \frac{\dfrac{1}{sC}}{\dfrac{1}{sC} + R} = \frac{1}{sCR + 1} E \tag{8.3.1}$$

Substituting $E = 1$ and $s = j\omega$

$$V_{out}(j\omega) = \frac{1}{j\omega CR + 1}$$

The absolute value is

$$|V_{out}| = \frac{1}{\sqrt{\omega^2 C^2 R^2 + 1}}$$

and the phase angle is

$$\phi = -\arctan \omega CR$$

We could now insert various values of C and R, select frequencies and evaluate the absolute value and phase, but for this simple network it is possible to introduce *normalization* and get a result which is valid for all such networks. Define the substitution

$$\Omega = \omega CR \tag{8.3.2}$$

The expressions simplify to

$$|V_{out}| = \frac{1}{\sqrt{\Omega^2 + 1}} \tag{8.3.3}$$

FIGURE 8.3.1   RC network.

and

$$\phi = -\arctan \Omega \qquad (8.3.4)$$

The table below gives results for a few values of Ω

Ω	$|V_{out}|$	$\phi°$
0	1	0
0.5	0.8944	−26.57
1	0.7071	−45.00
2	0.4470	−63.44
3	0.3162	−71.57
4	0.2425	−75.96
100	0.010	−89.43

The absolute value is plotted in Fig. 8.3.2a with respect to the *normalized* frequency Ω; this is the amplitude response of the network. The phase response is in Fig. 8.3.2b.

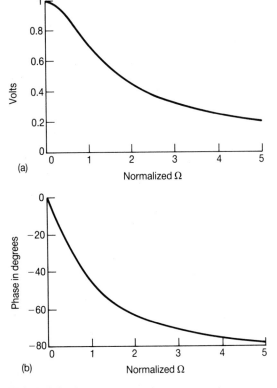

(a)

(b)

FIGURE 8.3.2   Responses of network in Fig. 8.3.1: (a) Amplitude response. (b) Phase response.

The second network we will study is in Fig. 8.3.3. Its output voltage is

$$V_{out}(s) = \frac{R}{\dfrac{1}{sC} + R} = \frac{sCR}{sCR + 1} E \qquad (8.3.5)$$

Substituting $E = 1$ and $s = j\omega$

$$V_{out}(j\omega) = \frac{j\omega CR}{j\omega CR + 1}$$

We can now introduce normalization Eq. (8.3.2)

$$V_{out}(j\Omega) = \frac{j\Omega}{j\Omega + 1}$$

and get

$$|V_{out}| = \frac{\Omega}{\sqrt{\Omega^2 + 1}} \qquad (8.3.6)$$

and

$$\phi = 90° - \arctan \Omega \qquad (8.3.7)$$

The table below gives results for a few values of Ω

Ω	$|V_{out}|$	$\phi°$
0	0	90
0.5	0.4472	63.43
1	0.7071	45.00
2	0.8944	26.57
3	0.9487	18.44
4	0.9701	14.04
100	0.9999	0.57

FIGURE 8.3.3   CR network.

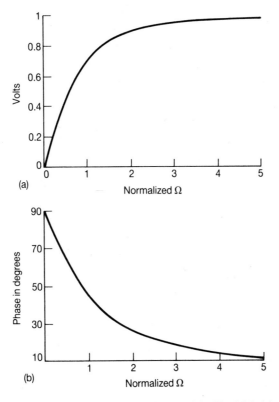

FIGURE 8.3.4   Responses of network in Fig. 8.3.3: (a) Amplitude response. (b) Phase response.

The amplitude response is plotted in Fig. 8.3.4a, the phase response in Fig. 8.3.4b.

Before discussing the results, we will consider two additional cases. The first one is in Fig. 8.3.5. The output is

$$V_{out}(s) = \frac{R}{sL + R} E = \frac{1}{s\dfrac{L}{R} + 1} E \tag{8.3.8}$$

Substituting $E = 1$ and $s = j\omega$ and defining

$$\Omega = \omega \frac{L}{R} \tag{8.3.9}$$

we get

$$|V_{out}| = \frac{1}{\sqrt{\Omega^2 + 1}}$$

Fig. 8.3.5

Fig. 8.3.6

FIGURE 8.3.5   LR network.

FIGURE 8.3.6   RL network.

and

$$\phi = -\arctan \Omega$$

These equations have *exactly* the same form as the equations describing the properties of the network in Fig. 8.3.1 and the curves in Fig. 8.3.2 are valid for the network in Fig. 8.3.3 as well, only the normalization variable is now Eq. (8.3.9) instead of Eq. (8.3.2).

The last network is in Fig. 8.3.6. Its output voltage is

$$V_{out}(s) = \frac{sL}{sL + R} = \frac{s\dfrac{L}{R}}{s\dfrac{L}{R} + 1} \tag{8.3.10}$$

Proceeding similarly as above we discover that properties of this network are *exactly* equivalent to the properties of the network in Fig. 8.3.3 and that the plots in Fig. 8.3.4 are valid for this network as well, only the normalization variable is now given by Eq. (8.3.9) instead of Eq. (8.3.2).

Let us now discuss the curves and how they can be used for similar networks with arbitrary component values. Consider, for instance, the network in Fig. 8.3.1 and let $R = 10^3 \, \Omega$ and $C = 10^{-6}$ F. We wish to know at what frequency will the response drop to 0.7071 of its value at dc. Reading from the curve (or from the table) we find that for this case $\Omega = \omega RC = 1$. Substituting the element values we have $\omega \times 10^3 \times 10^{-6} = 1$, from which $\omega = 10^3$ or $f = 1000/(2\pi) = 159$ Hz.

RC or RL networks may be more complicated than the above cases. Dependent sources may also be present and we will give two simple examples.

FIGURE 8.3.7   Network with one VC.

Example 1.

Consider the network in Fig. 8.3.7 and find the output as a function of the s variable. Let the phasor for the signal be $\mathbf{J} = 1$.

The nodal equations are

$$(1 + s)V_1 - sV_2 = 1$$
$$-sV_1 + (2 + s)V_2 - V_1 + V_2 = 0$$

As a matrix equation

$$\begin{bmatrix} 1 + s & -s \\ -1 - s & 3 + s \end{bmatrix} \begin{bmatrix} V_1 \\ V_2 \end{bmatrix} = \begin{bmatrix} 1 \\ 0 \end{bmatrix}$$

The determinant is $D = 3s + 3$, the numerator is $N_2 = s + 1$. The output voltage

$$V_2 = \frac{s + 1}{3s + 3} = \frac{1}{3}$$

In this case it so happens that the output is independent of frequency.

Example 2.

The network in Fig. 8.3.8 has a current-controlled current source. Find V_2 as a function of s and plot the amplitude response in absolute units. The source phasor is $\mathbf{J} = 1$.

FIGURE 8.3.8   Network with one CC.

The controlling current is $I = G_1 V_1$ and thus the controlled current is $\alpha I = 2G_1 V_1 = 4V_1$. The nodal equations are

$$\left(2 + \frac{1}{2s}\right)V_1 - \frac{1}{2s} V_2 = 1$$

$$-\frac{1}{2s} V_1 + \left(3 + \frac{1}{2s}\right)V_2 - 4V_1 = 0$$

In matrix form

$$\begin{bmatrix} \dfrac{4s+1}{2s} & -\dfrac{1}{2s} \\[2ex] \dfrac{-8s-1}{2s} & \dfrac{6s+1}{2s} \end{bmatrix} \begin{bmatrix} V_1 \\ V_2 \end{bmatrix} = \begin{bmatrix} 1 \\ 0 \end{bmatrix}$$

The determinant is $D = (12s + 1)/2s$, the numerator for the output is $N_2 = (8s + 1)/2s$ and the output voltage is $V_2 = (8s + 1)/(12s + 1)$. Inserting various values of $s = j\omega$ we can evaluate and plot the amplitude or phase response.

Since plotting without a computer is always a time-consuming task, it is useful to make an estimate of the frequency response at a few selected frequencies. The easiest is to find how the network will behave at $s = 0$ and at s approaching infinity. In the above example this will give $V_2(0) = 1$ and $V_2(\infty) = 8/12$.

Example 3.

Use NETOPT to find the frequency domain responses of the RL and RC networks solved above. If you have a computer with plotting facility, display the amplitude and phase responses. Also plot the real parts, imaginary parts and the dB option.

Use the symbolic option of NETOPT to get the functions in terms of s or in terms of s and up to five variable elements. Instructions how to do it are in Chapter 18 Section 5.

8.4. RLC NETWORKS

Networks consisting of one capacitor, one inductor and one or more resistors have important frequency domain properties which we will study in this section. The simplest ones, with just one resistor, are also called tuned circuits.

The series tuned circuit is shown in Fig. 8.4.1. Using mesh analysis and the phasor value $\mathbf{E} = E$ (corresponding to the input signal $E \cos \omega t$)

$$I = \frac{sC}{s^2 LC + sCR + 1} E = \frac{1}{L} \frac{s}{s^2 + s\dfrac{R}{L} + \dfrac{1}{LC}} E \qquad (8.4.1)$$

FIGURE 8.4.1   Series tuned circuit.

Consider first the special case when $R = 0$; the current becomes

$$I = \frac{1}{L} \frac{s}{s^2 + \dfrac{1}{LC}} E$$

If we now select $s = j\omega = j/\sqrt{LC}$, the denominator becomes zero and the current will be infinitely large. The voltage across the capacitor,

$$V_C = \frac{1}{sC} I$$

also becomes infinitely large for this frequency. The same will be true for the voltage across the inductor, because

$$V_L = sLI$$

Clearly, such frequency (in radians) represents a special property of the tuned circuit and is given a special name, the *resonant* frequency. It is commonly denoted by ω_0 and is given by

$$\omega_0 = \frac{1}{\sqrt{LC}} \tag{8.4.2}$$

If $R \neq 0$, then at the resonant frequency

$$I_{res} = \frac{1}{R} E$$

The current has the same phase as the input voltage (real value) and grows if R is made smaller. The voltage across the resistor will be

$$V_{R,res} = R \frac{E}{R} = E$$

exactly equal to the amplitude of the source signal. The voltage across the capacitor will become

$$V_{C,res} = \frac{1}{j\omega_0 C}\frac{E}{R} = -j\frac{1}{R}\left(\frac{L}{C}\right)^{1/2} E$$

and the voltage across the inductor will be

$$V_{L,res} = j\omega_0 L\frac{E}{R} = j\frac{1}{R}\left(\frac{L}{C}\right)^{1/2} E$$

Since the voltage source is expressed as a phasor, these voltages are also phasors and can be plotted in a diagram, Fig. 8.4.2. The voltage across the resistor is in phase with the current, the voltage across the capacitor is lagging behind the resistor voltage by 90° and the voltage across the inductor is leading the resistor voltage by 90°. The voltages across the capacitor and inductor have at resonance exactly opposite phases, have equal values and become larger if the resistor becomes smaller. Thus there is some kind of "quality" of the series tuned circuit; the smaller the resistor, the better is the quality of the tuned circuit. The concept of a *quality factor*, Q, is introduced,

$$Q = \frac{1}{R}\left(\frac{L}{C}\right)^{1/2} \tag{8.4.3}$$

We have now introduced two special names, the resonant frequency, ω_0, and the quality factor, Q. Their ratio

$$\frac{\omega_0}{Q} = \left(\frac{1}{LC}\right)^{1/2} R\left(\frac{C}{L}\right)^{1/2} = \frac{R}{L} \tag{8.4.4}$$

turns out to be exactly the middle coefficient of the denominator in Eq. (8.4.1). We can introduce both these new definitions into the second-order polynomial in

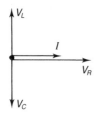

FIGURE 8.4.2   Phasors at resonance.

Eq. (8.4.1) and obtain

$$I = \frac{1}{L} \frac{s}{s^2 + s\dfrac{\omega_o}{Q} + \omega_0^2} E \tag{8.4.5}$$

We can also express the resonant voltages in terms of Q. The voltage across the capacitor becomes

$$V_{C,res} = -jQE \tag{8.4.6}$$

and across the inductor

$$V_{L,res} = jQE \tag{8.4.7}$$

It probably comes as a surprise that these voltages can be much larger than the voltage actually supplied by the independent voltage source.

Another possibility is a parallel tuned circuit shown in Fig. 8.4.3. If the input current is $j(t) = J \cos \omega t$, the phasor for the current is $\mathbf{J} = J$ and the voltage across the combination is

$$V = \frac{1}{C} \frac{s}{s^2 + s\dfrac{G}{C} + \dfrac{1}{LC}} J \tag{8.4.8}$$

Using our generalized definitions, we can identify

$$\omega_0^2 = \frac{1}{LC} \tag{8.4.9}$$

and

$$\frac{\omega_0}{Q} = \frac{G}{C}$$

FIGURE 8.4.3   Parallel tuned circuit.

from where we get

$$Q = \frac{\omega_0 C}{G} = \omega_0 C R = R\left(\frac{C}{L}\right)^{1/2} \qquad (8.4.10)$$

Note that this expression for the Q of a parallel tuned circuit differs from that of the series tuned circuit. Inserting Eq. (8.4.9) and Eq. (8.4.10) into Eq. (8.4.8)

$$V = \frac{1}{C} \frac{s}{s^2 + s\dfrac{\omega_0}{Q} + \omega_0^2} J \qquad (8.4.11)$$

At the resonant frequency the voltage

$$V_{res} = \frac{Q}{\omega_0 C} J = \frac{J}{G} = RJ \qquad (8.4.12)$$

is in phase with the current of the source and grows with the increase of R. If the resistor becomes very large, $R \to \infty$, then $Q \to \infty$ and the voltage becomes infinitely large.

It is advantageous to introduce normalized expressions because they can be used for any values of L and C. Divide the numerator and denominator of Eq. (8.4.8) by ω_0^2

$$V = \frac{1}{\omega_0 C} \frac{\dfrac{s}{\omega_0}}{\dfrac{s^2}{\omega_0^2} + \dfrac{s}{\omega_0}\dfrac{1}{Q} + 1} \qquad (8.4.13)$$

Substitute $s = j\omega$ and define *normalized* Ω,

$$\Omega = \frac{\omega}{\omega_0} \qquad (8.4.14)$$

This modifies Eq. (8.4.13) to

$$V = \frac{1}{\omega_0 C} \frac{j\Omega}{(1 - \Omega^2) + j\Omega\dfrac{1}{Q}} \qquad (8.4.15)$$

The product $\omega_0 C$ is a constant for any given network and the frequency-dependent properties are expressed by the second fraction only

$$S = \frac{j\Omega}{(1 - \Omega^2) + j\Omega\dfrac{1}{Q}} \tag{8.4.16}$$

The absolute value of S is

$$|S| = \frac{\Omega}{\left[(1 - \Omega^2)^2 + \dfrac{\Omega^2}{Q^2}\right]^{1/2}} \tag{8.4.17}$$

and the phase is

$$\phi = \arctan\frac{Q(1 - \Omega^2)}{\Omega} \tag{8.4.18}$$

Equation (8.4.17), sometimes called the *selectivity* curve, was evaluated for several Q and the plot is shown in Fig. 8.4.4. These normalized curves are valid for a parallel tuned circuit having the same Q and any values of L and C. They can be used to find suppression of signals away from resonance.

The concept of Q is extremely useful and we will derive a method how it can be determined from measurements. While the derivations may be skipped at first reading, the conclusions and the way how to measure Q should be remembered.

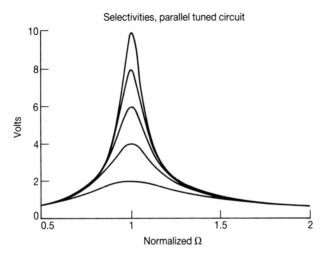

FIGURE 8.4.4   Selectivities for various Q.

Consider Eq. (8.4.11), insert $s = j\omega$ and take absolute value

$$|V(j\omega)| = \frac{1}{C} \frac{\omega}{\left[(\omega_0^2 - \omega^2)^2 + \frac{\omega^2 \omega_0^2}{Q^2} \right]^{1/2}} \tag{8.4.19}$$

Also obtain the value at resonance

$$|V(j\omega_0)| = \frac{Q}{\omega_0 C} \tag{8.4.20}$$

Form the ratio and make it equal to $\sqrt{2}$,

$$\frac{|V(j\omega_0)|}{|V(j\omega)|} = \sqrt{2} \tag{8.4.21}$$

This choice has the advantage that the derivations below become simple. Inserting from above and performing a few algebraic operations we arrive at

$$(\omega_0^2 - \omega^2)^2 = \frac{\omega^2 \omega_0^2}{Q^2} \tag{8.4.22}$$

Take the root

$$\omega_0^2 - \omega^2 = \pm \frac{\omega \omega_0}{Q}$$

and write

$$\omega^2 \pm \omega \frac{\omega_0}{Q} - \omega_0^2 = 0 \tag{8.4.23}$$

This represents two quadratic equations which must be considered separately. Taking first the positive sign in Eq. (8.4.23), we get two roots

$$\omega_{1,2} = -\frac{\omega_0}{2Q} \pm \omega_0 \left(\frac{1}{4Q^2} + 1 \right)^{1/2}$$

and only the positive one is meaningful (negative frequency does not exist)

$$\omega_A = -\frac{\omega_0}{2Q} + \omega_0 \left(\frac{1}{4Q^2} + 1 \right)^{1/2}$$

Taking next the negative sign in Eq. (8.4.23) we get another two roots

$$\omega_{3,4} = \frac{\omega_0}{2Q} \pm \omega_0\left(\frac{1}{4Q^2} + 1\right)^{1/2}$$

and only the positive root is meaningful,

$$\omega_B = \frac{\omega_0}{2Q} + \omega_0\left(\frac{1}{4Q^2} + 1\right)^{1/2}$$

Form the difference

$$\Delta\omega = \omega_B - \omega_A = \frac{\omega_0}{Q}$$

and rewrite as

$$Q = \frac{\omega_0}{\Delta\omega} = \frac{f_0}{\Delta f} \tag{8.4.24}$$

The result is interpreted as follows: Measure the voltage across the tuned circuit and find such frequency f_0 of the signal generator at which the voltage is maximum, V_{max}. Change the signal frequency below f_0 until the voltage across the tuned circuit drops to $V_{max}/\sqrt{2}$; this is f_A. Next change the signal frequency above f_0 until the voltage is again $V_{max}/\sqrt{2}$; this frequency is f_B. Insert into Eq. (8.4.24) to get the $Q = f_0/(f_B - f_A)$ of the tuned circuit. Application of the above is indicated schematically in Fig. 8.4.5.

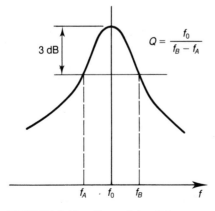

FIGURE 8.4.5   Determining Q from measurements.

The concept of Q and ω_0 is widely used in the literature and can be extended to any network function with a second-order polynomial and complex conjugate roots. We will demonstrate this generalization on the network shown in Fig. 8.4.6. Losses of inductors are usually modeled by a resistor in series with the inductance. Losses of capacitors are modeled by a resistor in parallel to the capacitance. Thus the network takes into account losses in both elements. We are interested in the output voltage. The analysis is simple, because the network is essentially a voltage divider in which the first impedance is

$$Z_1 = R + sL$$

and the second one is

$$Z_2 = \frac{1}{sC + G}$$

The transfer function is

$$V_{out} = \frac{Z_2}{Z_1 + Z_2} E = \frac{1}{s^2 LC + s(CR + LG) + (1 + GR)} E \qquad (8.4.25)$$

In order to get the form Eq. (8.4.5) we factor out the product LC

$$V_{out} = \frac{1}{LC} \frac{1}{s^2 + s\dfrac{CR + LG}{LC} + \dfrac{1 + GR}{LC}} E$$

Now we can identify

$$\omega_0^2 = \frac{1 + GR}{LC} \qquad (8.4.26)$$

and

$$\frac{\omega_0}{Q} = \frac{CR + LG}{LC}$$

FIGURE 8.4.6   Tuned circuit with losses.

from which we get

$$Q = \frac{\omega_0 LC}{CR + LG} = \frac{\sqrt{LC(1 + RG)}}{CR + LG} \qquad (8.4.27)$$

We will use similar generalized properties of the Q and ω_0 when we get to the analysis of active networks. We will also derive some additional formulas there. At this moment we remember the following steps:

1. Factor out the coefficient at s^2 so that the second-order polynomial is reduced to the form $s^2 + as + b$.
2. Then $\omega_0^2 = b$ and $\omega_0/Q = a$.

Example 1.

Analyze the network in Fig. 8.4.7 and find its ω_0 and Q. The nodal equations are

$$(s + 1)V_1 - sV_2 = J$$

$$-sV_1 + \left(s + \frac{1}{2s} + 1\right)V_2 + 2.5V_1 = 0$$

In matrix form

$$\begin{bmatrix} s + 1 & -s \\ 2.5 - s & \dfrac{2s^2 + 2s + 1}{2s} \end{bmatrix} \begin{bmatrix} V_1 \\ V_2 \end{bmatrix} = \begin{bmatrix} J \\ 0 \end{bmatrix}$$

The denominator is $D = (9s^2 + 3s + 1)/(2s)$, $N_2 = (s - 2.5)J$ and the output voltage is

$$V_2 = \frac{(s - 2.5)2s}{9s^2 + 3s + 1} J = \frac{1}{9} \frac{2s^2 - 5s}{s^2 + \dfrac{1}{3}s + \dfrac{1}{9}} J$$

FIGURE 8.4.7   RLC network with a VC.

Thus the resonant frequency is

$$\omega_0 = \frac{1}{3}$$

and Q is obtained by comparing the coefficient at the first power of s

$$\frac{\omega_0}{Q} = \frac{1}{3}$$

from which we obtain $Q = 1$.

Example 2.

Use NETOPT to analyze all networks of this section. Select any element values and study the plots of the amplitude response in magnitude and dB. Obtain symbolic functions in terms of s and also in terms of variable elements. Calculate Q and ω_0. Instructions are in Chapter 18.

8.5. PHASORS AND POWER

We introduced the concept of power in Chapter 1 and we stated, in Eq. (1.5.2), that it is equal to the product of the voltage and current,

$$p(t) = v(t)i(t) \tag{8.5.1}$$

The equation is valid for any signal at any instant of time and in this section we will apply it to the situations when the voltage and current are cosine functions. For generality we will take

$$v(t) = V \cos(\omega t + \alpha)$$
$$i(t) = I \cos(\omega t + \beta) \tag{8.5.2}$$

where V and I are *amplitudes* of the signals. The situation is sketched in Fig. 8.5.1 in the form of two phasors, stopped at one instant. Inserting Eq. (8.5.2) into Eq. (8.5.1) we get

$$p(t) = VI \cos(\omega t + \alpha) \cos(\omega t + \beta) \tag{8.5.3}$$

We will use several substitutions in order to get a more easily understandable result. Using the formula

$$\cos A \cos B = \frac{1}{2}[\cos(A - B) + \cos(A + B)]$$

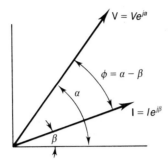

FIGURE 8.5.1   Phasors for voltage and current.

we rewrite Eq. (8.5.3) in the form

$$p(t) = \frac{VI}{2} \left[\cos(\omega t + \alpha - \omega t - \beta) + \cos(\omega t + \alpha + \omega t + \beta) \right]$$

$$= \frac{VI}{2} \cos(\alpha - \beta) + \frac{VI}{2} \cos(2\omega t + \alpha + \beta)$$

If we introduce the concept of phase difference of the two phasors, see Fig. 8.5.1,

$$\phi = \alpha - \beta \tag{8.5.4}$$

then the above can be further simplified to

$$p(t) = \frac{VI}{2} \cos \phi + \frac{VI}{2} \cos(2\omega t + 2\alpha - \phi)$$

For the complicated expression in the second cosine we can further apply the formula

$$\cos(A - B) = \cos A \cos B + \sin A \sin B$$

and obtain

$$p(t) = \frac{VI}{2} \cos \phi + \frac{VI}{2} \cos \phi \cos(2\omega t + 2\alpha) + \frac{VI}{2} \sin \phi \sin(2\omega t + 2\alpha)$$

Finally, since we know that the angle α is the result of our stopping the rotation, we can neglect it and get the final expression

$$p(t) = \frac{VI}{2} \cos \phi + \frac{VI}{2} \cos \phi \cos (2\omega t) + \frac{VI}{2} \sin \phi \sin (2\omega t) \qquad (8.5.5)$$

The first term represents a dc component of the power, the other two components are changing with twice the angular speed, are shifted 90° with respect to each other and have zero average value.

Let us now consider some special situations, the first being that the voltage and current are in phase, as they would be when $i(t)$ flows through a resistor and creates a voltage $v(t)$ across it. In such case $\phi = 0$ and Eq. (8.5.5) simplifies to

$$p(t) = \frac{VI}{2} + \frac{VI}{2} \cos (2\omega t)$$

The first component is dc power, consumed in the resistor, the second term is oscillating with zero average and thus the average power consumed by the resistor is equal to the dc component. The same power would be consumed by the same resistor if we applied a dc voltage having an *effective* value

$$V_{eff} = \frac{V}{\sqrt{2}} \qquad (8.5.6)$$

In such case the current would have an effective value

$$I_{eff} = \frac{I}{\sqrt{2}} \qquad (8.5.7)$$

Effective values are often used in equations but we will keep the absolute values.

If the element is a capacitor or inductor, then we know that the phase difference between the voltage and the current is $\pm 90°$ and Eq. (8.5.5) simplifies to

$$p(t) = \pm \frac{VI}{2} \sin (2\omega t)$$

This is an oscillating power with zero average value and thus no average power is lost in any of these two elements. If the element is composed of several simpler components, the angle between the voltage and current will differ from $0°$ and $\pm 90°$ and all three components of Eq. (8.5.5) will be nonzero.

Since the phasors were such a help in the analysis, it would be nice to use them for power calculations as well. This is indeed possible. The phasors for the signals

Eq. (8.5.2) are

$$V = Ve^{j\alpha}$$
$$I = Ie^{j\beta}$$

(8.5.8)

Form the product of the voltage phasor, the *complex conjugate* of the current phasor, and divide by 2

$$\mathbf{S} = \frac{1}{2}\mathbf{V}\overline{\mathbf{I}} = \frac{VI}{2}e^{j(\alpha-\beta)} = \frac{VI}{2}e^{j\phi}$$

or

$$\mathbf{S} = \frac{VI}{2}\cos\phi + j\frac{VI}{2}\sin\phi = P + jQ$$

(8.5.9)

This complex power determines the amplitudes of all three terms in Eq. (8.5.5) and is widely used. The first component expresses the real power and is measured in watts, W, the second component is measured in reactive volt-amperes, VAR, and their sum, **S**, is measured in volt-amperes, VA. Equation (8.5.9) gives us the possibility to use impedances and admittances to evaluate not only voltages and currents but also the components of the power, as we will show in several examples.

Example 1.

Consider a system with $f = 60$ Hz, which means that $\omega = 2\pi 60 = 377$ rad/s. The voltage $v(t) = 20 \cos \omega t$ is applied to a parallel combination of a conductance $G = 2$ and $C = 3/377$, as sketched in Fig. 8.5.2. Calculate the power and reactive power delivered into this combination.

The phasor of the voltage is $V = 20e^{j0} = 20$. Admittance of the element combination is, at the given frequency, $Y = G + j\omega C = 2 + j3$. The current is

$$\mathbf{I} = \mathbf{V}Y = 20(2 + j3) = 40 + j60$$

Inserting into Eq. (8.5.9) we get

$$\mathbf{S} = \frac{1}{2}\mathbf{V}\overline{\mathbf{I}} = \frac{1}{2}20(40 - j60) = 400 - j600$$

FIGURE 8.5.2   Simple GC network.

FIGURE 8.5.3   Simple RL network.

and thus

$$P = 400 \text{ W}$$

$$Q = -600 \text{ VAR}$$

Example 2.

Apply the same voltage as in Example 1 to the series combination in Fig. 8.5.3 and obtain the components of the complex power.

For the given frequency the impedance of this combination is

$$Z = 12 + j\omega L = 12 + j10$$

The current is

$$I = \frac{V}{Z} = \frac{20}{12 + j10} = 0.9836 - j0.8196$$

The complex power is

$$S = \frac{1}{2} V\bar{I} = \frac{1}{2} 20(0.9836 + j0.8196) = 9.836 + j8.196$$

and thus

$$P = 9.836 \text{ W}$$

$$Q = 8.196 \text{ VAR}$$

Example 3.

Calculate the complex power components for the network and signal in Fig. 8.5.4.

The phasor for the signal $2 \cos 3t$ is $V = 2$, with $\omega = 3$. Using standard mesh formulation

$$e(t) = 2 \cos 3t$$

FIGURE 8.5.4   Network with one inductor.

and substituting $s = j3$ we get

$$\begin{bmatrix} 5 + j6 & -j6 \\ -j6 & 3 + j6 \end{bmatrix} \begin{bmatrix} I_1 \\ I_2 \end{bmatrix} = \begin{bmatrix} 2 \\ 0 \end{bmatrix}$$

The determinant is $D = 15 + j48$, $N_1 = 6 + j12$ and the current

$$I_1 = \frac{6 + j12}{15 + j48}$$

To get a complex conjugate of I_1 change all signs of the imaginary parts. Thus the complex power is

$$\mathbf{S} = \frac{1}{2} 2 \frac{6 - j12}{15 - j48} = 0.2633 + j0.0427$$

The components are

$$P = 0.2633 \text{ W}$$

$$Q = 0.0427 \text{ VAR}$$

Example 4.

Find the power consumed by the network in Fig. 8.5.5.
    The signal phasor is $\mathbf{J} = 3e^{j0} = 3$ and $s = j5$. Nodal formulation provides

$$\begin{bmatrix} 1 + 3s & -2s \\ -2s & 2s + 3 \end{bmatrix} \begin{bmatrix} V_1 \\ V_2 \end{bmatrix} = \begin{bmatrix} 3 \\ 0 \end{bmatrix}$$

Inserting $s = j5$ we must solve

$$\begin{bmatrix} 1 + j15 & -j10 \\ -j10 & 3 + j10 \end{bmatrix} \begin{bmatrix} V_1 \\ V_2 \end{bmatrix} = \begin{bmatrix} 3 \\ 0 \end{bmatrix}$$

$j(t) = 3 \cos 5t$

FIGURE 8.5.5   Network with two capacitors.

The determinant is $D = -47 + j55$, the numerator is $N_1 = 9 + j30$, and the voltage

$$V_1 = \frac{9 + j30}{-47 + j10}$$

Since the phasor for the current does not have an imaginary component, the complex power is

$$S = \frac{1}{2} \frac{9 + j30}{-47 + j55} 3 = 0.3516 - j0.5459$$

and the power components are

$$P = 0.3516 \text{ W}$$

$$Q = -0.5459 \text{ VAR}$$

Example 5.

Suppose that we measured the voltage across a device and the current through it

$$v(t) = 10 \cos(60t + 70°)$$

$$i(t) = 5 \cos (60t - 50°)$$

Find the complex power components.
    The phasors are

$$\mathbf{V} = 10e^{j70}$$

$$\mathbf{I} = 5e^{-j50}$$

and the complex power is

$$\mathbf{S} = \frac{1}{2} 10e^{j70}5e^{j50} = 25e^{j120} = -12.5 + j21.65$$

Since the real component is negative, the device actually delivers power 12.5 W.

PROBLEMS CHAPTER 8

All problems of this Chapter can be solved by the program NETOPT. See Chapter 18 and problems P.8.28 through 31 for information.

P.8.1 Given are three complex numbers: $v_1 = 3 + j2$, $v_2 = -1 + j4$, $v_3 = -2 - j3$ and $v_4 = 4 - j2$. Perform the following operations:

(a) $v_1 + v_2$
(b) $v_1 - v_2 + v_3$
(c) $v_1 + v_3 - v_2 - v_4$
(d) $v_1 v_2$
(e) $\dfrac{v_1 v_3}{v_2}$
(f) $\dfrac{(v_1 - v_2)v_4}{v_3}$
(g) $v_1 v_2 v_3 v_4$
(h) $\dfrac{v_1 + v_2}{v_3 + v_4}$

P.8.2 Convert the following vectors in rectangular form into polar form:

(a) $v_1 = 3 + j5$
(b) $v_2 = -1 + j2$
(c) $v_3 = -2 - j4$
(d) $v_4 = 4 - j1$
(e) $v_5 = -4 - j3$
(f) $v_6 = 0 + j3$
(g) $v_7 = -1 - j0$
(h) $v_8 = 0 - j5$

P.8.3 Convert the following polar coordinates into rectangular form: The angles are given in degrees.

(a) $v_1 = 5e^{j60}$
(b) $v_2 = 2e^{j120}$
(c) $v_3 = 4e^{-j150}$
(d) $v_4 = 3e^{-j20}$
(e) $v_5 = 5e^{j150}$
(f) $v_6 = 4e^{j90}$
(g) $v_7 = 5e^{-j180}$
(h) $v_8 = 4e^{-j90}$

P.8.4 Write the phasor representations for the following functions:

(a) $3 \cos (2t + 40°)$
(b) $2 \cos (3t + 150°)$
(c) $4 \cos (4t - 30°)$
(d) $5 \sin \omega t$
(e) $4 \sin (\omega t + 20°)$
(f) $2 \sin (377t + 50°)$

P.8.5 Using phasors, find the resulting single cosine (or sine) functions.

(a) $4 \cos (2t + 40°) + 3 \cos (2t - 50°)$
(b) $\cos \omega t + \sin \omega t$
(c) $3 \cos (\omega t + 20°) + 2 \cos (\omega t + 40°)$
(d) $2 \cos (\omega t - 120°) - 3 \cos (\omega t + 140°)$
(e) $2 \cos \omega t + 4 \cos (\omega t + 45°)$
(f) $\cos (\omega t + 20°) + \cos (\omega t + 40°)$
$\qquad\qquad\qquad\qquad + \cos (\omega t + 60°)$
(g) $2 \cos (\omega t + 180°) + 4 \cos (\omega t + 90°)$
(h) $\sin (\omega t + 30°) - 2 \sin (\omega t - 30°)$
(i) $3 \cos (\omega t + 50°) + 2 \cos (\omega t + 150°)$
$\qquad\qquad\qquad\qquad + \cos (\omega t - 30°)$
(j) $2 \cos \omega t + 3 \cos (\omega t + 90°)$
$\qquad + \cos (\omega t + 180°) + 4 \cos (\omega t - 90°)$

P.8.6 Find the impedance and admittance for the series RL combination. In each case evaluate at $\omega = 1, 2, 3$.

P.8.7 Find the impedance and admittance for the series CR combination. In each case evaluate at $\omega = 1, 2, 3$.

P.8.8 Find the impedance and admittance for the parallel CR combination. In each case evaluate at $\omega = 1, 2, 3$.

P.8.9 Find the impedance and admittance for the parallel LR combination. In each case evaluate at $\omega = 1, 2, 3$.

P.8.10 Find the impedance of the series RCL combination at $\omega = 2$.

P.8.11 Find the input admittance of the CGRL combination at $\omega = 5$.

P.8.12 The input signal is $e(t) = 2 \cos (2t + 30°)$. Find the phasor of the current flowing into the network. Also express the result in terms of the input signal.

P.8.13 Find the output voltage for the current source being the function $j(t) = 3 \cos (3t + 25°)$. (a) Keep the variable s, get the result and then insert the phasor of the signal. (b) Substitute for $s = j3$ from the start and do the same.

P.8.14 Find the output voltage if the input signal is $j(t) = \cos \omega t$ and get the results for $\omega = 1, 2, 3$. Note that in this case it is to your advantage to keep the variable s, since you can reuse the solution step.

P.8.15 Do the same as in P.8.14.

P.8.16 Find the output voltage V_2 for $j(t) = \cos \omega t$ with $\omega = 1, 2$.

P.8.17 Using mesh formulation, obtain the output voltage and evaluate for $e(t) = \cos \omega t$ with $\omega = 1, 2$.

P.8.18 Find the output current for $e(t) = \cos \omega t$ with $\omega = 1$. Obtain the result by (a) keeping the variable s and (b) inserting $s = j1$ from the start.

P.8.19 Find the output voltage for $e(t) = \cos \omega t$ with $\omega = 1, 2$.

P.8.20 Calculate the complex power components for problem P.8.6 for (a) input voltage $e(t) = \cos \omega t$ and (b) input current $j(t) = \sin \omega t$, using $\omega = 1, 2$.

P.8.21 Do the same as in P.8.20 for the combination in P.8.9.

P.8.22 Do the same as in P.8.20 for the combination in P.8.10.

P.8.23 Do the same as in P.8.20 for the combination in P.8.11.

P.8.24 Find the complex power components consumed in the R, L part of the network in P.8.11. Let the source be a current source $j(t) = 2 \cos 2t$.

P.8.25 In problem P.8.14 calculate the complex power components delivered by the source. The input signal is $2 \cos 2t$.

P.8.26 Do the same as in P.8.25 for the network in P.8.15.

P.8.27 Do the same as in P.8.25 for the network in P.8.17.

P.8.28 You can use NETOPT to calculate input impedances by connecting a unit current source at the input and measuring the voltage across it. The sketch explains how to specify the input and output voltage. Confirm your results in P.8.6 through P.8.11 by running the program.

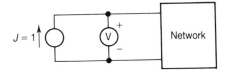

P.8.29 Frequency domain option (AC) in NETOPT assumes that the source is of the form $A \cos \omega t$ where A, the amplitude, is given as the value of the source. If you have several sources, all must have the same form (the program does not provide for various phase shifts). You can display (or print) the real and imaginary parts of the result or the absolute value and phase in degrees. Use NETOPT to confirm your results in the problems P.8.13 through P.8.19.

P.8.30 You can use NETOPT to obtain complex current through any part of the network by connecting there an ammeter. You can simultaneously obtain the voltage across the same combination by connecting a voltmeter. The current and the voltage can be obtained in the Re $+ j$ Im or in the $Ae^{j\phi}$ form. You can use these results to calculate complex power components. Use the program to obtain solutions for the problems P.8.20 through P.8.27.

P.8.31 Calculation of frequency domain responses was done in the above examples for at most three frequencies. Using NETOPT, calculate and display frequency domain responses for the above problems. Use 50 or more points.

P.8.32 The symbolic option in NETOPT can be used to obtain the desired output functions as ratios of two polynomials in the variable s. You have derived such functions above; check them by the program. Remember that the source in symbolic option is automatically assumed to have unit value and *only the first source* in the netlist is considered.

CHAPTER 9

Laplace Transformation

9.0. INTRODUCTION

The Laplace transformation is the most efficient method for solving linear networks. It was invented by Pierre Simon de Laplace in the eighteenth century but was widely accepted only in this century.

The purpose of a mathematical transformation is to introduce definitions which considerably simplify some complicated steps. In the case of the Laplace transform, differentiation is replaced by multiplication by the transform variable s. Integration is made even simpler, because it is replaced by division by s.

We will not give a fully comprehensive treatment of the Laplace transform theory and only those subjects which have direct relationship to linear networks will be selected. A detailed study is left to specialized mathematical courses.

9.1. DEFINITION OF THE LAPLACE TRANSFORM

The Laplace transform works with the complex variable

$$s = \sigma + j\omega \qquad (9.1.1)$$

The real part, σ, does not have any direct practical explanation but the imaginary part does. If we set $\sigma = 0$ and thus

$$s = j\omega = j2\pi f \tag{9.1.2}$$

then we can speak about the actual frequency, f, of a steady sinusoidal signal applied to the network. This was already explained in detail in previous chapters.

Let us have a function $f(t)$ of time t such that $f(t) = 0$ for $t < 0$. Then the (one-sided) Laplace transform is defined by

$$\mathscr{L}[f(t)] = F(s) = \int_{0^-}^{\infty} f(t)e^{-st}\,dt \tag{9.1.3}$$

It is common practice to use capital letters for the transformed variables (functions of s) and lower case letters for the functions of time, t.

The inverse transformation is defined by the integral

$$\mathscr{L}^{-1}[f(t)] = \frac{1}{2\pi j} \int_{c-j\infty}^{c+j\infty} F(s)e^{+st}\,ds \tag{9.1.4}$$

Note that the inverse transformation has the exponent with $+$ sign and the original transformation has $-$ sign. Also note the symbolic notation with $\mathscr{L}$ and $\mathscr{L}^{-1}$. The integration in Eq. (9.1.4) is along a line parallel with the imaginary axis.

Evaluations of Eq. (9.1.3) are relatively simple and require straightforward integration in the real variable, t. Evaluations of the inverse transform require considerably more mathematical background than can be expected from the student at this time. Fortunately, it is not necessary to go via Eq. (9.1.4). It is possible to get the transformations for a certain number of practical functions and prepare tables. The inverse transformations are then done by consulting the tables. This will be our way of handling the inverses.

Two fundamental rules of the Laplace transform will be needed in the following derivations. For a function $f(t)$, having the transform $F(s)$, we have

$$\mathscr{L}[af(t)] = aF(s) \tag{9.1.5}$$

This follows immediately from Eq. (9.1.3). The second rule considers addition. If we have two functions, $f_1(t)$ and $f_2(t)$ and their transforms are known to be $F_1(s)$ and $F_2(s)$, then

$$\mathscr{L}[f_1(t) + f_2(t)] = \int_{0^-}^{\infty} (f_1(t) + f_2(t))e^{-st}\,dt = \int_{0^-}^{\infty} f_1(t)e^{-st}\,dt + \int_{0^-}^{\infty} f_2(t)e^{-st}\,dt$$
$$= F_1(s) + F_2(s) \tag{9.1.6}$$

In other words, if an overall function is formed by the sum of simpler functions, then the same summation applies to the Laplace inversion.

9.2. SIMPLE FUNCTIONS AND THEIR TRANSFORMS

Finding the Laplace transform of more complicated functions may sometimes be achieved by applying various Laplace transform rules. We will introduce some of these rules later but here we will obtain all the necessary transforms by integrating Eq. (9.1.3) with selected functions of time.

The simplest and often used function is the unit step, see Fig. 9.2.1a. It is defined by

$$u(t) = 1 \text{ if } t \geq 0$$
$$= 0 \text{ if } t < 0$$

(9.2.1)

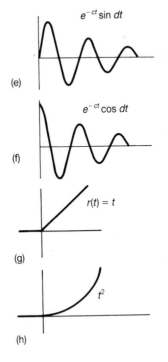

FIGURE 9.2.1   Typical time functions.

Inserting into Eq. (9.1.3), we must integrate

$$F_u(s) = \int_{0^-}^{\infty} 1 \cdot e^{-st}\, dt = \frac{-e^{-st}}{s}\bigg|_{0^-}^{\infty} = \frac{1}{s}$$

or

$$\mathscr{L}[u(t)] = F_u(s) = \frac{1}{s} \tag{9.2.2}$$

A rectangular pulse $p(t)$ having duration T and height $1/T$ is shown in Fig. 9.2.1b. Its transform is

$$F_p(s) = \frac{1}{T}\int_{0^-}^{T} e^{-st}\, dt = \frac{e^{-st}}{-sT}\bigg|_{0^-}^{T} = \frac{1 - e^{-sT}}{sT} \tag{9.2.3}$$

The definition is such that the area of the function is equal to one, no matter what T we select. This function is used sometimes but we will need it to define the Dirac impulse.

A Dirac impulse, $\delta(t)$, is the limit of the pulse $p(t)$ for $T \to 0$. This strange function, sketched in Fig. 9.2.1c, is simultaneously infinitely high and infinitely narrow, but still having an area equal to 1. Its Laplace transform is found by applying the L'Hospital limiting rule to Eq. (9.2.3)

$$\mathscr{L}[\delta(t)] = F_\delta(s) = \lim_{T \to 0} \frac{1 - e^{-sT}}{sT} = 1 \tag{9.2.4}$$

The function is extensively used in theoretical studies. As will be shown later, it is the derivative of $u(t)$.

Another commonly occuring function is the exponential,

$$f(t) = e^{-pt} \tag{9.2.5}$$

It is sketched in Fig. 9.2.1d for real p, although p can also be complex. The transform is

$$\mathscr{L}[e^{-pt}] = \int_{0^-}^{\infty} e^{-pt} e^{-st}\, dt = \frac{e^{-(s+p)t}}{-(s+p)}\bigg|_{0^-}^{\infty} = \frac{1}{s+p} \tag{9.2.6}$$

A sinusoid multiplied by an exponential is another important function

$$f(t) = e^{-ct} \sin dt \tag{9.2.7}$$

It is sketched in Fig. 9.2.1e for positive real c and d. The integral to be solved is

$$F(s) = \int_{0^-}^{\infty} e^{-ct} \sin dt \, e^{-st} \, dt = \int_{0^-}^{\infty} e^{-(s+c)t} \sin dt \, dt$$

Integration is done by parts using the formula $\int u'v = uv - \int uv'$. Define $u' = e^{-(s+c)t}$ and $v = \sin dt$. Then $u = -e^{-(s+c)t}/(s+c)$, $v' = d \cos dt$ and

$$F(s) = \frac{-e^{-(s+c)t} \sin dt}{s+c} \bigg|_{0^-}^{\infty} + \frac{d}{s+c} \int_{0^-}^{\infty} e^{-(s+c)t} \cos dt \, dt$$

Inserting the limits we get zero for the first term. The second term is integrated by parts using the same formula, selecting u and u' as above but setting $v = \cos dt$ and $v' = -d \sin dt$. This leads to

$$F(s) = \frac{d}{s+c} \left[\frac{-e^{-(s+c)t} \cos dt}{s+c} \bigg|_{0^-}^{\infty} - \frac{d}{s+c} \int_{0^-}^{\infty} e^{-(s+c)t} \sin (dt) \, dt \right]$$

Inserting the limits again and realizing that the integral on the right side is nothing but $F(s)$, we obtain

$$F(s) = \frac{d}{s+c} \left[\frac{1}{s+c} - \frac{d}{s+c} F(s) \right]$$

Transferring terms with $F(s)$ to the left side we get

$$\mathscr{L}[e^{-ct} \sin dt] = \frac{d}{(s+c)^2 + d^2} \tag{9.2.8}$$

Proceeding similarly we can also find

$$\mathscr{L}[e^{-ct} \cos dt] = \frac{s+c}{(s+c)^2 + d^2} \tag{9.2.9}$$

This function is sketched in Fig. 9.2.1f.

If the function does not have the exponential component, we set $c = 0$ in the function as well as in the Laplace transform and obtain

$$\mathscr{L}[\sin dt] = \frac{d}{s^2 + d^2} \tag{9.2.10}$$

$$\mathscr{L}[\cos dt] = \frac{s}{s^2 + d^2} \tag{9.2.11}$$

Finally, realizing that $\sin (dt + \phi) = \sin dt \cos \phi + \cos dt \sin \phi$ and taking into account the fact that the Laplace transform is a linear operation, see Eq. (9.1.6),

$$\mathscr{L}[e^{-ct} \sin (dt + \phi)] = \frac{d \cos \phi + (s + c) \sin \phi}{(s + c)^2 + d^2} \tag{9.2.12}$$

Should $c = 0$, then

$$\mathscr{L}[\sin (dt + \phi)] = \frac{d \cos \phi + s \sin \phi}{s^2 + d^2} \tag{9.2.13}$$

Another function which can have some practical application is the ramp function, $r(t) = kt$ where k is the slope. It is sketched in Fig. 9.2.1g. If $k = 1$, the ramp function is clearly the integral of the unit step. Its transform is

$$F_r(s) = \int_{0^-}^{\infty} te^{-st} \, dt$$

Setting $u' = e^{-st}$ and $v = t$ we find for $k = 1$

$$\mathscr{L}[t] = \frac{1}{s^2} \tag{9.2.14}$$

We can go further and consider higher powers; t^2 is sketched in Fig. 9.2.1h. Inserting into Eq. (9.1.3) and integrating by parts, we get after the first integration

$$\mathscr{L}[t^2] = -\frac{t^2 e^{-st}}{s} \Big|_{0^-}^{\infty} + \frac{2}{s} \int_{0^-}^{\infty} te^{-st} \, dt$$

The first term is zero after inserting the limits and the integral has just been solved in Eq. (9.2.14). Using this result we get

$$\mathscr{L}[t^2] = \frac{2}{s^3} \tag{9.2.15}$$

We can continue similarly to higher powers and obtain, by induction,

$$\mathscr{L}[t^n] = \frac{n!}{s^{n+1}} \tag{9.2.16}$$

The transformations are summarized in Table 9.2.1 and we will be referring to it frequently. Most of the terms are multiplied by $u(t)$ to indicate, in a formally correct way, that the functions all start at $t = 0$. The table contains all the transformations which can reasonably appear in our studies.

TABLE 9.2.1   Laplace transform pairs

No.	$F(s)$	$f(t)$	Remark
1	1	$\delta(t)$	Dirac impulse
2	$\dfrac{1}{s}$	$u(t)$	unit step
3	$\dfrac{1}{s^2}$	$tu(t)$	ramp function
4	$\dfrac{1}{s^n}$	$\dfrac{t^{n-1}}{(n-1)!}u(t)$	$0! = 1$
5	$\dfrac{1}{s+p}$	$e^{-pt}u(t)$	p real or complex
6	$\dfrac{1}{(s+p)^n}$	$\dfrac{t^{n-1}e^{-pt}}{(n-1)!}u(t)$	p real or complex
7	$\dfrac{d}{s^2+d^2}$	$\sin dt\, u(t)$	d real
8	$\dfrac{s}{s^2+d^2}$	$\cos dt\, u(t)$	d real
9	$\dfrac{d\cos\phi + s\sin\phi}{s^2+d^2}$	$\sin(dt+\phi)u(t)$	d, ϕ real
10	$\dfrac{d}{(s+c)^2+d^2}$	$e^{-ct}\sin dt\, u(t)$	c, d real
11	$\dfrac{s+c}{(s+c)^2+d^2}$	$e^{-ct}\cos dt\, u(t)$	c, d real
12	$\dfrac{As+B}{(s+c)^2+d^2}$	$e^{-ct}\left[A\cos dt + \dfrac{B-Ac}{d}\sin dt\right]u(t)$	A, B, c, d real
13	$\dfrac{A+jB}{s+c+jd} + \dfrac{A-jB}{s+c-jd}$	$2e^{-ct}[A\cos dt + B\sin dt]u(t)$	A, B, c, d real

9.3. RATIONAL FUNCTIONS WITH SIMPLE POLES

Laplace domain functions which describe properties of linear time-independent networks are rational functions in the s variable. We have already established this in some of our previous examples. Such functions can be written in the following forms:

$$F(s) = \frac{N(s)}{D(s)} = \frac{\displaystyle\sum_{0}^{M} a_i s^i}{\displaystyle\sum_{i=0}^{N} b_i s^i} = K_0 \frac{\displaystyle\prod_{i=1}^{M}(s - z_i)}{\displaystyle\prod_{1}^{N}(s - p_i)} \tag{9.3.1}$$

where the coefficient in K_0 is equal to

$$K_0 = \frac{a_M}{b_N}$$

In the above equation, $N(s)$ and $D(s)$ are polynomials. The roots of the numerator, $N(s)$, are called the zeros, z_i, the roots of the denominator, $D(s)$, are called poles, p_i. For inversion of the Laplace transform we *must* know the poles. Knowledge of the zeros is not a necessity but is useful in most cases. This section will cover cases where all poles are *distinct*, which means that there are no cases of the type $(s - p_i)^m$ where m is larger than one. Problems with multiple poles, $m > 1$, will be handled in the next section.

We start our explanations with two examples:

$$F_1(s) = \frac{3s^2 + 27}{2s^2 + 4s + 4} = \frac{3(s^2 + 9)}{2(s^2 + 2s + 2)} = \frac{3}{2} \frac{(s + j3)(s - j3)}{(s + 1 + j)(s + 1 - j)} \qquad (9.3.2)$$

and

$$F_2(s) = \frac{3s^2 + 27}{2s^3 + 6s^2 + 8s + 4} = \frac{3}{2} \frac{s^2 + 9}{s^3 + 3s^2 + 4s + 2}$$

$$= \frac{3}{2} \frac{(s + j3)(s - j3)}{(s + 1)(s + 1 + j)(s + 1 - j)} \qquad (9.3.3)$$

In Eq. (9.3.2), it was easy to find both the poles and the zeros by a simple and well known formula for quadratic polynomials. In most situations we are not so lucky.

Very often, students do not clearly understand what is involved in the step from the polynomial to its roots. Let us state here that finding the roots is simple for a polynomial of degree two, but not much can be done without a computer for polynomials of higher degrees. Theoretically, formulas exist to find roots of polynomials of third and fourth degree, but such formulas are difficult to apply. Should the polynomials be in symbolic form (with letters describing the types of elements having yet unknown values), then it will be difficult to find the roots of polynomials of degree as low as three. Our suggestion is not even to try in the case of a symbolic function. In numerical cases use the computer or a suitable calculator.

Recall the functions in Section 2 or look at Table 9.2.1. The Laplace transforms there have, in the denominator, either a term of the form $s + p$, or a polynomial of the type $(s + c)^2 + d^2$, which has complex conjugate roots. If we can break a more complicated rational function into a sum of such simple rational functions, we can then use Table 9.2.1 for each of them and find the inverse transformation as a sum of simple time functions. The steps are called *partial fraction expansion* and will be the method that we will use.

Decomposition can be done by several methods which we will explain. We will also remember the following rule: If the numerator has the same or higher degree than the denominator, then we must first reduce the numerator degree by division and make it one less than the denominator. For instance, should we have the function

$$F(s) = \frac{s^3 + 4s^2 + 5s + 1}{s^2 + s + 1}$$

then we first apply long division to get

$$F(s) = s + 3 + \frac{s - 2}{s^2 + s + 1}$$

and only then do we look for a decomposition of the remaining rational function. The numerator is usually not higher than the denominator but we will have cases when the degrees are equal. In such cases we apply one step of such division and obtain a constant. Should its value be k then the inverse transformation will have a term $k\delta(t)$ as can be seen from the first pair in Table 9.2.1.

9.3.1. Decomposition Using Real Arithmetic

We will first introduce a method which operates with real numbers. It is based on the decomposition $K_i/(s + p_i)$ for each real pole and on the decomposition $(A_i s + B_i)/[(s + c_i)^2 + d_i^2]$ for each complex conjugate pair.

Before we proceed we derive a useful formula for the inversion. Consider the following sequence of equivalent operations

$$\frac{As + B}{(s + c)^2 + d^2} = \frac{(As + Ac) + (B - Ac)}{(s + c)^2 + d^2} = A\frac{s + c}{(s + c)^2 + d^2} + \frac{B - Ac}{d}\frac{d}{(s + c)^2 + d^2}$$

Comparing the last two terms with formulas 10 and 11 in Table 9.2.1 we see that

$$\mathscr{L}^{-1}\left[\frac{As + B}{(s + c)^2 + d^2}\right] = e^{-ct}\left[A \cos dt + \frac{B - Ac}{d} \sin dt\right]u(t) \qquad (9.3.4)$$

This is formula 12 in Table 9.2.1.

Function (9.3.2) has the same degree in the numerator and in the denominator. The degree of the numerator is first lowered by division

$$F_1(s) = \frac{3}{2}\left[\frac{s^2 + 9}{s^2 + 2s + 2}\right] = \frac{3}{2}\left[1 + \frac{-2s + 7}{(s + 1)^2 + 1}\right]$$

Comparing with the above derived formula, we identify $A = -2$, $B = 7$, $c = 1$ and $d = 1$ and the inversion is

$$f_1(t) = \frac{3}{2}\delta(t) + 3e^{-t}\left(-\cos t + \frac{9}{2}\sin t\right)u(t) \qquad (9.3.5)$$

The second example already has the degree of the numerator less than the degree of the denominator and no division is needed. We use the decomposition

$$F_2(s) = \frac{3}{2}\left[\frac{s^2 + 9}{(s + 1)(s^2 + 2s + 2)}\right] = \frac{3}{2}\left[\frac{K}{s + 1} + \frac{As + B}{s^2 + 2s + 2}\right]$$

Cross-multiply the terms on the right to get a single fraction and compare the resulting numerator with the numerator of the left side. This leads to

$$s^2 + 9 = Ks^2 + 2Ks + 2K + As^2 + As + Bs + B$$

The equation can be satisfied only if coefficients at equal powers are equal. This leads to the system

$$K + A = 1$$
$$2K + A + B = 0 \qquad (9.3.6)$$
$$2K + B = 9$$

Its solution is $K = 10$, $A = -9$ and $B = -11$ and

$$F_2(s) = \frac{3}{2}\left[\frac{10}{s + 1} + \frac{-9s - 11}{(s + 1)^2 + 1}\right] \qquad (9.3.7)$$

Identify $A = -9$, $B = -11$, $c = 1$, $d = 1$ and $(B - Ac)/d = -2$. Using the table, the inversion is

$$f_2(t) = \frac{3}{2}[10e^{-t} - 9e^{-t}\cos t - 2e^{-t}\sin t]u(t)$$

$$= \frac{3}{2}e^{-t}[10 - 9\cos t - 2\sin t]u(t) \qquad (9.3.8)$$

Summarizing the method, assume the decomposition $K_i/(s - p_i)$ for each real simple pole and the decomposition $(A_i s + B_i)/[(s - c_i)^2 + d_i^2]$ for each pair of complex conjugate poles. Express the sum of fractions as one fraction by cross-multiplying the terms and compare coefficients with the numerator of the given function. This provides a set of linear equations which are solved for the unknown constants. Then consult Table 9.2.1 for the inverse functions.

Example 1.

Find the time domain response of the function

$$F(s) = \frac{s^2 + 4}{(s^2 + 9)(s^2 + 4s + 5)}$$

The function has complex conjugate poles $p_{1,2} = \pm j3$ and $p_{3,4} = -2 \pm j1$. We need the decomposition in the form

$$\frac{s^2 + 4}{(s^2 + 9)(s^2 + 4s + 5)} = \frac{A_1 s + B_1}{s^2 + 9} + \frac{A_2 s + B_2}{s^2 + 4s + 5}$$

Cross-multiplying on the right side we get the equation for the numerators

$$s^2 + 4 = A_1 s^3 + A_1 4s^2 + A_1 5s + B_1 s^2 + B_1 4s$$
$$+ B_1 5 + A_2 s^3 + A_2 9s + B_2 s^2 + B_2 9$$

and equating coefficients at the same powers of s we get the system

$$A_1 + A_2 = 0$$
$$4A_1 + B_1 + B_2 = 1$$
$$5A_1 + 4B_1 + 9A_2 = 0$$
$$5B_1 + 9B_2 = 4$$

Since we see that $A_2 = -A_1$, we can reduce the size of the problem by substituting into the last three equations. This leads to

$$\begin{bmatrix} 4 & 1 & 1 \\ -4 & 4 & 0 \\ 0 & 5 & 9 \end{bmatrix} \begin{bmatrix} A_1 \\ B_1 \\ B_2 \end{bmatrix} = \begin{bmatrix} 1 \\ 0 \\ 4 \end{bmatrix}$$

The solution is $D = 160$, $A_1 = 1/8$, $B_1 = 1/8$, $A_2 = -1/8$ and $B_2 = 3/8$. The problem is decomposed into

$$\frac{s^2 + 4}{(s^2 + 9)(s^2 + 4s + 5)} = \frac{\frac{1}{8}s + \frac{1}{8}}{s^2 + 9} + \frac{-\frac{1}{8}s + \frac{3}{8}}{(s + 2)^2 + 1}$$

At this point we can use formula 12 in Table 9.2.1, identify $A_1 = 1/8$, $B_1 = 1/8$, $c_1 = 0$, $d_1 = 3$ and calculate $(B_1 - A_1 c_1)/d_1 = 1/24$. For the second term identify $A_2 = -1/8$,

$B_2 = 3/8$, $c_2 = 2$, $d_2 = 1$ and calculate $(B_2 - A_2c_2)/d_2 = 5/8$. The inverted function is

$$f(t) = \left\{ \frac{1}{8} \cos 3t + \frac{1}{24} \sin 3t + e^{-2t} \left[-\frac{1}{8} \cos t + \frac{5}{8} \sin t \right] \right\} u(t)$$

9.3.2. Decomposition Using Complex Arithmetic

Instead of using second-order polynomials as above, we can break the denominator polynomial into the product $(s + c + jd)(s + c - jd)$ and treat each term as a simple pole using Eq. (9.2.6). Since for any complex pole there must be another complex conjugate pole, it will *always* be true that the decomposition constants will also be complex conjugate.

Similarly as above we will first derive a useful formula. Consider a complex conjugate pair with complex conjugate numerator constants and use Eq. (9.2.6) for inversion

$$\mathscr{L}^{-1}\left[\frac{A + jB}{s + c + jd} + \frac{A - jB}{s + c - jd} \right] = (A + jB)e^{(-c+jd)t} + (A - jB)e^{(-c-jd)t}$$

$$= e^{-ct}\left[2A \frac{e^{jdt} + e^{-jdt}}{2} - 2B \frac{e^{jdt} - e^{-jdt}}{2j} \right]$$

This leads to the formula

$$\mathscr{L}^{-1}\left[\frac{A + jB}{s + c + jd} + \frac{A - jB}{s + c - jd} \right] = 2e^{-ct}[A \cos dt - B \sin dt] \qquad (9.3.9)$$

which is also in Table 9.2.1, formula 13.

For instance, in Eq. (9.3.1) we could use the decomposition

$$\frac{3}{2} \frac{s^2 + 9}{s^2 + 2s + 2} = \frac{3}{2}\left[1 + \frac{A + jB}{s + 1 + j} + \frac{A - jB}{s + 1 - j} \right]$$

Proceeding similarly as above, we cross-multiply the terms on the right to get the equation

$$s^2 + 9 = s^2 + 2s + 2 + (A + jB)(s + 1 - j) + (A - jB)(s + 1 + j)$$

After the multiplications we obtain

$$s^2 + 9 = s^2 + s(2 + 2A) + 2 + 2A + 2B$$

Comparing terms at equal powers of s we get the system

$$2 + 2A = 0$$

$$2 + 2A + 2B = 9$$

with the solution $A = -1$, $B = 9/2$. The decomposition is

$$F_1(s) = \frac{3}{2}\left[1 + \frac{-1 - j9/2}{s + 1 - j} + \frac{-1 + j9/2}{s + 1 + j} \right] \tag{9.3.10}$$

Consulting Table 9.2.1, formula 13 we get the final result

$$f_1(t) = \frac{3}{2}\delta(t) + 3e^{-t}\left(-\cos t + \frac{9}{2}\sin t \right)u(t) \tag{9.3.11}$$

which is the same as obtained in Eq. (9.3.5).

9.3.3. Residues of Simple Poles

In this section, we cover the best method for evaluating the constants of the partial fractions under the condition that *all poles are simple*. The method is suitable for programming as well as hand calculations, even if the poles are complex.

Let us have a rational function in which we know the poles. The numerator may (but need not) be in factored form but the denominator must be. Thus we assume that the function has already one of the forms

$$F(s) = \frac{\displaystyle\sum_0^M a_i s^i}{\displaystyle\prod_1^N (s - p_i)} = K_0 \frac{\displaystyle\prod_{i=1}^M (s - z_i)}{\displaystyle\prod_1^N (s - p_i)} \tag{9.3.12}$$

In the first expression we assume that all numerator coefficients have been divided by a possible term $b_N \neq 1$. In the second expression

$$K_0 = \frac{a_M}{b_N}$$

Since we assumed that the poles are simple, we know from the previous discussion that we can decompose this function into two forms. If $M < N$, then

$$F(s) = \sum_{i=1}^N \frac{K_i}{s - p_i} \tag{9.3.13}$$

if $M = N$, then

$$F(s) = K_0 + \sum_{i=1}^{N} \frac{K_i}{s - p_i} \qquad (9.3.14)$$

The numbers K_i, $i = 1, 2, \ldots N$, are called the *residues* of the function. The steps are easiest to explain by taking an example. Consider

$$F(s) = \frac{(s + 2)(s + 4)}{(s + 1)(s + 3)(s + 5)} = \frac{K_1}{s + 1} + \frac{K_2}{s + 3} + \frac{K_3}{s + 5} \qquad (9.3.15)$$

and let us be interested in K_1. Multiply the equation by the term $s + 1$. This results in

$$(s + 1)F(s) = \frac{(s + 2)(s + 4)}{(s + 3)(s + 5)} = K_1 + K_2 \frac{s + 1}{s + 3} + K_3 \frac{s + 1}{s + 5}$$

Note that in the middle expression the multiplication *removed* the term $s + 1$. If we now substitute $s = -1$, we will get

$$[(s + 1)F(s)]_{s=-1} = \frac{(-1 + 2)(-1 + 4)}{(-1 + 3)(-1 + 5)} = K_1 + 0 + 0$$

or

$$K_1 = \frac{3}{8}$$

For the second pole we multiply Eq. (9.3.15) by $s + 3$ to get

$$[(s + 3)F(s)] = \frac{(s + 2)(s + 4)}{(s + 1)(s + 5)} = K_1 \frac{s + 3}{s + 1} + K_2 + K_3 \frac{s + 3}{s + 5}$$

and substituting $s = -3$ in the whole equation will result in

$$[(s + 3)F(s)]_{s=-3} = \frac{1}{4} = K_2$$

Similarly, for the last pole multiply Eq. (9.3.15) by $s + 5$

$$[(s + 5)F(s)] = \frac{(s + 2)(s + 4)}{(s + 1)(s + 3)} = K_1 \frac{s + 5}{s + 1} + K_2 \frac{s + 5}{s + 3} + K_3$$

Substituting $s = -5$ results in

$$[(s + 5)F(s)]_{s=-5} = \frac{3}{8} = K_3$$

and the decomposition is finished

$$F(s) = \frac{3}{8}\frac{1}{s+1} + \frac{1}{4}\frac{1}{s+3} + \frac{3}{8}\frac{1}{s+5}$$

Using the table we get the inversion

$$f(t) = \left[\frac{3}{8}e^{-t} + \frac{1}{4}e^{-3t} + \frac{3}{8}e^{-5t}\right]u(t) \qquad (9.3.16)$$

The above steps can be written as a formula

$$K_i = [(s - p_i)F(s)]_{s=p_i} \qquad (9.3.17)$$

where the symbols on the right side mean that we

1. first cancel the term $s - p_i$ against the same term in the denominator of $F(s)$ and
2. only then substitute $s = p_i$.

This method works equally well for complex conjugate poles, with the additional advantage that we need not calculate the residue of the complex conjugate pole because we know that its residue is complex conjugate. Considering, for instance, our example in Eq. (9.3.3), we have

$$F_2(s) = \frac{3}{2}\frac{s^2+9}{(s+1)(s+1+j)(s+1-j)}$$

$$= \frac{K_1}{s+1} + \frac{K_2}{s+1+j} + \frac{\overline{K_2}}{s+1-j} \qquad (9.3.18)$$

Then

$$K_1 = [(s+1)F(s)]_{s=-1} = \frac{3}{2}\frac{(-1)^2+9}{(-1+1+j)(-1-1+j)} = 15$$

and

$$K_2 = [(s+1+j)F(s)]_{s=-1-j} = \frac{3}{2}\frac{(-1+j)^2+9}{(-1-j+1)(-1-j+1-j)} = -\frac{27}{4} - j\frac{3}{2}$$

If we are confident that we did not make a mistake in the evaluations, the decomposition is finished because the complex conjugate pole will have a complex conjugate residue:

$$F_2(s) = \frac{3}{2} \frac{s^2 + 9}{(s + 1)(s + 1 + j)(s + 1 - j)}$$

$$= \frac{15}{s + 1} + \frac{-\dfrac{27}{4} - j\dfrac{3}{2}}{s + 1 + j} + \frac{-\dfrac{27}{4} + j\dfrac{3}{2}}{s + 1 - j} \qquad (9.3.19)$$

We do not even need to write the complete decomposition. We can use Eq. (9.3.9) for the pair of complex conjugate poles, identify $A = -27/4$, $B = -3/2$, $c = 1$, $d = 1$ and write the inversion

$$f(t) = \left[15e^{-t} + 2e^{-t}\left(-\frac{27}{4}\cos t - \frac{3}{2}\sin t \right) \right]u(t) \qquad (9.3.20)$$

The result corresponds to Eq. (9.3.8).

Example 2.

Find the time domain response of the function

$$F(s) = \frac{s^2 + 4}{(s^2 + 9)(s^2 + 4s + 5)}$$

by using the residue calculus. The function has complex conjugate poles $p_{1,2} = \pm j3$ and $p_{3,4} = -2 \pm j1$.

The function can be rewritten in the form

$$F(s) = \frac{s^2 + 4}{(s + j3)(s - j3)(s + 2 + j1)(s + 2 - j1)}$$

The residue at $s = -j3$ is

$$K_1 = \frac{(-j3)^2 + 4}{(-j3 - j3)(-j3 + 2 + j1)(-j3 + 2 - j1)} = \frac{3 + j1}{48}$$

The residue at $s = -2 - j1$ is

$$K_2 = \frac{(-2 - j1)^2 + 4}{(-2 - j1 + j3)(-2 - j1 - j3)(-2 - j1 + 2 - j1)} = \frac{-1 + j5}{16}$$

The above calculations may be fairly lengthy if we do them by hand, but many calculators have complex arithmetic and make the evaluation quite easy. We can now identify the terms for Eq. (9.3.9). For the first pole, $-j3$, we have $A_1 = 1/16$, $B_1 = 1/48$, $c_1 = 0$, $d_1 = 3$. For the second pole, $-2 - j1$, we have $A_2 = -1/16$, $B_2 = 5/16$, $c_2 = 2$, $d_2 = 1$. Inserting into Eq. (9.3.9),

$$f(t) = \left\{ 2\left[\frac{1}{16}\cos 3t + \frac{1}{48}\sin 3t \right] + 2e^{-2t}\left[-\frac{1}{16}\cos t + \frac{5}{16}\sin t \right] \right\}u(t)$$

9.4. DECOMPOSITION WITH MULTIPLE POLES

Before we go into details of decompositions with multiple poles, we will discuss some practical points. Multiple poles are important theoretically, simply because they can exist, but finding them in higher-order polynomials is not a trivial problem. If the multiplicity is high, we may not even be able to find the multiple root correctly. Fortunately, when we look at the situation from a practical point of view, we discover that multiple poles are extremely rare. In network applications, if we have the choice, we will *always* avoid situations when they could occur. Responses of networks with multiple poles are inferior to responses where we are free to assign the poles as simple ones. The only situation of some practical importance is the case of the ramp function, see Table 9.2.1, but even this function is somewhat artificial. As a result of these *practical* considerations we feel that the student can skip this section in his first reading and return to it when needed.

If the function has multiple poles of the type $(s - p_i)^m$, $m > 1$, then the decomposition is done differently and an example will show best what has to be done. Let the function have one pole of multiplicity $m = 3$ and some other simple poles, let us say two. Then we must assume the decomposition in the form

$$F(s) = \frac{N(s)}{(s - p_1)^3(s - p_2)(s - p_3)}$$

$$= \frac{K_{1,3}}{(s - p_1)^3} + \frac{K_{1,2}}{(s - p_1)^2} + \frac{K_{1,1}}{(s - p_1)^1} + \frac{K_{2,1}}{s - p_2} + \frac{K_{3,1}}{s - p_3}$$

(9.4.1)

The first subscript corresponds to the subscript of the pole and the second indicates the power of the term $s - p_1$. For each pole of multiplicity m we must assume m terms with all the powers from m to 1.

The first step to find the coefficients is to multiply Eq. (9.4.1) by the highest power of the pole we are considering. In the above case, multiply by $(s - p_1)^3$ to get

$$[(s - p_1)^3 F(s)] = \frac{N(s)}{(s - p_2)(s - p_3)} = K_{1,3} + K_{1,2}(s - p_1) + K_{1,1}(s - p_1)^2$$

$$+ K_{2,1}\frac{(s - p_1)^3}{s - p_2} + K_{3,1}\frac{(s - p_1)^3}{s - p_3}$$

(9.4.2)

If we now substitute $s = p_1$, we get

$$[(s - p_1)^3 F(s)]_{s=p_1} = \frac{N(p_1)}{(p_1 - p_2)(p_1 - p_3)} = K_{1,3}$$

(9.4.3)

and the remaining terms disappear because their numerators are zero.

In order to get $K_{1,2}$, we differentiate Eq. (9.4.2) with respect to s:

$$\frac{d}{ds}[(s-p_1)^3 F(s)] = \frac{d}{ds}\left[\frac{N(s)}{(s-p_2)(s-p_3)}\right] = K_{1,2} + 2K_{1,1}(s-p_1)$$
$$+ [terms \ having \ (s-p_1) \ in \ the \ numerator] \tag{9.4.4}$$

Check that after the differentiation we have, indeed, all the remaining terms multiplied by $s - p_1$. If we now substitute $s = p_1$, the result will be

$$\frac{d}{ds}[(s-p_1)^3 F(s)]_{s=p_1} = K_{1,2} \tag{9.4.5}$$

Finally, to obtain $K_{1,1}$ we differentiate Eq. (9.4.4) to get

$$\frac{d^2}{ds^2}[(s-p_1)^3 F(s)] = 2K_{1,1} + [terms \ having \ (s-p_1) \ in \ the \ numerator]$$

Substituting $s = p_1$ provides

$$\frac{d^2}{ds^2}[(s-p_1)^3 F(s)]_{s=p_1} = 2K_{1,1}$$

or

$$K_{1,1} = \frac{1}{2}\frac{d^2}{ds^2}[(s-p_1)^3 F(s)]_{s=p_1}$$

The residues of the simple poles are obtained by the method in Section 9.3.3.

In general, if the ith pole has multiplicity m, then the decomposition coefficient $K_{i,j}$, $m \geq j \geq 1$, will be

$$K_{i,j} = \frac{1}{(m-j)!}\frac{d^{m-j}}{ds^{m-j}}[(s-p_i)^m F(s)]_{s=p_i} \tag{9.4.6}$$

with the understanding that we

1. first cancel the term with which we multiply against the same term in the denominator,
2. then obtain the necessary derivative and
3. only then substitute the pole p_i for s.

As an example, consider the function

$$F(s) = \frac{1}{(s-1)^2 s^3}$$

with the poles $p_1 = 1$ and $p_2 = 0$. The decomposition will have the form

$$F(s) = \frac{K_{1,2}}{(s-1)^2} + \frac{K_{1,1}}{s-1} + \frac{K_{2,3}}{s^3} + \frac{K_{2,2}}{s^2} + \frac{K_{2,1}}{s} \qquad (9.4.7)$$

For the first pole we have

$$(s-1)^2 F(s) = s^{-3} \qquad (9.4.8)$$

Thus

$$K_{1,2} = \frac{1}{(2-2)!} [(s-1)^2 F(s)]_{s=1} = s^{-3}|_{s=1} = 1$$

$$K_{1,1} = \frac{1}{(2-1)!} \frac{d}{ds} [(s-1)^2 F(s)]_{s=1} = -3s^{-4}|_{s=1} = -3$$

For the second pole we have

$$s^3 F(s) = (s-1)^{-2} \qquad (9.4.9)$$

$$K_{2,3} = \frac{1}{(3-3)!} [s^3 F(s)]_{s=0} = [(s-1)^{-2}]_{s=0} = 1$$

$$K_{2,2} = \frac{1}{(3-2)!} \frac{d}{ds} [s^{-3} F(s)]_{s=0} = [-2(s-1)^{-3}]_{s=0} = 2$$

$$K_{2,1} = \frac{1}{(3-1)!} \frac{d^2}{ds^2} [s^{-3} F(s)]_{s=0} = \frac{1}{2} [6(s-1)^{-4}]_{s=0} = 3$$

The complete decomposition is

$$F(s) = \frac{1}{(s-1)^2 s^3} = \frac{1}{(s-1)^2} + \frac{-3}{s-1} + \frac{1}{s^3} + \frac{2}{s^2} + \frac{3}{s}$$

The above steps are equally valid for complex conjugate poles. The decomposition coefficient of the conjugate pole will be the conjugate of the coefficient of the other pole. Once we have the term $K_{i,j}$, then the time domain function corresponding to this term is

$$\mathcal{L}^{-1} \left[\frac{K_{i,j}}{(s-p_i)^j} \right] = K_{ij} \frac{t^{j-1} e^{p_i t}}{(j-1)!} \qquad (9.4.10)$$

9.5. LAPLACE TRANSFORM OPERATIONS

Operations with time functions are usually complicated while operations in the Laplace domain are almost always much simpler. In order to be able to take advantage of these simplifications, we now transform several mathematical operations into the Laplace domain. The transformations are summarized in Table 9.5.1.

9.5.1. Differentiation

Differentiation of time functions is expressed in the Laplace domain by multiplication by the s variable. For our network theory this is the most important operation.

Consider $f(t)$ for which we know the Laplace transform $F(s)$. We wish to express differentiation of the time function in terms of $F(s)$; this means that we are looking for

$$\mathscr{L}[f'(t)] = \int_{0_-}^{\infty} f'(t)e^{-st}\, dt \tag{9.5.1}$$

TABLE 9.5.1   Laplace transform operations

No.	$f(t)$	$F(s)$
1	$a \cdot f(t)$	$a \cdot F(s)$
2	$\dfrac{d}{dt} f(t)$	$s \cdot F(s) - f(0)$
3	$\dfrac{d^n}{dt^n} f(t)$	$s^n F(s) - \sum_{i=1}^{n} s^{n-i} f^{(i-1)}(0)$ where $f^{(k)}(0) = \lim_{t \to 0} \dfrac{d^k}{dt^k} f(t)$
4	$\displaystyle\int_0^t f(z)\, dz$	$\dfrac{F(s)}{s} + \dfrac{f^{(-1)}(0)}{s}$ where $f^{(-1)}(0) = \lim_{t \to 0} \displaystyle\int_0^t f(z)\, dz$
5	$f(kt)$	$\dfrac{1}{k} F\!\left(\dfrac{s}{k}\right)$
6	$f(t - T)$	$e^{-sT} F(s)$
7	$\displaystyle\int_0^t f_1(t-z) f_2(z)\, dz$	$F_1(s) \cdot F_2(s)$
8	$\lim_{t \to 0} f(t)$	$\lim_{s \to \infty} s \cdot F(s)$
9	$\lim_{t \to \infty} f(t)$	$\lim_{s \to 0} s \cdot F(s)$

The integral is solved by selecting $u' = f'$, $u = f$, $v = e^{-st}$, $v' = -se^{-st}$:

$$\mathscr{L}[f'(t)] = f(t)e^{-st}\Big|_{0^-}^{\infty} + s\int_{0^-}^{\infty} f(t)e^{-st}\,dt$$

or

$$\mathscr{L}[f'(t)] = sF(s) - f(0) \tag{9.5.2}$$

where $f(0)$ is the value of the function at $t = 0$.

The Laplace transform of the second derivative can be obtained similarly but it is easier to apply Eq. (9.5.2):

$$\mathscr{L}[f''(t)] = s[sF(s) - f(0)] - f'(0)$$

or

$$\mathscr{L}[f''(t)] = s^2 F(s) - sf(0) - f'(0) \tag{9.5.3}$$

Continuing to higher-order derivatives, we obtain a general formula

$$\mathscr{L}[f^{(n)}(t)] = s^n F(s) - s^{n-1}f(0) - s^{n-2}f'(0) - \ldots$$
$$\ldots - sf^{(n-2)}(0) - f^{(n-1)}(0) \tag{9.5.4}$$

Let us now see how this is used in applications. The most important case is that of an initial voltage on the capacitor or an initial current through the inductor. The current flowing through a capacitor is given by

$$i = C\frac{dv}{dt}$$

Denote by I the Laplace transform of the current i and by V the Laplace transform of the voltage v. Let the initial voltage across the capacitor be E_0. The situation is sketched in Fig. 9.5.1. Using Eq. (9.5.2)

$$I = C[sV - E_0]$$

or

$$I = sCV - CE_0 \tag{9.5.5}$$

Should the initial voltage change its polarity with respect to the $+$ and $-$ signs indicated at the terminals, the sign in Eq. (9.5.5) will change to $+$. In the Laplace transform, the product CE_0 represents the Dirac impulse with an area CE_0 instead of 1.

The voltage across the inductor is given by

$$v = L\frac{di}{dt}$$

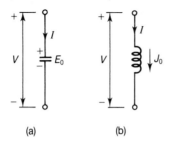

FIGURE 9.5.1   Initial conditions for: (a) Capacitor. (b) Inductor.

Let the initial current be J_0 with the direction shown in Fig. 9.5.1. Then in the Laplace domain

$$V = L[sI - J_0] = sLI - LJ_0 \qquad (9.5.6)$$

The product LJ_0 represents a Dirac impulse with the area LJ_0 instead of 1. Should the initial current flow in the opposite direction with respect to the $+$ and $-$ sign at the terminals, the sign in the formula is changed to $+$.

If there is no initial voltage on the capacitor or no initial current through the inductor, E_0 or J_0 are zero and

$$I = sCV$$
$$V = sLI \qquad (9.5.7)$$

We have already used these expressions in Chapter 7, but now we can see the theoretical connection. We will return to the above derivations in more detail in the next chapter.

9.5.2. Integration

Formulas for integration are derived similarly as in the case of differentiation

$$\mathscr{L}\left[\int_{0^-}^{t} f(\tau)\, d\tau\right] = \int_{0^-}^{\infty} \left(\int_{0}^{t} f(\tau)\, d\tau\right) e^{-st}\, dt$$

Integrating by parts

$$\mathscr{L}\left[\int_{0^-}^{t} f(\tau)\, d\tau\right] = \frac{-e^{-st}}{s} \int_{0^-}^{t} f(\tau)\, d\tau \bigg|_{0^-}^{\infty} + \frac{1}{s}\int_{0^-}^{\infty} f(t) e^{-st}\, dt$$

or

$$\mathscr{L}\left[\int_{0^-}^{t} f(\tau)\, d\tau\right] = \frac{F(s)}{s} + \frac{f^{(-1)}(0)}{s} \tag{9.5.8}$$

The symbol $f^{(-1)}(0)$ denotes the integral of the function from 0^- to 0^+. This integral will normally be zero, except in the case of a Dirac impulse for which we have derived in Section 9.2 that it has zero duration but, nevertheless, has an area equal to one. This is the case of the initial voltage on a capacitor or the initial current through the inductor. For the capacitor, we have in the time domain

$$v(t) = \frac{1}{C} \int_{0^-}^{t} i(t)\, dt$$

In the Laplace domain this will become

$$V = \frac{1}{C}\left[\frac{I}{s} + \frac{CE_0}{s}\right]$$

or

$$V = \frac{I}{sC} + \frac{E_0}{s} \tag{9.5.9}$$

The reader should note that this could be obtained simply by separating algebraically V in Eq. (9.5.5). There is, however, one important point to remember: the influence of the initial voltage across the capacitor is now expressed by E_0/s, which is equivalent to a *step* of voltage having a height E_0, see Table 9.2.1, formula 2.

The expression for inductor current can be found similarly but we will proceed differently here. We start with Eq. (9.5.6) and separate I algebraically:

$$I = \frac{V}{sL} + \frac{J_0}{s} \tag{9.5.10}$$

The initial condition is now taken into account by a step function having the amplitude J_0. If the initial voltage across the capacitor and the initial current through the inductor are zero, then the expressions simplify to

$$V = \frac{I}{sC}$$

$$I = \frac{V}{sL} \tag{9.5.11}$$

These expressions were already introduced in Chapter 7.

9.5.3. Initial and Final Value Theorems

It is sometimes advantageous to know how the function $f(t)$ behaves at times $t \to 0$ and $t \to \infty$. Formulas for both cases are derived by starting with Eq. (9.5.2):

$$\mathscr{L}[f'(t)] = \int_{0^-}^{\infty} f'(t)e^{-st}\, dt = sF(s) - f(0)$$

Take the limit $s \to 0$:

$$\lim_{s \to 0} \int_{0^-}^{\infty} f'(t)e^{-st}\, dt = \lim_{s \to 0} [sF(s) - f(0)]$$

Since $e^0 = 1$, on the left side there remains only $f'(t)$ under the integral sign and the solution is $f(t)$; inserting the limits

$$\lim_{t \to \infty} [f(t) - f(0)] = \lim_{s \to 0} [sF(s) - f(0)]$$

Canceling $f(0)$ on both sides leads to the *final value theorem*:

$$\lim_{t \to \infty} f(t) = \lim_{s \to 0} sF(s) \qquad\qquad (9.5.12)$$

Another formula is obtained by starting again with Eq. (9.5.2) but assuming that the limit is $s \to \infty$:

$$\lim_{s \to \infty} \int_{0^-}^{\infty} f'(t)e^{-st}\, dt = \lim_{s \to \infty} [sF(s) - f(0)]$$

Since $e^{-\infty} = 0$, the integral is zero and the expression reduces to

$$0 = \lim_{s \to \infty} [sF(s) - f(0)]$$

Transferring $f(0)$ on the left side and realizing that this may be written as a limit of $f(t)$ for $t \to 0$, we obtain the *initial value theorem*:

$$\lim_{t \to 0} f(t) = \lim_{s \to \infty} sF(s) \qquad\qquad (9.5.13)$$

Note that Eq. (9.5.12) and Eq. (9.5.13) have the same form, except for the limits. If the limit of t is 0, the limit of s is ∞ and vice versa.

Both theorems can give considerable insight into the behavior of functions. As an example, consider a Laplace domain function

$$F(s) = \frac{N(s)}{D(s)} = \frac{s^2 + \cdots}{s^5 + \cdots}$$

where the dots indicate that there are other terms with powers of s less than the one indicated. The initial value theorem indicates that

$$\lim_{t \to 0} f(t) = \lim_{s \to \infty} \frac{s(s^2 + \cdots)}{s^5 + \cdots} = 0$$

since the highest power in the numerator is less than the highest power in the denominator. How will the derivative of $f(t)$ behave at $t \to 0$? We already know that the derivative is obtained by multiplying the Laplace domain function by s. Therefore

$$f'(0) = \lim_{s \to \infty} \frac{s \cdot s(s^2 + \cdots)}{s^5 + \cdots}$$

Here one multiplication by s comes from the formula, the second from the differentiation. The numerator has the highest power 4, the denominator 5, and for extremely large s the limit will be zero. What about the second derivative? Multiply again by s. Now we have

$$f''(0) = \lim_{s \to \infty} \frac{s^2 \cdot s(s^2 + \cdots)}{s^5 + \cdots}$$

The highest powers in the numerator and denominator are equal, resulting in a nonzero limit of the second-order derivative at $t \to 0$. We know that the time domain response starts at the origin and, in addition, has the first derivative equal to zero; this means that it must start horizontally from the same point. We cannot say anything else about the form of the curve, but even this gives some insight into the time domain behavior of the function. Note that all we need to know about the function are the highest powers of s in the numerator and denominator.

9.5.4. Scaling in the Time Domain

Assume that we have the function $f(t)$ and know its Laplace transform $F(s)$. We wish to know how scaling of time will influence the transform. We are looking for a formula for the scaled situation $f(kt)$. Using the definition

$$\mathscr{L}[f(kt)] = \int_{0^-}^{\infty} f(kt)e^{-st}\, dt$$

and changing the variable by selecting

$$kt = \tau$$

$$dt = \frac{d\tau}{k}$$

we arrive at

$$\mathscr{L}[f(kt)] = \frac{1}{k} \int_{0^-}^{\infty} f(\tau)e^{-(s/k)\tau}\, d\tau = \frac{1}{k} \int_{0^-}^{\infty} f(t)e^{-(s/k)t}\, dt$$

In the last step we returned to the original time variable because the use of a different letter cannot change the integral. The expression is similar to Eq. (9.1.3) with the exception that we now have s/k instead of s. This means that we can write the formula

$$\mathscr{L}[f(kt)] = \frac{1}{k} F\left(\frac{s}{k}\right) \tag{9.5.14}$$

9.5.5. Delay

One of the important questions is: What will happen with the transform $F(s)$ if instead of $f(t)$ we consider $f(t - T)$, a function delayed (shifted to the right) by the amount of T. Using Eq. (9.1.3), we are looking for

$$\mathscr{L}[f(t - T)] = \int_{0^-}^{\infty} f(t - T)e^{-st}\, dt$$

Substitute a new variable

$$\tau = t - T$$

Then

$$\mathscr{L}[f(t - T)] = \int_{-T}^{\infty} f(\tau)e^{-s(\tau + T)}\, d\tau$$

Recalling that all functions start, by definition, at $t = 0$, we can modify the lower limit to 0 and write

$$\mathscr{L}[f(t - T)] = e^{-sT} \int_{0^-}^{\infty} f(\tau)e^{-s\tau}\, d\tau$$

To write the final formula, we replace τ by t, because the change of the letter cannot change the value of the integral. But then the integral is nothing but $F(s)$,

Eq. (9.1.3), and

$$\mathscr{L}[f(t - T)] = e^{-sT}F(s) \qquad (9.5.15)$$

A few explanations will clarify the importance of this formula. The symbol $f(t - T)$ means that in the time domain every occurrence of t is replaced by $(t - T)$. For instance, let the parabola

$$
\begin{aligned}
f(t) &= t^2 \qquad \text{for } t \geq 0 \\
&= 0 \qquad \text{for } t < 0
\end{aligned}
\qquad (9.5.16)
$$

be shifted to the right by $T = 5$. Then the delayed parabola is defined by

$$
\begin{aligned}
f(t - T) &= (t - T)^2 \qquad \text{for } t - 5 \geq 0 \\
&= 0 \qquad\qquad \text{for } t - 5 < 0
\end{aligned}
\qquad (9.5.17)
$$

As we know from Section 9.2, the Laplace transform of Eq. (9.5.16) is $2/s^3$. According to the formula we just derived, the Laplace transform of Eq. (9.5.17) is $e^{-5s}2/s^3$.

In most application-oriented studies, only a few functions are needed: the Dirac impulse, the unit step, the ramp function, the sine and cosine functions, possibly multiplied by e^{-ct}. Many other functions can be composed of several of the above functions, properly shifted in time. Figure 9.5.2 gives some of them.

Linear networks have the superposition property: if several signals are applied to the network, we can consider each of them separately, solve for the response of the network to each of them and add the responses to get the final answer. For this reason, it is not even necessary to consider shifts of input signals. Take, for instance, the trapezoidal impulse in Fig. 9.5.2, composed of four equal ramp functions. All we have to do is solve for one of them and plot the result. The responses to the other three ramp functions will be the same curves, only shifted to the right by appropriate amounts and either added or subtracted. This will give the response to the trapezoidal pulse.

9.6. SIGNALS AND NETWORKS

The most important property of the Laplace transform is the ease with which we can calculate the response of any linear network to any given input signal. Recall what we learned in Chapter 7. We introduced the impedances and admittances, we stated that the independent signal will be a symbol, to be specified later, and we learned how to get the solution and the output. Except for the case of initial conditions, which will be covered in the next chapter, we already know all the steps to find the output in terms of the network elements and the as-yet unspecified input signal.

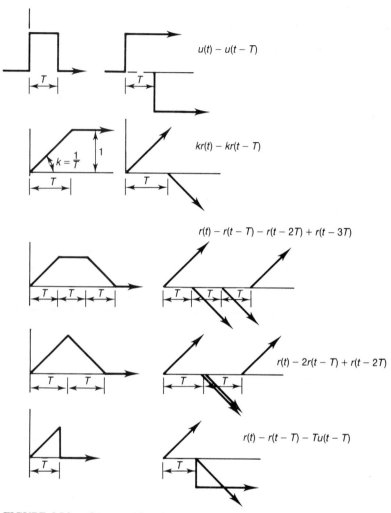

FIGURE 9.5.2   Composed functions.

For time domain evaluation, all we have to do is replace the symbol of the independent source by *the Laplace transform of the signal* actually applied. Afterwards we must go ahead with the decompositions and inversions as we have learned in this chapter. Since the whole of Chapter 10 is dedicated to time domain responses, we will use only one small example here.

Example 1.

Consider the RL voltage divider in Fig. 9.6.1. Using E as a symbol, we get the output

$$V_{out}(s) = \frac{sL}{sL + R} E = \frac{s}{s + 1} E$$

If the signal is a unit step, then we consult Table 9.2.1 and read that $E = 1/s$. Inserting for the symbol of the independent source we get the expression

$$V_{out,step}(s) = \frac{s}{s + 1}\frac{1}{s} = \frac{1}{s + 1}$$

The inverse is taken from Table 9.2.1, formula 5, and is

$$v_{out,step}(t) = e^{-t}u(t)$$

Suppose that instead of the unit step we wish to apply a cosine function, $e(t) = 2 \cos 2t$. Consulting Table 9.2.1 we see that $E = 2s/(s^2 + 4)$. Substituting for the source symbol,

$$V_{out,cos}(s) = \frac{s}{s + 1}\frac{2s}{s^2 + 4}$$

To get the time domain response we must proceed with the decomposition

$$\frac{2s^2}{(s + 1)(s^2 + 4)} = \frac{K}{s + 1} + \frac{As + B}{s^2 + 4}$$

Cross-multiplication provides the equation for the numerators:

$$2s^2 = Ks^2 + 4K + As^2 + As + Bs + B$$

FIGURE 9.6.1   *RL* voltage divider.

which leads to the system of equations

$$K + A = 2$$

$$A + B = 0$$

$$4K + B = 0$$

The solution is $A = 8/5$, $B = -8/5$, $K = 2/5$ and thus the decomposition is

$$V_{out, cos}(s) = \frac{\frac{2}{5}}{s + 1} + \frac{\frac{8}{5}s - \frac{8}{5}}{s^2 + 4}$$

To be able to use formula 12 from Table 9.2.1 we identify, in addition, $c = 0$ and $d = 2$, and get the time domain response

$$v_{out, cos}(t) = \left[\frac{2}{5}e^{-t} + \frac{8}{5}\cos 2t - \frac{8}{5}\sin 2t \right] u(t)$$

PROBLEMS CHAPTER 9

The program described in Appendix B calculates partial fraction decomposition for functions with *simple* poles and plots time responses of rational functions to the Dirac impulse, unit step, sin ωt and cos ωt with your choice of ω. You can use the program to get the poles and residues by asking for the Dirac impulse response. To obtain time domain functions from the residues printed by the program use formulas in Table 9.2.1.

P.9.1 Find the partial fraction decompositions and get the time domain responses for the following functions:

(a) $F(s) = \dfrac{s + 12}{s(s + 4)}$

(b) $F(s) = \dfrac{3s + 2}{s^2 + s}$

(c) $F(s) = \dfrac{6s^2 + 10s + 1}{s(s^2 + 4s + 13)}$

(d) $F(s) = \dfrac{2s}{s + 2}$

(e) $F(s) = \dfrac{2(s^2 + 2s + 3)}{(s + 1)(s^2 + 1)}$

(f) $F(s) = \dfrac{2s + 1}{s + 3}$

(g) $F(s) = \dfrac{s + 1}{(s + 1 + j)(s + 1 - j)}$

(h) $F(s) = \dfrac{(s + 3)(s + 5)}{(s + 1)(s^2 + 2s + 2)}$

(i) $F(s) = \dfrac{(s + 3)(s + 5)}{(s^2 + 2s + 2)(s^2 + 1)}$

(j) $F(s) = \dfrac{s^2 + 3s + 1}{s(s^2 + 2s + 5)}$

(k) $F(s) = \dfrac{(s + 1)(s + 3)(s + 5)}{s(s + 2)(s + 4)(s + 6)}$

(l) $F(s) = \dfrac{s^2 + 3s + 5}{(s + 1)(s + 2)(s + 3)}$

(m) $F(s) = \dfrac{s^3 + 3s + 7}{(s + 1)(s^2 + 4s + 8)}$

(n) $F(s) = \dfrac{5s^2 + 1}{s(s^2 + 1)}$

P.9.2 Functions of time, obtained by the Laplace transform, start at $t = 0$. This is formally expressed by multiplication of the function by $u(t)$, the unit step. For instance, $F(s) = 4/(s + 2)$ leads to $f(t) = 4e^{-2t}u(t)$. If we differentiate the time function, we must treat the expression as a product of two functions and the derivative is

$$f'(t) = -8e^{-2t}u(t) + 4e^{-2t}\delta(t)$$

Any function multiplied by $\delta(t)$ results in the Dirac impulse being multiplied by the value of the function at $t = 0$. In the above case

$$f'(t) = -8e^{-2t}u(t) + 4\delta(t)$$

Using the formula for differentiation we also obtain

$$sF(s) = \frac{4s}{s + 2} = 4 - \frac{8}{s + 2}$$

and the time function is

$$f'(t) = 4\delta(t) - 8e^{-2t}u(t)$$

the same result. Using the above explanation find the inverse of each of the functions below and differentiate the time functions with respect to t. Afterwards form the product $sF(s)$ and find the inverse.

(a) $F(s) = \dfrac{s + 12}{s(s + 4)}$

(b) $F(s) = \dfrac{2}{(s + 1)(s + 2)}$

(c) $F(s) = \dfrac{1}{(s + 2)(s + 3)(s + 4)}$

(d) $F(s) = \dfrac{s + 1}{s(s^2 + 2s + 2)}$

P.9.3 Using the limit theorems, find where will the time domain functions of problems in P.9.1 start at $t = 0$ and end at $t \to \infty$.

P.9.4 In the problems P.9.1 multiply the functions by $1/s$, calculate the residues and find time domain responses. Skip problems (a), (b), (c), (j), (k) and (n), because they lead to a double pole at the origin. Check your results by running the program from Appendix B.

P.9.5 Divide the functions (a), (b), (c), (j) and (n) in P.9.1 by s and find the partial fraction decompositions and the time domain responses. In problem P.9.4 you skipped problems which lead to a double pole at the origin. Find the partial fraction decompositions and the inverses for these functions.

P.9.6 The following problems have multiple poles. Using the theory of Section 4, find the partial fraction decompositions and the time domain responses.

(a) $F(s) = \dfrac{1}{(s + 1)^3(s + 2)^2}$

(b) $F(s) = \dfrac{s^2 + 1}{(s + 1)^3(s + 2)}$

(c) $F(s) = \dfrac{7s^2 + 14s + 9}{(s + 1)^2(s + 2)^2}$

CHAPTER 10

Time Domain

10.0. INTRODUCTION

This chapter will cover all the theory necessary for the calculation of time domain responses of linear networks. It is based on the Laplace transform and has the advantage that there are almost no differences between the frequency and time domain methods. In fact, the only difference in the analysis is the introduction of special independent sources corresponding to the initial voltages on capacitors and initial currents through inductors. Thus everything we have learned so far can be used for time domain analysis as well.

We start separately with networks composed of resistors and capacitors, first without and then with initial conditions. Next we cover networks with resistors and inductors, also first without and then with initial conditions. Finally we go to networks which have all the linear components we have already discussed in this book.

The program NETOPT can calculate time domain responses of linear networks to Dirac impulses, steps and pulses with given duration. Instructions how to use it are in Chapter 18, Section 9. In addition, Appendix B describes how to use another program in which the input is not a network, but a rational function.

10.1. NETWORKS WITH CAPACITORS

We have discussed capacitors and their properties in several earlier chapters, but let us repeat some of the basic conclusions.

A capacitor is an element which can store electrons and in such cases we say that it has an initial voltage, E_0. The current, $i_C(t)$, flowing through the linear capacitor, and the voltage, $v_C(t)$, appearing across it, are related by the equation

$$i_C(t) = C \frac{dv_C(t)}{dt} \tag{10.1.1}$$

We also derived in Chapter 9, Eq. (9.5.2), that the derivative of a time function, $f(t)$, is expressed in the Laplace domain by

$$\mathscr{L}\left[\frac{df(t)}{dt}\right] = sF(s) - f(0)$$

Since Eq. (10.1.1) asks for the derivative of the voltage and since our initial voltage is E_0, the derivative of the voltage across the capacitor will be, in the Laplace domain,

$$\mathscr{L}\left[\frac{dv(t)}{dt}\right] = sV - E_0$$

Using Eq. (10.1.1) in the Laplace domain we get

$$IC = C[sV_C - E_0]$$

and finally

$$I_C = sCV_C - CE_0 \tag{10.1.2}$$

This fundamental expression was already derived in Eq. (9.5.5). The next two subsections will consider separately the cases when the capacitor is without and with the initial voltage (initial condition).

10.1.1. Capacitors without Initial Conditions

If the capacitor has no initial voltage, then $E_0 = 0$ and Eq. (10.1.2) simplifies to

$$I_C = sCV_C \tag{10.1.3}$$

This was already introduced in Chapter 7 without giving any reasons. We also used a special name, *admittance*, for the product

$$Y_C = sC \tag{10.1.4}$$

We can invert Eq. (10.1.3) and obtain

$$V_C = \frac{1}{sC} I_C \tag{10.1.5}$$

The name *impedance* was introduced for the expression

$$Z_C = \frac{1}{sC} \tag{10.1.6}$$

It is worth noting again that at zero frequency, $s = 0$, the impedance becomes infinitely large and the capacitor behaves as an open circuit for dc. At infinitely high frequency it behaves as a short circuit. We have used these expressions many times before. We also stated that E (for an independent voltage source), or J (for an independent current source) will be symbols, to be replaced later by an appropriate expression. In the time domain, the expression is the Laplace transform of the signal to be applied, and all the necessary transformations have been collected in Table 9.2.1. We will now use the previous derivations to obtain time domain responses of various networks.

We will start with the RC and RL dividers, already analyzed in Section 8.3. The first is shown in Fig. 10.1.1. The output voltage was derived in Eq. (8.3.1) and selecting $R = 1\,\Omega$, and $C = 1\,\text{F}$ we simplify it to

$$V_{out}(s) = \frac{1}{s + 1} E$$

We can now study the time domain responses by substituting the Laplace transform of the signal for the symbol E. Suppose that we wish to know the response to the Dirac impulse. Consulting Table 9.2.1 we find in formula 1 that $\mathcal{L}[\delta(t)] = 1$ and the output becomes

$$V_{out,\,Dirac}(s) = \frac{1}{s + 1}$$

No partial fraction decomposition is needed and formula 5 of Table 9.2.1 gives the answer

$$v_{out,\,Dirac}(t) = e^{-t}u(t)$$

FIGURE 10.1.1   RC voltage divider.

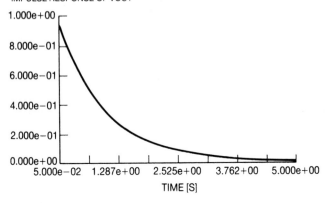

NETWORK FIGURE 10.1.1
TIME DOMAIN ANALYSIS (0/2 APPROXIMATION)
IMPULSE RESPONSE OF VOUT

FIGURE 10.1.2   Impulse response of network in Fig. 10.1.1.

where multiplication by $u(t)$ formally makes the signal zero for $t < 0$. The response, done by NETOPT, is plotted in Fig. 10.1.2. Consult Chapter 18, Section 9 for instructions how to use the program. Suppose that instead of the Dirac impulse we wish to use the unit step. Table 9.2.1 tells us (formula 2) that $\mathscr{L}[u(t)] = 1/s$ and thus

$$V_{out,step}(s) = \frac{1}{s+1}\frac{1}{s}$$

Using the residue method of Section 9.3.3 we obtain the decomposition

$$V_{out,step}(s) = \frac{1}{s} - \frac{1}{s+1}$$

and the inverse is again found by using Table 9.2.1:

$$v_{out,step}(t) = [1 - e^{-t}]u(t)$$

This response, obtained by NETOPT, is plotted in Fig. 10.1.3.

The other RC voltage divider considered in Chapter 8 is redrawn in Fig. 10.1.4. We derived for it Eq. (8.3.5), and for the indicated elements it simplifies to

$$V_{out}(s) = \frac{s}{s+1} E$$

Suppose that we want to find the time domain response for the Dirac impulse.

NETWORK FIGURE 10.1.1
TIME DOMAIN ANALYSIS (0/2 APPROXIMATION)
STEP RESPONSE OF VOUT

FIGURE 10.1.3   Step response of network in Fig. 10.1.1.

FIGURE 10.1.4   CR voltage divider.

Substituting $E = 1$ we have

$$V_{out,\,Dirac}(s) = \frac{s}{s + 1}$$

Since the numerator and denominator have the same degrees, we must first divide the denominator into the numerator to obtain

$$V_{out,\,Dirac}(s) = 1 - \frac{1}{s + 1}$$

For the inversion we consult Table 9.2.1 where we find that

$$v_{out,\,Dirac}(t) = \delta(t) - e^{-t}u(t)$$

The plot is given in Fig. 10.1.5, but the Dirac impulse at the time origin cannot

NETWORK FIGURE 10.1.4
TIME DOMAIN ANALYSIS (0/2 APPROXIMATION)
IMPULSE RESPONSE OF VOUT

FIGURE 10.1.5   Impulse response of network in Fig. 10.1.4.

be plotted. In addition, the response starts from the time $t = 0.05$. This is due to the fact that NETOPT uses a special integration method which gives the response at the first step, but not at $t = 0$. Suppose that we would like to know the step response of this network. Substituting $E = 1/s$ we get

$$V_{out,\,step}(s) = \frac{s}{s+1}\frac{1}{s} = \frac{1}{s+1}$$

and the time domain response is

$$v_{out,\,step}(s) = e^{-t}u(t)$$

which happens to be the same function as plotted in Fig. 10.1.2.

As we already know, the admittance form is suitable for nodal analysis and the impedance form for mesh analysis. We now consider three somewhat more complicated networks and find their time responses.

Example 1.

Find the output voltage V_2 for the network in Fig. 10.1.6 by keeping the elements and the source as variables. Afterwards substitute unit values for all elements and find the step response. Using admittances and the symbol J for the independent current source, we get the nodal equations

$$(G_1 + sC)V_1 - sCV_2 = J$$

$$-sCV_1 + (G_2 + sC)V_2 = 0$$

FIGURE 10.1.6   Network with one capacitor.

In matrix form

$$\begin{bmatrix} G_1 + sC & -sC \\ -sC & G_2 + sC \end{bmatrix}\begin{bmatrix} V_1 \\ V_2 \end{bmatrix} = \begin{bmatrix} J \\ 0 \end{bmatrix}$$

The determinant is equal to

$$D = (G_1 + sC)(G_2 + sC) - s^2C^2 = sC(G_1 + G_2) + G_1G_2$$

and, applying Cramer's rule,

$$V_2 = \frac{sC}{sC(G_1 + G_2) + G_1G_2}\, J(s)$$

Substituting unit values for the elements,

$$V_2(s) = \frac{s}{2s + 1}\, J$$

For the unit step this changes to

$$V_{2,step}(s) = \frac{1}{2(s + 0.5)}$$

and the time response is

$$v_{2,step}(t) = \frac{1}{2}e^{-0.5t}u(t)$$

This is one exponential and the plot will have a similar shape as in Fig. 10.1.2.

Example 2.

Use the mesh method for the network in Fig. 10.1.7 and find the voltage across the capacitor C_2 in terms of the input voltage E. Afterwards find the Dirac impulse response.

FIGURE 10.1.7   Network with two meshes.

The mesh equations are

$$\left(R_1 + \frac{1}{sC_1}\right)I_1 - R_1I_2 = E$$

$$-R_1I_1 + \left(R_1 + R_2 + \frac{1}{sC_2}\right)I_2 = 0$$

In matrix form

$$\begin{bmatrix} R_1 + \dfrac{1}{sC_1} & -R_1 \\ \\ -R_1 & R_1 + R_2 + \dfrac{1}{sC_2} \end{bmatrix} \begin{bmatrix} I_1 \\ I_2 \end{bmatrix} = \begin{bmatrix} E \\ 0 \end{bmatrix}$$

Inserting numerical values the matrix becomes

$$\begin{bmatrix} 1 + \dfrac{1}{2s} & -1 \\ \\ -1 & 2 + \dfrac{1}{s} \end{bmatrix} \begin{bmatrix} I_1 \\ I_2 \end{bmatrix} = \begin{bmatrix} E \\ 0 \end{bmatrix}$$

The determinant is

$$D = \left(1 + \frac{1}{2s}\right)\left(2 + \frac{1}{s}\right) - 1 = 1 + \frac{2}{s} + \frac{1}{2s^2}$$

and the current I_2 is

$$I_2 = \frac{1}{1 + \dfrac{2}{s} + \dfrac{1}{2s^2}} E$$

To simplify the result we multiply both the numerator and denominator by $2s^2$ to get

$$I_2 = \frac{2s^2}{2s^2 + 4s + 1} E$$

The output voltage will be

$$V_{out} = \frac{1}{sC_2} I_2 = \frac{2s}{2s^2 + 4s + 1} E$$

We will return to this result in Example 3.

Example 3.

Consider again the network in Fig. 10.1.7, but apply first the source transformation and then use nodal analysis. Find the output voltage and response to the Dirac impulse.
The transformation is shown in Fig. 10.1.8. The nodal equations are

$$(sC_1 + G_1 + G_2)V_1 - G_2V_2 = sC_1E$$
$$-G_2V_1 + (G_2 + sC_2)V_2 = 0$$

Substituting numerical values and writing in matrix form,

$$\begin{bmatrix} 2s + 2 & -1 \\ -1 & 1 + s \end{bmatrix} \begin{bmatrix} V_1 \\ V_2 \end{bmatrix} = \begin{bmatrix} 2sE \\ 0 \end{bmatrix}$$

The determinant is

$$D = (2s + 2)(1 + s) - 1 = 2s^2 + 4s + 1$$

and the output voltage is

$$V_2 = \frac{2s}{2s^2 + 4s + 1} E$$

FIGURE 10.1.8   Network Fig. 10.1.8 modified for nodal formulation.

NETWORK FIGURE 10.1.7
TIME DOMAIN ANALYSIS (0/2 APPROXIMATION)
IMPULSE RESPONSE OF VOUT

FIGURE 10.1.9   Impulse response of network in Fig. 10.1.7.

The result is the same as in Example 2. Continuing for Dirac impulse, we set $E = 1$ and

$$V_{2,Dirac}(s) = \frac{2s}{2s^2 + 4s + 1} = \frac{s}{s^2 + 2s + 0.5}$$

The decomposition is

$$V_{2,Dirac}(s) = \frac{1.20711}{s + 1.7011} - \frac{0.20711}{s + 0.29289}$$

and the time domain response is

$$v_{2,Dirac}(t) = \{1.20711e^{-1.7011t} - 0.20711e^{-0.29289t}\}u(t)$$

The response is plotted in Fig. 10.1.9, obtained by NETOPT.

Let us now summarize the results of this section. In the derivations keep the independent source as a symbol E or J. For the capacitor with no initial charge ($E_0 = 0$) use its admittance, $Y_C = sC$, or its impedance, $Z_C = 1/sC$, depending on the formulation method. Derive the desired output and only then substitute the Laplace transform of the input signal for the symbol E or J. Find the partial fraction expansion and use Table 9.2.1 to obtain the time domain response.

10.1.2. Capacitors with Initial Conditions

If the capacitor has an initial voltage, the situation is as shown in Fig. 10.1.10a. The initial voltage, E_0, can have signs either as shown, or reversed, depending on

$$i(t) = C \frac{dv(t)}{dt}$$

$$v(t) = E_0 + \frac{1}{C_0} \int_0^t i(t)\, dt$$

$$I = sCV - CE_0$$

$$V = \frac{I}{sC} + \frac{E_0}{s}$$

(a)   (b)   (c)

FIGURE 10.1.10   Capacitor with initial voltage: (a) Symbol. (b) Equivalent for nodal formulation. (c) Equivalent for mesh formulation.

how we charged the capacitor before we connected it into the network. Due to the influences coming from the network there will be, in addition, a voltage, $v(t)$, across it and a current, $i(t)$, flowing through it. In agreement with our general definition, this current is taken as positive when it flows from + to −, as in the figure.

We derived in Eq. (10.1.2) how a capacitor with an initial voltage is described in the Laplace domain. We also defined its admittance, $Y_C = sC$. Substitute the admittance into Eq. (10.1.2) and rewrite in the form

$$I_C + CE_0 = Y_C V_C \qquad (10.1.7)$$

This is an equation in currents so every term in it must be a current. Without the initial voltage $I_C = Y_C V_C$. With the initial condition we have another component, CE_0, flowing into the capacitor as well. This can be interpreted as shown in Fig. 10.1.10b. The additional current acts as a special independent current source. Recall what we said in Chapter 9 about the Dirac impulse, $\delta(t)$. It is a strange impulse, simultaneously infinitely large and infinitely narrow, but, curiously, having an area equal to 1. We derived for it that $\mathscr{L}[\delta(t)] = 1$. There is no s in the transform, similarly as in our current source due to the initial condition. The only difference is that this source is CE_0 times larger, or has an area CE_0. We thus conclude that the initial voltage across the capacitor, E_0, is expressed in the Laplace domain by a current source in parallel with the capacitor and its signal is a Dirac impulse with an area CE_0. Should the + and − sign of the initial voltage be reversed, the direction of the equivalent source is reversed as well.

Let us now return to Eq. (10.1.2). Simple algebraic operations get us

$$V_C = \frac{1}{sC} I_C + \frac{E_0}{s} \qquad (10.1.8)$$

We can modify this expression by using the impedance, $Z_C = 1/sC$,

$$V_C = Z_C I_C + \frac{E_0}{s}$$

This is an equation in voltages and every term must describe a voltage. The overall voltage across the capacitor is equal to the sum of two components, one due to the current through it and one due to the initial condition. The orientations are as shown in Fig. 10.1.10c. Going back to the Laplace transform theory, we see that the additional voltage is a *step source* having an amplitude E_0.

The figures greatly simplify the way we take initial conditions into consideration. If the capacitor has an initial voltage, all we have to do is draw an additional independent source into the network. If we are using nodal equations, we take advantage of Fig. 10.1.10b, because the initial condition is in the form of an independent current source. If we use mesh formulation, we use the form with the independent voltage source, Fig. 10.1.10c. The Kirchhoff voltage and current laws apply the same way as before. It is also becoming clear why we have used the letter E_0 to indicate the initial condition: In the Laplace transform, this initial condition forms an independent source, and as such must be transferred to the right-hand side of the equation before we can proceed with the solution.

We will now take examples with initial conditions and set up the equations. Consider first the RC voltage divider, now with an initial condition as shown in Fig. 10.1.11a. Using the results summarized in Fig. 10.1.10 we can redraw this network either as shown in Fig. 10.1.11b, suitable for mesh formulation, or as shown in Fig. 10.1.11c, suitable for nodal formulation. From now on the analysis proceeds exactly the same way as before. For mesh analysis we write

$$I\left(R + \frac{1}{sC}\right) = E - \frac{E_0}{s}$$

or

$$I = \frac{sC}{sCR + 1}\left(E - \frac{E_0}{s}\right) = \frac{sCE - CE_0}{sCR + 1}$$

(a)   (b)   (c)

FIGURE 10.1.11   RC network with initial condition: (a) Original. (b) Equivalent for mesh formulation. (c) Equivalent for nodal formulation.

The output voltage is

$$V_{out}(s) = E - RI = E - R\frac{sCE - CE_0}{sCR + 1}$$

If we substitute the numerical values we get

$$V_{out}(s) = \frac{E + E_0}{s + 1}$$

At this point we must decide what we will use for the external voltage source. If we select a unit step, the result is

$$V_{out, step}(s) = \frac{1}{s(s + 1)} + \frac{E_0}{s + 1}$$

Partial fraction decomposition of the first term provides

$$V_{out, step}(s) = \frac{1}{s} - \frac{1}{s + 1} + \frac{E_0}{s + 1}$$

and for $E_0 = 2$ the final result is

$$V_{out, step}(s) = \frac{1}{s} + \frac{1}{s + 1}$$

Consulting Table 9.2.1 we get the inverse

$$v_{out, step}(t) = (1 + e^{-t})u(t)$$

This result is plotted in Fig. 10.1.12, obtained by NETOPT. Let us state again that the program plots the first point at the first step, in this case at $t = 0.05$.

If we consider nodal formulation, then we first transform the external source into a current source, as shown in Fig. 10.1.11c, and write one nodal equation

$$V(G + sC) = EG + CE_0$$

which, after the substitution of element values, results in

$$V = \frac{E + E_0}{s + 1}$$

the same as above.

NETWORK FIGURE 10.1.11
TIME DOMAIN ANALYSIS (0/2 APPROXIMATION)
STEP RESPONSE OF VOUT

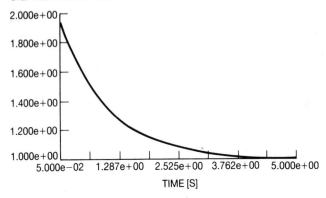

FIGURE 10.1.12   Step response of network in Fig. 10.1.11.

Our next problem will be the other possible RC divider, as sketched in Fig. 10.1.13a. If we consider first the mesh formulation, the network is redrawn as in Fig. 10.1.13b, for which the single mesh equation is

$$\left(R + \frac{1}{sC}\right)I = E - \frac{E_0}{s}$$

the current is

$$I = \frac{sC}{sCR + 1}\left(E - \frac{E_0}{s}\right)$$

and the output voltage is

$$V_{out} = RI = \frac{sCR}{sCR + 1}\left(E - \frac{E_0}{s}\right)$$

Inserting element values

$$V_{out} = \frac{sE - E_0}{s + 1}$$

For the unit step $E = 1/s$, and since $E_0 = 2$, we get

$$V_{out, step} = \frac{-1}{s + 1}$$

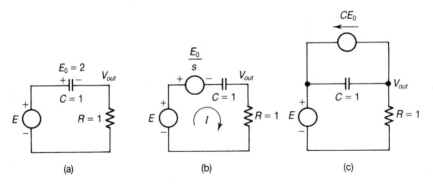

FIGURE 10.1.13   CR network with initial condition: (a) Original. (b) Equivalent for mesh formulation. (c) Equivalent with impulse source.

and the time domain response is

$$v_{out, step}(t) = -e^{-t}u(t)$$

If we wish to use the current source equivalent for the initial condition, then we have to analyze the network in Fig. 10.1.13c, but here we have a dilemma: we cannot transform the voltage source on the left into a current source, because the capacitor is now with its own current source. However, with a bit of thinking there is always a way. What we can do is take the network in Fig. 10.1.13b, and transform both sources simultaneously into one current source. This is shown in Fig. 10.1.14, for which we now write

$$(G + sC)V = sC\left(E - \frac{E_0}{s}\right)$$

with the result

$$V = \frac{sE - E_0}{s + \dfrac{G}{C}}$$

the same as above.

FIGURE 10.1.14   Transformation of network in Fig. 10.1.13(c).

The above networks have shown that there is nothing new in terms of network analysis. There are some additional independent sources, due to the initial conditions, but all the methods we learned in the previous chapters remain valid. What is even more important, the analysis remains exactly the same if we move to more complicated networks. One of them will be analyzed now.

Example 4.

Consider the network in Fig. 10.1.15a. It has no independent external source but all three capacitors have initial conditions. The network is redrawn in b with equivalent current sources, suitable for nodal formulation, and in c with voltage sources, suitable for mesh formulation. We will use both. For Fig. 10.1.15b the nodal equations are

$$(sC_1 + G_1)V_1 - G_1V_2 + 0V_3 = C_1E_{01}$$

$$-G_1V_1 + (G_1 + sC_2 + sC_3)V_2 - sC_3V_3 = -C_2E_{02} + C_3E_{03}$$

$$0V_1 - sC_3V_2 + (sC_3 + G_2)V_3 = -C_3E_{03}$$

In matrix form

$$\begin{bmatrix} sC_1 + G_1 & -G_1 & 0 \\ -G_1 & G_1 + sC_2 + sC_3 & -sC_3 \\ 0 & -sC_3 & sC_3 + G_2 \end{bmatrix} \begin{bmatrix} V_1 \\ V_2 \\ V_3 \end{bmatrix} = \begin{bmatrix} C_1E_{01} \\ -C_2E_{02} + C_3E_{03} \\ -C_3E_{03} \end{bmatrix}$$

To continue we insert numerical values:

$$\begin{bmatrix} s+1 & -1 & 0 \\ -1 & 3s+1 & -s \\ 0 & -s & s+1 \end{bmatrix} \begin{bmatrix} V_1 \\ V_2 \\ V_3 \end{bmatrix} = \begin{bmatrix} 1 \\ -1 \\ -3 \end{bmatrix}$$

After a few intermediate steps, we find the determinant of the matrix to be

$$D = s(2s^2 + 6s + 4)$$

Using Cramer's rule for V_3, we find, after a few steps, that the numerator is

$$N_3 = s(-10s - 12)$$

We can cancel $2s$ in the numerator against the same term in the denominator and get

$$V_3 = \frac{-5s - 6}{s^2 + 3s + 2}$$

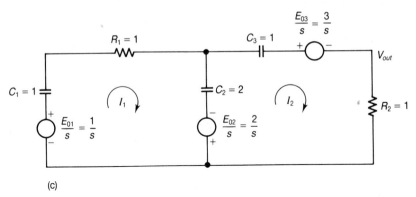

FIGURE 10.1.15   Network with initial conditions: (a) Original. (b) Transformed for nodal formulation. (c) Transformed for mesh formulation.

The same network can also be solved by the mesh formulation and the other representation of the initial voltages, as shown in Fig. 10.1.15c. The equations are

$$\left(\frac{1}{sC_1} + R_1 + \frac{1}{sC_2}\right)I_1 - \frac{1}{sC_2}I_2 = \frac{E_{01} + E_{02}}{s}$$

$$-\frac{1}{sC_2}I_1 + \left(\frac{1}{sC_2} + \frac{1}{sC_3} + R_2\right)I_2 = \frac{-E_{02} - E_{03}}{s}$$

Using numerical values

$$\begin{bmatrix} \dfrac{2s+3}{2s} & -\dfrac{1}{2s} \\[4mm] -\dfrac{1}{2s} & \dfrac{2s+3}{2s} \end{bmatrix} \begin{bmatrix} I_1 \\[4mm] I_2 \end{bmatrix} = \begin{bmatrix} \dfrac{3}{s} \\[4mm] -\dfrac{5}{s} \end{bmatrix}$$

The determinant of the matrix is

$$D = \frac{s^2 + 3s + 2}{s^2}$$

and, using Cramer's rule, the numerator for I_2 is

$$N_2 = \frac{-5s - 6}{s^2}$$

As a result, the current of the second mesh is

$$I_2 = \frac{-5s - 6}{s^2 + 3s + 2}$$

The output voltage, V_{out}, is taken across the resistor R_2 and is numerically equal to the current I_2. This is in agreement with our previous solution, obtained by the nodal method. The function can be decomposed into

$$V_3 = -\frac{1}{s+1} - \frac{4}{s+2}$$

and consulting Table 9.2.1 the inverse is

$$v_3(t) = \{-1e^{-t} - 4e^{-2t}\}u(t)$$

This is plotted in Fig. 10.1.16. The first plotted point is at $t = 0.05$. The response at $t = 0$ is -5.

Example 5.

Consider the network in Fig. 10.1.17a, and its Laplace domain equivalent in b. Find the response of the network for $j(t) = \sin t$. The nodal equations are

$$(G_1 + sC)V_1 - sCV_2 = J - CE_0$$
$$-sCV_1 + (sC + G_2)V_2 = CE_0$$

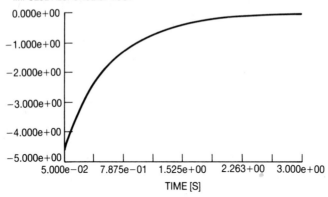

NETWORK FIGURE 10.1.15
TIME DOMAIN ANALYSIS (0/2 APPROXIMATION)
IMPULSE RESPONSE OF VOUT

FIGURE 10.1.16   Response of network in Fig. 10.1.15.

In matrix form

$$\begin{bmatrix} 1+s & -s \\ -s & 2+s \end{bmatrix} \begin{bmatrix} V_1 \\ V_2 \end{bmatrix} = \begin{bmatrix} J-2 \\ 2 \end{bmatrix}$$

The denominator is $D = 3s + 2$, $N_2 = Js + 2$ and

$$V_2 = \frac{Js + 2}{3s + 2}$$

The Laplace transform of the input signal is $J = 1/(s^2 + 1)$ and thus

$$V_2 = \frac{\dfrac{2}{3}s^2 + \dfrac{1}{3}s + \dfrac{2}{3}}{(s^2 + 1)\left(s + \dfrac{2}{3}\right)}$$

We need the decomposition

$$\frac{\dfrac{2}{3}s^2 + \dfrac{1}{3}s + \dfrac{2}{3}}{(s^2 + 1)\left(s + \dfrac{2}{3}\right)} = \frac{K}{s + \dfrac{2}{3}} + \frac{As + B}{s^2 + 1}$$

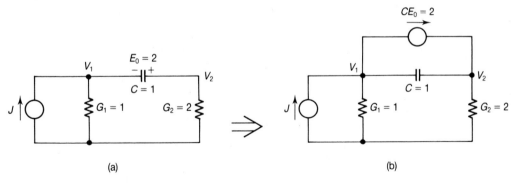

FIGURE 10.1.17   Network with initial voltage: (a) Original. (b) Nodal formulation equivalent.

Cross-multiplying and comparing coefficients at equal powers, we get the system

$$K + A = \frac{2}{3}$$

$$\frac{2}{3}A + B = \frac{1}{3}$$

$$K + \frac{2}{3}B = \frac{2}{3}$$

Its solution is $K = 20/39$, $A = 6/39$ and $B = 9/39$. The decomposition is

$$V_2 = \frac{20}{39} \frac{1}{s + \frac{2}{3}} + \frac{\frac{6}{39}s + \frac{9}{39}}{s^2 + 1}$$

Using formulas from Table 9.2.1 we get the time domain response

$$v_2(t) = \left[\frac{20}{39} e^{-2t/3} + \frac{6}{39} \cos t + \frac{9}{39} \sin t \right] u(t)$$

It is plotted in Fig. 10.1.18.

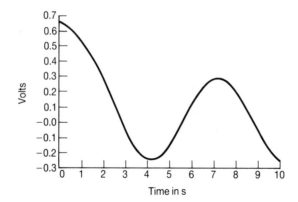

FIGURE 10.1.18   Response of network Fig. 10.1.17 to $j(t) = \sin t$.

10.2. NETWORKS WITH INDUCTORS

The voltage across an inductor and the current flowing through it are coupled by the fundamental equation

$$v_L(t) = L\frac{di_L(t)}{dt} \tag{10.2.1}$$

The current creates a magnetic field which stores energy and which tries to continue the flow of current even if the source of current has changed. We speak about initial current through the inductor and denote it by J_0. If we proceed similarly as in the case of the capacitor and use the expression for the Laplace transform of the derivative, we get

$$V_L = sLI_L - LJ_0 \tag{10.2.2}$$

This fundamental expression relates, in the Laplace transform, the voltage across the inductor, V_L, the current through the inductor, I_L, and the initial current, J_0. It was already derived in Eq. (9.5.6). We will consider separately the cases when the inductor is without and with the initial current.

10.2.1. Inductors without Initial Conditions

If no initial current flows through the inductor, $J_0 = 0$ and Eq. (10.2.2) simplifies to

$$V_L = sLI_L \tag{10.2.3}$$

The product sL is an *impedance*

$$Z_L = sL \qquad (10.2.4)$$

If we invert Eq. (10.2.3), we obtain

$$I_L = \frac{1}{sL} V_L \qquad (10.2.5)$$

and the word *admittance* is used for the ratio

$$Y_L = \frac{1}{sL} \qquad (10.2.6)$$

We use the impedance when writing mesh equations and the admittance in nodal equations. If $s = 0$ then the impedance is zero and the inductor acts as a short circuit at dc. If s is very large, approaching infinity, then the impedance becomes very large and the inductor acts as an open circuit.

Since RL dividers lead to the same equations as the RC dividers (see our derivations in Section 8.3), we will not consider them here. Instead we will take somewhat more complicated networks.

Example 1.

Analyze the network in Fig. 10.2.1 using the mesh and nodal equations. Find the time domain response for a Dirac impulse and unit step.

The mesh equations are

$$(R_1 + sL_1)I_1 - R_1 I_2 = E$$
$$-R_1 I_1 + (R_1 + R_2 + sL_2)I_2 = 0$$

Using numerical values, the matrix equation is

$$\begin{bmatrix} 1+s & -1 \\ -1 & 2+2s \end{bmatrix} \begin{bmatrix} I_1 \\ I_2 \end{bmatrix} = \begin{bmatrix} E \\ 0 \end{bmatrix}$$

FIGURE 10.2.1   Network with two inductors.

The determinant is

$$D = 2s^2 + 4s + 1$$

and

$$I_2 = \frac{1}{2s^2 + 4s + 1} E$$

For the output voltage across the inductor we must multiply the current I_2 by the impedance $Z_L = sL_2 = 2s$. The result is

$$V_{out}(s) = sLI_2 = \frac{2s}{2s^2 + 4s + 1} E$$

Finding the roots of the denominator, substituting for the Dirac impulse $E = 1$ and finding the residues:

$$V_{out, Dirac}(s) = \frac{1.20711}{s + 1.70711} - \frac{0.20711}{s + 0.29289}$$

and the time domain response is

$$v_{out, Dirac} = [1.20711e^{-1.70711t} - 0.20711e^{-0.29289t}]u(t)$$

For unit step we substitute $E = 1/s$ and the function becomes

$$V_{out, step} = \frac{2s}{2s^2 + 4s + 1} \frac{1}{s} = \frac{1}{2s^2 + 4s + 1} = \frac{0.70711}{s + 0.29289} - \frac{0.70711}{s + 1.70711}$$

The time domain response is

$$v_{out, step} = [0.70711e^{0.29289t} - 0.70711e^{-1.70711t}]u(t)$$

For nodal formulation we first transform the voltage source into a current source as shown in Fig. 10.2.2. The equations are

$$\left(G_1 + G_2 + \frac{1}{sL_1}\right)V_1 - G_2V_2 = \frac{E}{sL_1}$$

$$-G_2V_1 + \left(G_2 + \frac{1}{sL_2}\right)V_2 = 0$$

FIGURE 10.2.2   Network Fig. 10.2.1 prepared for nodal formulation.

Substituting numerical values and writing in matrix form

$$
\begin{bmatrix} 2 + \dfrac{1}{s} & -1 \\[3mm] -1 & 1 + \dfrac{1}{2s} \end{bmatrix} \begin{bmatrix} V_1 \\[2mm] V_2 \end{bmatrix} = \begin{bmatrix} \dfrac{E}{s} \\[2mm] 0 \end{bmatrix}
$$

The determinant is $D = (2s^2 + 4s + 1)/2s^2$ and the numerator is $N_2 = E/s$. The output voltage, V_2, is the same as obtained above.

Example 2.

Dependent sources are handled the same way as before, as seen from the analysis of the network in Fig. 10.2.3.
   The equations are

$$
\left(1 + \frac{1}{2s}\right)V_1 - \frac{1}{2s}V_2 = J
$$

$$
-\frac{1}{2s}V_1 + \left(\frac{1}{2s} + 2\right)V_2 + 1.5V_1 = 0
$$

and in matrix form

$$
\begin{bmatrix} \dfrac{2s + 1}{2s} & -\dfrac{1}{2s} \\[4mm] \dfrac{3s - 1}{2s} & \dfrac{4s + 1}{2s} \end{bmatrix} \begin{bmatrix} V_1 \\[2mm] V_2 \end{bmatrix} = \begin{bmatrix} J \\[2mm] 0 \end{bmatrix}
$$

The determinant is $D = (8s + 9)/4s$, $N_2 = [(1 - 3s)/2s]J$ and the output voltage is

$$
V_{out} = \frac{-6s + 2}{8s + 9} J
$$

FIGURE 10.2.3   Network with a VC.

If the source is a Dirac impulse, then we must first divide the two polynomials to obtain

$$V_{out, Dirac} = -0.75 + \frac{1.09375}{s + 1.125}$$

The time domain response is

$$v_{out, Dirac} = -0.75\delta(t) + 1.09375e^{-1.125t}u(t)$$

10.2.2. Inductors with Initial Conditions

An inductor may have an initial current, J_0, in addition to a given voltage $v(t)$ across it and a current $i(t)$ flowing through it. As always, the positive direction of current $i(t)$ is from $+$ to $-$ but initial current can flow in either direction, depending on the previous state of the network. Our reference direction is as shown in Fig. 10.2.4a.

In the Laplace domain, an inductor with an initial current is described by Eq. (10.2.2). If we substitute $Z_L = sL$, then

$$V_L = Z_L I_L - LJ_0 \tag{10.2.7}$$

This indicates that the voltage across the inductor is formed from two components, one due to the current flowing through the impedance and another due to the initial current J_0. This is interpreted as shown in Fig. 10.2.4b. The voltage source

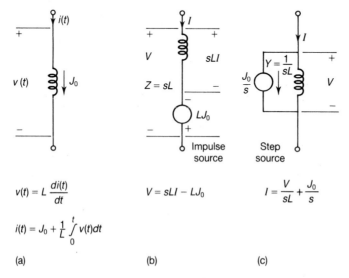

(a)   (b)   (c)

FIGURE 10.2.4   Inductor with initial current: (a) Symbol. (b) Equivalent for mesh formulation. (c) Equivalent for nodal formulation.

is a Dirac impulse having an area LJ_0. Should the direction of the initial current J_0 be reversed, the $+$ and $-$ signs of the equivalent voltage source will be reversed as well.

Let us now return to Eq. (10.2.7) and separate the current on the left side of the equation

$$I_L = \frac{1}{sL} V_L + \frac{J_0}{s} \qquad (10.2.8)$$

If we now insert $Y_L = 1/sL$ we can write

$$I_L - \frac{J_0}{s} = Y_L V_L$$

The equation indicates that two currents are flowing into admittance, one due to the voltage V_L across the inductor and one due to the initial current, J_0. This is interpreted as shown in Fig. 10.2.4c. Going back to the Laplace transform theory, we see that the source is a step having an amplitude J_0.

Similarly as in the case of the capacitor, the figures considerably simplify our work. All we have to do is redraw the network using one of the two possible representations. For mesh equations we use the form with the voltage source, for nodal equations the form with the current source and apply KVL or KCL for the modified network. It is also becoming clear why we have used the letter J_0 to indicate the initial condition: In the Laplace transform, this initial condition forms an independent source and in the equations all such sources must go to the right-hand side before we can go ahead with the solution.

We will use examples to clarify application.

Example 3.

Find the unit step response of the network in Fig. 10.2.5a using the mesh and nodal formulation.

For mesh formulation we redraw the network as shown in Fig. 10.2.5b and obtain

$$\begin{bmatrix} 2 + 2s & -2s \\ -2s & 1 + 2s \end{bmatrix} \begin{bmatrix} I_1 \\ I_2 \end{bmatrix} = \begin{bmatrix} E + 4 \\ -4 \end{bmatrix}$$

The determinant is $D = 6s + 2$, $N_2 = 2(Es - 4)$ and

$$I_2 = \frac{Es - 4}{3s + 1}$$

The output voltage is $V_{out} = R_2 I_2$ and since $R_2 = 1$ the above result applies for V_{out} as well.

For nodal formulation we redraw the network as in Fig. 10.2.5c. Only one equation is

(a)   (b)

(c)

FIGURE 10.2.5   Network with initial current: (a) Original. (b) Transformed for mesh formulation. (c) Transformed for nodal formulation.

needed

$$\left(\frac{1}{2} + \frac{1}{2s} + 1\right)V = \frac{E}{2} - \frac{2}{s}$$

and the output voltage is

$$V_{out} = \frac{Es - 4}{3s + 1}$$

the same as above. For unit step we substitute $E = 1/s$ and

$$V_{out,step} = -\frac{3}{3s + 1} = -\frac{1}{s + \frac{1}{3}}$$

The time domain response is

$$v_{out,step} = -e^{-0.33333t}u(t)$$

Example 4.

Consider Fig. 10.2.6a with three inductors, each having an initial current. The network has no independent external source. Solve using both mesh and nodal methods.

FIGURE 10.2.6   Network with initial currents: (a) Schematic. (b) Transformed for mesh formulation. (c) Transformed for nodal formulation.

For mesh analysis we redraw the network as shown in Fig. 10.2.6b. The equations are

$$(sL_1 + sL_2 + R_1)I_1 - sL_2I_2 = -L_1J_{01} - L_2J_{02}$$
$$-sL_2I_2 + (sL_2 + sL_3 + R_2)I_2 = L_2J_{02} - L_3J_{03}$$

Inserting numerical values and writing in matrix form

$$\begin{bmatrix} 3s + 1 & -2s \\ -2s & 3s + 1 \end{bmatrix} \begin{bmatrix} I_1 \\ I_2 \end{bmatrix} = \begin{bmatrix} -5 \\ 1 \end{bmatrix}$$

The determinant of the matrix is

$$D = 5s^2 + 6s + 1$$

Using Cramer's rule we find

$$I_2 = \frac{-7s + 1}{5s^2 + 6s + 1}$$

The output voltage is

$$V_{out} = R_2 I_2 = \frac{-7s + 1}{5s^2 + 6s + 1}$$

Finding the roots and the partial fraction decomposition

$$V_{out} = \frac{0.6}{s + 0.2} - \frac{2}{s + 1}$$

and the time domain output is

$$v_{out} = [0.6e^{-0.2t} - 2e^{-t}]u(t)$$

If we use the representation with current sources, then the same network can be redrawn as shown in Fig. 10.2.6c. Applying the nodal equations

$$\left(\frac{1}{sL_1} + G_1\right)V_1 - G_1 V_2 + 0V_3 = -\frac{J_{01}}{s}$$

$$-G_1 V_1 + \left(G_1 + \frac{1}{sL_2} + \frac{1}{sL_3}\right)V_2 - \frac{1}{sL_3} V_3 = \frac{J_{02}}{s} + \frac{J_{03}}{s}$$

$$0V_1 - \frac{1}{sL_3} V_2 + \left(\frac{1}{sL_3} + G_2\right)V_3 = -\frac{J_{03}}{s}$$

If we substitute numerical values, then the system will be

$$\begin{bmatrix} \dfrac{s+1}{s} & -1 & 0 \\[2mm] -1 & \dfrac{2s+3}{2s} & -\dfrac{1}{s} \\[2mm] 0 & -\dfrac{1}{s} & \dfrac{s+1}{s} \end{bmatrix} \begin{bmatrix} V_1 \\[2mm] V_2 \\[2mm] V_3 \end{bmatrix} = \begin{bmatrix} -\dfrac{1}{s} \\[2mm] \dfrac{5}{s} \\[2mm] -\dfrac{3}{s} \end{bmatrix}$$

The determinant of the matrix is

$$D = \frac{5s^2 + 6s + 1}{2s^3}$$

If we wish to get only the output voltage, $V_{out} = V_3$, then the numerator is

$$N_3 = \frac{-7s + 1}{2s^3}$$

and the output voltage is

$$V_{out} = \frac{-7s + 1}{5s^2 + 6s + 1}$$

the same as above.

The examples have shown that proper choice of the equivalent source can save us a lot of work; Fig. 10.2.6b leads to only two equations, Fig. 10.2.6c leads to three. If the network has only inductors and resistors, it will be generally advantageous to select mesh formulation with voltage sources because we will avoid fractions of the type $1/s$. It can pay to first decide which of the methods leads to a smaller set of equations.

Another fact which is good to remember (although it was not derived here) is that for networks combined of only resistors and capacitors, or resistors and inductors, without any dependent sources, all poles are real, irrespective of the size of the network.

We will now make several conclusions and recommendations:

1. When writing the system equations, keep the symbols for external sources, E and/or J, throughout all calculations, up to the point when partial fraction decomposition can be applied. Only then substitute the actual Laplace domain expression for the input signal.
2. Decomposition requires factorization of the denominator polynomial. For polynomials of first and second order, this can be done symbolically, with elements kept as variables. If the polynomial is of order 3 or more, only numerical solutions are feasible and a computer should be used to find them.
3. Responses are composed of two types of components. Those which die out with time are called *transients*. If the input is a periodic signal (sine or cosine), then the output contains a component of the same periodic function, modified in amplitude and in phase. This is called the *steady state* component.

We also make a few observations.

a. The unit step seems to be the most useful signal; it is simple and smaller networks can be handled by hand.
b. The Dirac impulse is even simpler, but is an artificial signal. It is the derivative of the unit step and therefore the response of the network to the Dirac impulse will be the derivative of the response to a unit step.
c. Finding the time domain response of a network to a periodic input signal (sine or cosine) is a laborious process even for small networks.

10.3. RLC CIRCUITS

In this section we will study time domain responses of series and parallel tuned circuits. In order to simplify the derivations, we will set fixed values for the capacitor and inductor; only the resistor will be changing.

For the series tuned circuit, Fig. 10.3.1, we have

$$\left(R + \frac{1}{sC} + sL\right)I = E$$

and

$$I = \frac{1}{L}\frac{s}{s^2 + s\frac{R}{L} + \frac{1}{LC}}E$$

If we insert the element values $L = 1\,\text{H}$, $C = 1\,\text{F}$, select unit step $E = 1/s$ and calculate the voltage across the capacitor, we get

$$V_{C,step} = \frac{1}{sC}I = \frac{1}{s(s^2 + sR + 1)}$$

In dealing with time responses it is always useful to find out beforehand where the response will be at $t = 0$ and where it will be for $t \to \infty$. For this we use the limiting theorems derived in Eq. (9.5.12) and Eq. (9.5.13). At the time $t = 0$

$$v_{C,step}(0) = \lim_{s \to \infty} sV = \lim_{s \to \infty} \frac{1}{s^2 + sR + 1} = 0$$

At $t \to \infty$ we use the other formula

$$v_{C,step}(\infty) = \lim_{s \to 0} sV = \lim_{s=0} \frac{1}{s^2 + sR + 1} = 1$$

and the response will start at 0 and end at 1.

FIGURE 10.3.1   Series tuned circuit with $L = 1\,\text{H}$ and $C = 1\,\text{F}$.

The roots of the quadratic polynomial are

$$p_{1,2} = -\frac{R}{2} \pm \left[\frac{R^2}{4} - 1\right]^{1/2}$$

Depending on the value of the discriminant (expression in the square brackets), these roots can be either distinct real, or coincide as one double real root, or they can be complex conjugate. We will consider all three cases.

a. If $R^2/4 - 1 > 0$, the polynomial has two real roots. Selecting, for instance, $R = 4$, the roots become

$$p_1 = -0.268$$
$$p_2 = -3.732$$

and

$$V_{C,step} = \frac{1}{s(s - p_1)(s - p_2)} = \frac{K_1}{s} + \frac{K_2}{s - p_1} + \frac{K_3}{s - p_2}$$

where

$$K_1 = 1$$
$$K_2 = -1.07735$$
$$K_3 = 0.07735$$

The time response is

$$v_{C,step}(t) = [1 - 1.07735e^{-0.268t} + 0.07735e^{-3.732t}]u(t)$$

and is plotted in Fig. 10.3.2, full curve.

b. If $R = 2$, then

$$V_{C,step} = \frac{1}{s(s + 1)^2} = \frac{1}{s} - \frac{1}{(s + 1)^2} - \frac{1}{s + 1}$$

The decomposition for the double root is obtained by the theory explained in Section 9.4. The inverse for the double pole is in formula 6, Table 9.2.1 and the time domain response is

$$v_{C,step}(t) = [1 - te^{-t} - e^{-t}]u(t)$$

It is plotted in Fig. 10.3.2, dotted curve.

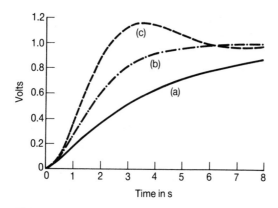

FIGURE 10.3.2   Unit step responses of the circuit in Fig. 10.3.1 for: (a) $R = 4\,\Omega$ (full curve). (b) $R = 2\,\Omega$ (dotted curve). (c) $R = 1\,\Omega$ (dashed curve).

c. If $R < 2$, then

$$V_{C,step} = \frac{1}{s(s^2 + sR + 1)} = \frac{1}{s} - \frac{s + R}{\left(s + \dfrac{R}{2}\right)^2 + d^2}$$

where

$$d = \left[1 - \frac{R^2}{4}\right]^{1/2}$$

We identify for formula 12 in Table 9.2.1 $A = 1$, $B = R$, $c = R/2$, d as given above and $(B - Ac)/d = R/2d$. The inverse is

$$v_{C,step}(t) = [1 - e^{-Rt/2}\cos dt - \frac{R}{2d}e^{-Rt/2}\sin dt]u(t)$$

If we select $R = 1$, then $c = 0.5$, $d = 0.866$ and

$$v_{C,step}(t) = [1 - e^{-0.5t}(\cos 0.866t + 0.577\sin 0.866t)]u(t)$$

see the dashed curve in Fig. 10.3.2.

The shape of the curves depends on the value of the resistor; the smaller the resistor, the more pronounced is the oscillatory behavior of the network. As a matter of interest, if we select $R = 0.1$, then the time domain response will be as shown in Fig. 10.3.3. If we make the resistor zero, the expression for the voltage

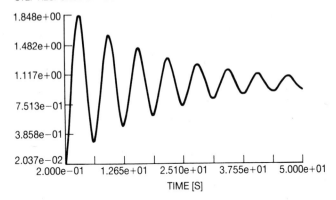

UNIT STEP RESPONSE OF THE NETWORK IN FIGURE 10.3.1 WITH R = 0.1
TIME DOMAIN ANALYSIS (0/2 APPROXIMATION)
STEP RESPONSE OF VOUT

FIGURE 10.3.3   Step response of network Fig. 10.3.1 for $R = 0.1\,\Omega$.

across the capacitor reduces to

$$V_{C,step} = \frac{1}{s(s^2 + 1)} = \frac{1}{s} - \frac{s}{s^2 + 1}$$

the inverse will be

$$v_{C,step}(t) = 1 - \cos t$$

and will oscillate permanently.

Another practical case is a parallel tuned circuit shown in Fig. 10.3.4. The output voltage is given by

$$V = \frac{1}{C} \frac{s}{s^2 + s\dfrac{1}{RC} + \dfrac{1}{LC}} J$$

FIGURE 10.3.4   Parallel tuned circuit.

The equation is very similar to the equations discussed above. If we use $C = 1\,\mathrm{F}$ and $L = 1\,\mathrm{H}$, it simplifies to

$$V = \frac{s}{s^2 + s\dfrac{1}{R} + 1}\, J$$

The difference from the previous case is that the middle term of the denominator is now divided by R, instead of multiplied.

10.4. SWITCHED NETWORKS

Switched networks are typical examples where we deal with initial conditions. An ideal switch is an element which is equivalent to a short circuit when closed and to an open circuit when open. A nonideal switch can be modeled by means of an ideal switch and some other network components. Figure 10.4.1 shows, for instance, an ideal switch with two resistors, one of which may be large and one small. This combination can simulate a switch which has a small resistance when closed and a large resistance when open.

We will use two networks to explain how one must proceed in the case of switched networks. The first one is in Fig. 10.4.2a. We start with the left switch closed and the right switch open; the network is an RC divider with a step source, as shown in Fig. 10.4.2b. We calculate the time response of the voltage across the capacitor. At some later instant, the first switch opens and the second switch closes, changing the network to that in Fig. 10.4.2c. There is now an initial voltage, E_0, on the capacitor, equal to the voltage at the moment of switching. The initial condition is handled by an equivalent current source in parallel with the capacitor.

At the beginning the voltage across the capacitor is

$$V = \frac{1}{sCR + 1}\, E \qquad\qquad (10.4.1)$$

FIGURE 10.4.1   Possible modeling of a nonideal switch.

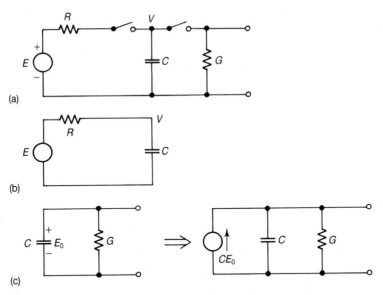

FIGURE 10.4.2   Network with two switches: (a) Schematic. (b) Left switch closed, right open until $t = 1$. (c) Left switch open, right closed for $t > 1$.

If the voltage source is a unit step, then

$$V = \frac{1}{RC} \frac{1}{s + \dfrac{1}{RC}} \frac{1}{s} = \frac{1}{s} - \frac{1}{s + \dfrac{1}{RC}}$$

It is convenient to define the time constant

$$\tau_1 = RC \tag{10.4.2}$$

and rewrite Eq. (10.4.1) as

$$V = \frac{1}{s} - \frac{1}{s + \dfrac{1}{\tau_1}}$$

The time response is

$$v(t) = [1 - e^{-t/\tau_1}]u(t) \tag{10.4.3}$$

Suppose that the switches change their positions at $t = \tau_1$. At this instant, the voltage across the capacitor is

$$v(\tau_1) = E_0 = 0.6321 \tag{10.4.4}$$

equivalent to the initial voltage on the capacitor. Analyzing next the network in Fig. 10.4.2c we have

$$V = \frac{0.6321}{s + \dfrac{1}{\tau_2}} \tag{10.4.5}$$

where we have defined

$$\tau_2 = \frac{C}{G} \tag{10.4.6}$$

The time response is

$$v(t) = [0.6321e^{-t/\tau_2}]u(t) \tag{10.4.7}$$

and is measured from the instant when we switched from the first network to the second. If we select, for instance, $\tau_1 = \tau_2 = 1$, then the response will be as shown in Fig. 10.4.3.

The responses in Eq. (10.4.3) and Eq. (10.4.7) start at different times, the first one at $t = 0$ and the second at $t = \tau_1$. A verbal explanation is sufficient for understanding this, but we also need a formal way of expressing these facts mathematically. Multiplication of the first expression by $u(t)$ guarantees that the response is zero for $t < 0$, because $u(t)$ is zero for such times. The response in Eq. (10.4.7) is shifted by the amount $t = \tau_1$, which we indicate by writing $t - \tau_1$ instead

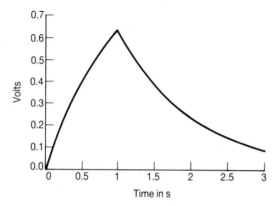

FIGURE 10.4.3   Step response of network Fig. 10.4.2.

of t. The overall response of the network in Fig. 10.4.2 is thus expressed by two equations

$$v(t) = u(t)(1 - e^{-t/\tau_1}) \qquad\qquad t \leq \tau_1$$
$$v(t) = u(t - \tau_1)0.6321e^{-(t-\tau_1)/\tau_2} \qquad t > \tau_1 \tag{10.4.8}$$

The next example is in Fig. 10.4.4a. We are interested in the voltage across the resistor. The switching is done by one toggle switch, first connected to the source, as shown in Fig. 10.4.4b. We must calculate both the current flowing through the inductor (to obtain the initial current for the inductor after switching), as well as the desired output voltage. Using mesh analysis

$$I = \frac{1}{sL + R}\frac{1}{s}$$

Introducing the time constant

$$\tau = \frac{L}{R} \tag{10.4.9}$$

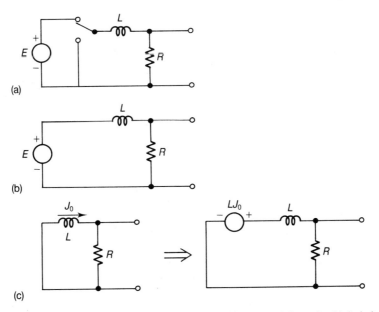

(a)

(b)

(c)

FIGURE 10.4.4   LR network with a toggle switch: (a) Schematic. (b) Switch in upper position until $t = 1$. (c) Switch in lower position, for $t > 1$.

we modify the expression to

$$I = \frac{1}{R}\left[\frac{1}{s} - \frac{1}{s + \frac{1}{\tau}}\right]$$

The current waveform is

$$i(t) = \frac{1}{R}(1 - e^{-t/\tau}) \qquad (10.4.10)$$

and the output voltage is

$$v(t) = i(t)R = \{1 - e^{-t/\tau}\}u(t) \qquad (10.4.11)$$

If we switch to the second position at the time $t = \tau$, the current flowing through the inductor will be

$$i(\tau) = J_0 = \frac{0.6321}{R} \qquad (10.4.12)$$

and we must analyze the network in Fig. 10.4.4c. We obtain

$$I = \frac{J_0}{s + \frac{1}{\tau}} = \frac{0.6321}{R} \frac{1}{s + \frac{1}{\tau}}$$

The output voltage is

$$v = RI = \frac{0.6321}{s + \frac{1}{\tau}} \qquad (10.4.13)$$

and the time response of the output is

$$v(t) = \{0.6321e^{-t/\tau}\}u(t) \qquad (10.4.14)$$

measured from the instant of switching. We have chosen the values such that the equations are exactly the same as for the first example; Fig. 10.4.3 and Eq. (10.4.8) apply for this case as well.

The two examples show that in switched networks we must calculate currents through all inductors, voltages across all capacitors, plus the desired output of the network. Only then do we have all the information needed for the analysis of the network after switching.

PROBLEMS CHAPTER 10

Use the program from Appendix B to obtain partial fraction decompositions for your derivations.

You can display responses of networks with initial conditions by using NETOPT. You can also use NETOPT and its symbolic function method, but do remember that only the first source in the netlist is taken into account and the value is automatically set equal to one. You can use the superposition principle to solve the network for each source separately, but you must edit the netlist for each run and multiply each result by the true value of the source. See Chapter 18 for additional information.

P.10.1 Analyze the network by first keeping all elements and the source as variables and obtain the voltage V_2. Afterwards substitute numerical values for elements and find the time responses for (a) the Dirac impulse and (b) for unit step input.

P.10.2 Do the same as in P.10.1.

P.10.3 Obtain the output voltage by keeping the source as a variable. Obtain time domain responses for input being (a) Dirac impulse, (b) unit step, (c) $\sin 2t$, and (d) $\cos 2t$.

P.10.4 Do the same as in P.10.3.

P.10.5 Do the same as in P.10.3.

P.10.6 Use mesh formulation to calculate the output voltage in terms of E, then let the input be the Dirac impulse, the unit step and $\sin 2t$.

P.10.7 Use mesh formulation and the voltage-source equivalent for the initial condition. Calculate the output voltage for the source being (a) Dirac impulse, (b) unit step, (c) sin 2t. Also transform the voltage source into a current source, use the current source representation for the initial condition and do the same in nodal analysis.

P.10.8 Do the same as in P.10.7.

P.10.9 Find the voltage V_2 for the source being (a) Dirac impulse and (b) unit step.

P.10.10 Do the same as in P.10.9.

P.10.11 Do the same as in P.10.9.

P.10.12 Do the same as in P.10.9.

P.10.13 Use mesh analysis to obtain the voltage across the output resistor. Use sin 3t for the input.

P.10.14 Do the same as in P.10.13.

P.10.15 Use nodal formulation to obtain the output response for (a) no source, (b) $j(t) = \delta(t)$, (c) $j(t) = u(t)$.

P.10.19 Find the output response for unit step. Mesh analysis is advantageous in this network.

P.10.16 Find the output voltage using mesh formulation. Let the source be (a) no source, (b) $e(t) = \delta(t)$.

P.10.20 Find the output voltage for (a) no source, (b) $e(t) = \delta(t)$. It is advantageous to use nodal formulation.

P.10.17 Use nodal formulation and calculate the output voltage for (a) no input, (b) $j(t) = \delta(t)$.

P.10.21 Find the output voltage by mesh and nodal formulation. Use two cases: (a) no source, (b) $e(t) = u(t)$.

P.10.18 Derive the expression for the output voltage. This is a third-order network with a real root at $s = -1$. Find (a) the Dirac impulse and (b) unit step response.

P.10.22 Find the time domain response of this network at the output node.

P.10.23 Find the output voltage for (a) no source, (b) $e(t) = u(t)$. Proceed with both mesh and nodal formulations.

P.10.24 Find the output voltage for (a) no source, (b) $e(t) = \delta(t)$.

P.10.25 Do the same as in P.10.24.

P.10.26 Keep both E and E_0 as variables and find the output voltage. Afterwards solve for (a) $E = 0$ and $E_0 = 2$, (b) $E = 1/s$ and $E_0 = 0$, (c) $E = 1/s$ and $E_0 = 2$. Note that nodal formulation is advantageous in this case.

P.10.27 The dc current source has been in position 1 until the steady state has been reached. Use this information to find initial conditions. Find the output voltage response when the switch moves to position 2.

P.10.28 The switch is in position 1 and unit step is applied at the input. Find the response at the output. At $t = 1$ the switch changes its position. Find the initial conditions and the response at the output for $t > 1$.

P.10.29 Do the same as in P.10.28.

P.10.30 Do the same as in P.10.28.

P.10.31 Do the same as in P.10.28.

P.10.32 Use NETOPT to display the response of the network to a Dirac impulse and to a unit step.

P.10.33 Do the same as in P.10.32.

CHAPTER 11

Network Functions

11.0. INTRODUCTION

The concept of network functions is one of the most practical ones in the theory of linear networks. Network functions are expressed in terms of the Laplace domain variable and contain all information about the investigated output of the network. It is possible to go from there to the frequency or time domain. In addition, we can establish stability properties and also derive some other fundamental rules, provided we know the decomposition of the network function into its poles and zeros.

While NETOPT can provide all the functions from the network definition and can plot the responses, it is always beneficial to have the possibility of studying responses from given network functions or from the positions of poles and zeros. Appendix B describes how to use a program which can start either from the network function or from given poles and zeros. The program calculates frequency responses (amplitude in magnitude or dB, phase, group delay) or time responses for a Dirac impulse, unit step, sine and cosine. It also plots the responses on the screen if the computer has the appropriate hardware.

11.1. DEFINITION OF NETWORK FUNCTIONS

Network functions are usually considered in cases where the network has one independent source, E or J, and one output. The functions are obtained in the Laplace domain and initial conditions are ignored.

In order to introduce the concept, we take the simplest possible network, an R–L voltage divider shown in Fig. 11.1.1. The output voltage is equal to

$$V_{out} = \frac{2}{s+2} E$$

where E can be any input signal. If we do not take a specific signal but rather divide the equation by E, we get a transfer function

$$T = \frac{V_{out}}{E} = \frac{2}{s+2}$$

This ratio is one of the possible network functions; specifically, for this example, it is a voltage transfer function.

Had we chosen a larger network, the output would be in the form

$$V_{out} = \frac{N}{D} E \qquad (11.1.1)$$

where N and D are polynomials in the variable s. Dividing by E we get the network function

$$T = \frac{V_{out}}{E} = \frac{N}{D} \qquad (11.1.2)$$

One network can have many network functions, depending on where we apply the independent source and where we take the output. In order to be able to define the function, we must satisfy two conditions:

1. The network must be linear.
2. There are zero initial conditions on capacitors and inductors.

If these conditions are satisfied, then we define the network function as the ratio

$$Network\ function = \frac{output}{input} \qquad (11.1.3)$$

FIGURE 11.1.1   LR voltage divider.

The output variable, V or I, is *always* in the numerator while the input variable, E or J, is in the denominator, similarly as in Eq. (11.1.2).

Depending on the type of the input source (voltage or current) and the type of the output (voltage or current), we can define the following network functions:

1. Voltage transfer functions:

$$T_V = \frac{V_{out}}{E} \tag{11.1.4}$$

The source is an independent voltage source, the output is either the voltage of one node or the difference of two nodal voltages. The voltage transfer function is a dimensionless quantity.

2. Current transfer function:

$$T_I = \frac{I_{out}}{J} \tag{11.1.5}$$

The source is an independent current source, the output is the current through any element. The current transfer function is a dimensionless quantity.

3. Transfer impedance:

$$Z_{TR} = \frac{V_{out}}{J} \tag{11.1.6}$$

The source is an independent current source, the output is the voltage of any node or the difference of two nodal voltages. The transfer impedance has the dimension of an impedance.

4. Transfer admittance:

$$Y_{TR} = \frac{I_{out}}{E} \tag{11.1.7}$$

The source is an independent voltage source, the output is the current through any element. Transfer admittance has the dimension of an admittance.

5. Input impedance:

$$Z_{in} = \frac{V_{in}}{J} \tag{11.1.8}$$

The output variable is the voltage across the source J. The input impedance is an impedance.

6. Input admittance:

$$Y_{in} = \frac{I_{in}}{E}$$
(11.1.9)

The output variable is the current flowing through the voltage source. The input admittance is an admittance.

The input impedance and input admittance are *always* related by the expression

$$Z_{in} = \frac{1}{Y_{in}}$$
(11.1.10)

However, such relation *does not* apply for the transfer impedance and transfer admittance. The reader should remember that

$$Z_{TR} \ \ IS \ \ NOT \ \ \frac{1}{Y_{TR}}$$

because they use different variables, see Eq. (11.1.6) and Eq. (11.1.7).

Let us now look again at the definition of the network function and give it still another explanation. Suppose that instead of dividing Eq. (11.1.1) by E we leave the source on the right but substitute instead $E = 1$. The result will be the same. However, we learned that the Laplace transform of the Dirac impulse is equal to 1 and we can thus say that the network function describes simultaneously the response of the network to the Dirac impulse. Once we have the network function, we can always get the output by multiplying the network function by E or J and use it for frequency domain or time domain analysis. We have covered a similar subject already in Chapter 7, only now we give it the above additional meaning. Several examples will provide some additional practice.

Example 1.

Find Z_{TR} and T_V for the network in Fig. 11.1.2.

The nodal equations are

$$(G_1 + sC_1)V_1 - sC_1 V_2 = J$$
$$-(sC_1 + g)V_1 + (sC_1 + sC_2 + G_2)V_2 = 0$$

Although it should already be clear, the dependent source transconductance must stay on the left and the independent source on the right. In matrix form

$$\begin{bmatrix} (G_1 + sC_1) & -sC_1 \\ -(g + sC_1) & (sC_1 + sC_2 + G_2) \end{bmatrix} \begin{bmatrix} V_1 \\ V_2 \end{bmatrix} = \begin{bmatrix} J \\ 0 \end{bmatrix}$$

FIGURE 11.1.2   Network with a CC.

The determinant of the system is

$$D = s^2 C_1 C_2 + s(C_1 G_1 + C_1 G_2 + C_2 G_1 - C_1 g) + G_1 G_2$$

Using Cramer's rule we evaluate V_2 and also define $I_{out} = sC_2 V_2$. The transfer functions are

$$Z_{TR} = \frac{V_2}{J} = \frac{sC_1 + g}{D}$$

$$T_I = \frac{I_{out}}{J} = \frac{s^2 C_1 C_2 + sC_2 g}{D}$$

At this point we make a very important observation: All network functions of any given network have the same denominator. This clearly follows from the solution by Cramer's rule where the determinant of the matrix turns out to be the denominator of every network function.

Example 2.

Find V_{out} and I_{out} for the network in Fig. 11.1.3.

This example will show that the numerator and denominator may be functions of both s and $1/s$. Although in this case it would be easier to transform the voltage source into a current source and apply nodal formulation, we will use the mesh method. The system matrix is

$$\begin{bmatrix} \left(R_1 + \dfrac{1}{sC_1}\right) & -\dfrac{1}{sC_1} \\ -\dfrac{1}{sC_1} & \left(R_2 + \dfrac{1}{sC_1} + \dfrac{1}{sC_2}\right) \end{bmatrix} \begin{bmatrix} I_1 \\ I_2 \end{bmatrix} = \begin{bmatrix} E \\ 0 \end{bmatrix}$$

FIGURE 11.1.3   Network with two capacitors.

The determinant of the matrix is

$$D = \frac{1}{s^2 C_1 C_2} + \frac{1}{s}\left(\frac{R_1}{C_1} + \frac{R_1}{C_2} + \frac{R_2}{C_1}\right) + R_1 R_1$$

$$= \frac{s^2 C_1 C_2 R_1 R_2 + s(R_1 C_1 + R_1 C_2 + R_2 C_2) + 1}{s^2 C_1 C_2}$$

The output current is $I_{out} = I_2$ and the output voltage is $V_{out} = I_2/sC_2$. Applying Cramer's rule for I_2 we obtain

$$I_2 = I_{out} = \frac{\dfrac{E}{sC_1}}{D}$$

The output current is

$$I_{out} = I_2 = \frac{sC_2 E}{s^2 C_1 C_2 R_1 R_2 + s(R_1 C_1 + R_1 C_2 + R_2 C_2) + 1}$$

and the network functions are

$$Y_{TR} = \frac{I_{out}}{E} = \frac{I_2}{E} = \frac{sC_2}{s^2 C_1 C_2 R_1 R_2 + s(R_1 C_1 + R_1 C_2 + R_2 C_2) + 1}$$

$$T_V = \frac{V_{out}}{E} = \frac{I_2}{sC_2} = \frac{1}{s^2 C_1 C_2 R_1 R_2 + s(R_1 C_1 + R_1 C_2 + R_2 C_2) + 1}$$

Example 3.

For the network in Fig. 11.1.4 find the indicated output voltage and current.
    The nodal matrix equation can be written by inspection:

$$\begin{bmatrix} (G_1 + G_2 + sC) & -G_2 \\ -G_2 & \left(G_2 + \dfrac{1}{sL}\right) \end{bmatrix} \begin{bmatrix} V_1 \\ V_2 \end{bmatrix} = \begin{bmatrix} J \\ 0 \end{bmatrix}$$

FIGURE 11.1.4   GLC network.

The denominator of the transfer functions is

$$D = sCG_2 + \left(G_1G_2 + \frac{C}{L}\right) + \frac{G_1 + G_2}{sL}$$

Using Cramer's rule, we obtain

$$V_{out} = \frac{G_2J}{sCG_2 + \left(G_1G_2 + \dfrac{C}{L}\right) + \dfrac{G_1 + G_2}{sL}}$$

and since $I_{out} = V_{out}/sL$, we also get

$$I_{out} = \frac{\dfrac{G_2}{sL}}{sCG_2 + \left(G_1G_2 + \dfrac{C}{L}\right) + \dfrac{G_1 + G_2}{sL}} J$$

Similarly as we have done before, we remove the fractions by multiplying the numerator and denominator by sL. The results are

$$Z_{TR} = \frac{sLG_2}{s^2LCG_2 + s(LG_1G_2 + C) + G_1 + G_2}$$

$$T_I = \frac{G_2}{s^2LCG_2 + s(LG_1G_2 + C) + G_1 + G_2}$$

Example 4.

Find the transfer function T_V for the network in Fig. 11.1.5. Use mesh analysis and numerical values of the elements.

Since the direction of the driving current of the CV is the same as the direction of I_1,

FIGURE 11.1.5   Network with a CV.

it must be true that

$$I_{R2} = I_1 - I_2$$

The mesh equations are

$$(3 + s)I_1 - (2 + s)I_2 = E$$

$$-(2 + s)I_1 + (2 + 3s)I_2 + r(I_1 - I_2) = 0$$

Here we used the letter r of the CV to clarify the equation. Substituting $r = 1$ and writing in matrix form

$$\begin{bmatrix} (3 + s) & -(2 + s) \\ -(1 + s) & (1 + 3s) \end{bmatrix} \begin{bmatrix} I_1 \\ I_2 \end{bmatrix} = \begin{bmatrix} E \\ 0 \end{bmatrix}$$

The determinant of the system is

$$D = 2s^2 + 7s + 1$$

Since in this example the output voltage is given by $V_{out} = r(I_1 - I_2) = I_1 - I_2$, we must find both I_1 and I_2:

$$I_1 = \frac{1 + 3s}{D} E$$

$$I_2 = \frac{1 + s}{D} E$$

Then

$$V_{out} = \frac{2s}{D} E$$

and the transfer function is

$$T_V = \frac{V_{out}}{E} = \frac{2s}{2s^2 + 7s + 1}$$

Let us summarize what we have learned in this section.

1. The network function is defined only for linear networks with zero initial conditions.
2. To obtain the network function, connect the required source at the input, but specify it only as E or J.
3. Solve the network to find the desired output and divide by E or J to obtain the network function. The input variable is always in the denominator.

4. Should the numerator and/or denominator have fractions of the type $1/s$, remove them by simultaneously multiplying D and N by an appropriate factor.
5. The network function can be used to obtain frequency domain response: replace the source symbol by the source phasor and set $s = j\omega$ where ω is the angular frequency of the signal.
6. The network function can be used to obtain the time domain response: first multiply the equation by the source symbol, replace the source symbol by the Laplace transform of the actual input signal, find the poles and get the partial fraction expansion. For the inverse we use Table 9.2.1.

You can obtain all the above network functions by running the symbolic option in NETOPT, see Chapter 18.

11.2. POLES, ZEROS AND STABILITY

In the first section we introduced the concept of the network function. In all cases, the final results were rational functions in s. If the elements were kept as variables, then each power of s was accompanied by some combination of the elements. When the network elements were given numerically, then the final network function was a ratio of two polynomials in s with constant coefficients. Roots of the denominator are the *poles*, roots of the numerator are the *zeros*. We also discovered that although a network may have many network functions, *all* will have the same denominator. The numerator depends on the type of the input (E or J), on the place where the source is applied and where we take the output.

From this discussion we see that the denominator expresses some fundamental properties of the network whereas the numerator is related to a particular input and output. A network function will have the general form

$$F(s) = \frac{N(s)}{D(s)} = \frac{A_M s^M + A_{M-1} s^{M-1} + \cdots + A_1 s^1 + A_0}{B_N s^N + B_{N-1} s^{N-1} + \cdots + B_1 s^1 + B_0} = \frac{\sum\limits_{i=0}^{M} A_i s^i}{\sum\limits_{i=0}^{N} B_i s^i} \qquad (11.2.1)$$

Since it is known from mathematics that a polynomial of order n has exactly n roots, we can also rewrite the network function in the form

$$F(s) = K_0 \frac{(s - z_1)(s - z_2)\cdots(s - z_M)}{(s - p_1)(s - p_2)\cdots(s - p_N)} = K_0 \frac{\prod\limits_{i=1}^{M} (s - z_i)}{\prod\limits_{i=1}^{N} (s - p_i)} \qquad (11.2.2)$$

where

$$K_0 = \frac{A_M}{B_N} \tag{11.2.3}$$

is equal to the ratio of the coefficients at the highest powers of s. Most of the poles and zeros of practical networks are complex,

$$z_i = \alpha_i + j\beta_i$$
$$p_i = \gamma_i + j\delta_i \tag{11.2.4}$$

with $j = \sqrt{-1}$.

The step from Eq. (11.2.1) to Eq. (11.2.2) may be difficult, but the two forms are equivalent, at least theoretically. If the coefficients of the polynomials are numbers, we can find the roots by numerical methods. If the coefficients are functions of the element variables, then the situation is much more difficult and we will remember the following *practical* advice:

1. The root of the first order polynomial $P(s) = as + b = 0$ is $s = -b/a$.
2. The roots of a second order polynomial $P(s) = as^2 + bs + c = 0$ are found by means of the formula

$$s_{1,2} = \frac{-b \pm \sqrt{b^2 - 4ac}}{2a}$$

The coefficients may be either numbers or element symbols.
3. For polynomials with constant coefficients of order 3 or more, use a root-finding computer program.
4. Do not try to find the roots of polynomials of order 3 or more if the coefficients are not numerically given.

In these few sentences are hidden many problems of network design: only very few practical networks can be fully solved by hand. The rest must use special computer methods which are the subject of advanced courses[1] and which are used in NETOPT.

The positions of the zeros and poles can be plotted in the complex plane. Zeros are marked as small circles, poles as crosses. Thus the transfer function

$$F = 5 \frac{s^2 + 4}{s^3 + 4s^2 + 6s + 4} = 5 \frac{(s + j2)(s - j2)}{(s + 2)(s + 1 + j1)(s + 1 - j1)} \tag{11.2.5}$$

[1] J. Vlach and K. Singhal: *Computer Methods for Circuit Analysis and Design* Van Nostrand Reinhold 1983.

FIGURE 11.2.1   Pole-zero plot.

has the plot shown in Fig. 11.2.1. The multiplicative constant 5 cannot be determined from the pole-zero plot. If the polynomial of the numerator or denominator has a complex root, then there exists *always* another, complex conjugate root, having the same real part but opposite sign of the imaginary part.

Suppose that instead of the function Eq. (11.2.5) we have another function with multiple poles, for instance

$$F = 5 \frac{s^2 + 4}{s^4 + 6s^3 + 14s^2 + 16s + 8} = 5 \frac{(s + j2)(s - j2)}{(s + 2)^2(s + 1 + j1)(s + 1 - j1)} \quad (11.2.6)$$

The pole $s = -2$ has multiplicity 2 but the positions of the poles are the same as in the previous example. To indicate that the pole has multiplicity n, we write a small n next to this pole. The same will apply to multiple zeros.

The concept of poles and zeros can also be related to matrix algebra. Since the denominator is found as the determinant of the system matrix, and the pole is its root, then any value of s which makes the determinant equal to zero is the *pole*. The numerator polynomial is obtained by Cramer's rule by replacing one column of the matrix by the right-hand side. Since the numerator polynomial is found as the determinant of this modified matrix, then any value of s which makes the determinant of the modified matrix zero is also the *zero* of the transfer function.

Poles and zeros are mathematical abstractions, valid only for linear networks. They cannot be easily related to elements or how the network is constructed. The poles are clearly more fundamental than the zeros because all network functions of any given network will have the same poles. For this reason, they are sometimes called *natural frequencies* of the network. The zeros depend on the place where the source is connected to the network and on the place where we take the output. If the zeros are purely imaginary (lying on the vertical axis), then they do have a clear practical meaning: they completely suppress a signal occuring at their frequency.

Importance of the poles can be best understood by considering the Dirac impulse as the input. This signal has finite area and thus delivers a finite amount of energy to the network.

We will say that the network is stable if its response to the Dirac impulse, a signal with finite energy, does not grow for large t. Should it grow, then we will say that the network is unstable. Networks composed of passive components like R, L, C cannot be unstable but stability problems can occur in networks with sources of energy, like dependent sources or amplifiers.

Assume that we know the poles. If all poles are simple, then the network function can be decomposed into partial fractions

$$F(s) = \sum_{i=1}^{N} \frac{K_i}{s - p_i} \tag{11.2.7}$$

and the Laplace transform inverse is

$$f(t) = \sum_{k=1}^{N} K_i e^{p_i t} \tag{11.2.8}$$

Consider now just one term in Eq. (11.2.7) and $s = p$. If the pole is negative real, then it lies on the negative real axis, to the left of the origin. The time response of such term is an exponential, $f(t) = e^{-pt}$, and decreases with time. It will become practically zero for large t. This is a *stable* situation. On the other hand, if the pole lies on the positive real axis, the response will be $f(t) = e^{pt}$, an exponential which grows very rapidly; we say that such term is unstable.

For a complex conjugate pair the exact response depends on the residue, but the type of the function will be $f(t) = e^{\gamma t} \cos(\delta t + \phi)$. If the real part of the pole, γ, is negative, we will have a decaying oscillatory behavior which will become practically zero for large t. If the real part is positive, the oscillations will grow without bound. We can also have a special case when the real part of the pole is zero. In such case we will get oscillations which neither grow nor decay. Such situation can occur either if there are no resistors in the network or if all the losses in the resistors are somehow restored from other sources of energy (battery).

If the poles are multiple, the functions become more complicated but the real part must be negative in order to get a response which decreases with time. A multiple pole on the imaginary axis will always lead to an unstable behavior.

We can now summarize:

1. A stable network must have all poles (simple or multiple) in the left half plane.
2. Only simple poles may lie on the imaginary axis; in such a case the network will oscillate with constant amplitude.

The positions of the zeros did not come into our stability considerations. They do not influence stability but they do influence amplitudes of the exponentials and phase shifts of the sinusoids.

Example 1.

Find the poles and zeros of the function

$$T = \frac{2s}{2s^2 + 7s + 1}$$

and determine whether the function is stable or not.
 The poles are the roots of the denominator,

$$2s^2 + 7s + 1 = 0$$

Using the formula for a quadratic polynomial we find that the roots are $p_1 = -0.14922$ and $p_2 = -3.3508$. They are real, in the left half plane and thus the network function is stable. The function can be rewritten as

$$T = \frac{2s}{2(s + 0.14922)(s + 3.3508)}$$

Note that we must not forget the coefficient 2 which multiplied the highest power of s in the denominator.

Example 2.

The network in Fig. 11.2.2a has all elements given by their values, except for the voltage controlled voltage source which has gain A. Find the value of A for which the network will become unstable.
 The network is redrawn in Fig. 11.2.2b with transformed sources so that we can use

(a)

(b) $V_{out} = AV_2$

FIGURE 11.2.2   Network with a VV: (a) Original. (b) Transformed for nodal formulation.

nodal analysis. The equations are

$$(2 + s)V_1 - V_2 = E + AsV_2$$

$$-V_1 + (1 + s)V_2 = 0$$

Transferring the term with V_2 to the left and leaving E on the right leads to the following matrix equation

$$\begin{bmatrix} 2 + s & -(1 + sA) \\ -1 & 1 + s \end{bmatrix}\begin{bmatrix} V_1 \\ V_2 \end{bmatrix} = \begin{bmatrix} E \\ 0 \end{bmatrix}$$

The determinant of the system matrix is $D = s^2 + s(3 - A) + 1$. Normally, to find poles, we set the polynomial equal to zero and find the roots. In this problem we only have to determine the value of A for which the poles will move on the imaginary axis. This will happen when the middle term of the polynomial becomes zero, $A = 3$. Then the equation reduces to

$$s^2 + 1 = 0$$

and the poles are $p_{1,2} = \pm j1$. The network will be stable for $A < 3$ and unstable for $A \geq 3$.

11.3. FREQUENCY RESPONSES FROM POLES AND ZEROS

If the network function poles and zeros are known, it is very easy to evaluate various frequency domain responses. We will derive the necessary expressions.
The network function in terms of poles and zeros has the form

$$F = K_0 \frac{(s - z_1)(s - z_2)\cdots(s - z_M)}{(s - p_1)(p - p_2)\cdots(s - p_N)} \tag{11.3.1}$$

where

$$z_i = \alpha_i + j\beta_i$$
$$p_i = \gamma_i + j\delta_i \tag{11.3.2}$$

For evaluation in the frequency domain we have always substituted $s = j\omega$, which is a point on the imaginary axis. A zero, $\alpha + j\beta$, is also a point in the complex plane and both points are shown in Fig. 11.3.1. Considering one term in Eq. (11.3.1), we must form the difference

$$(s - z) = (0 + j\omega) - (\alpha + j\beta) \tag{11.3.3}$$

To subtract two vectors graphically we first copy the vector to be subtracted in the other direction (point $-\alpha - j\beta$) and then form a parallelogram with the unchanged vector (point $0, j\omega$), see Fig. 11.3.1. Note that the vector of the difference

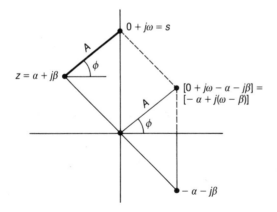

FIGURE 11.3.1   Subtraction of two vectors.

and its angle is replicated by connecting the point $z = \alpha + j\beta$ with the point $j\omega$ directly (thick line). The absolute value

$$A = \sqrt{\alpha^2 + (\omega - \beta)^2} \tag{11.3.4}$$

represents the distance of the point z from the point $s = j\omega$. We can also calculate the angle ϕ

$$\phi = \arctan \frac{\omega - \beta}{-\alpha} \tag{11.3.5}$$

Thus for one term in Eq. (11.3.1) we can write the equivalence

$$s - z = A e^{j\phi} \tag{11.3.6}$$

The same steps apply for the denominator terms $s - p$. For the amplitude response, the absolute values from zeros will be in the numerator and absolute values from the poles in the denominator. The formula becomes

$$|F(j\omega)| = K_0 \frac{A_{z,1} A_{z,2} \cdots A_{z,M}}{A_{p,1} A_{p,2} \cdots A_{p,N}} \tag{11.3.7}$$

The angles appear as exponents of e. The exponents are added separately in the numerator and in the denominator, but those in the denominator can be brought to the numerator and by changing their signs. Thus the overall phase will be

$$\Phi = (\phi_{z,1} + \phi_{z,2} + \cdots \phi_{z,M}) - (\phi_{p,1} + \phi_{p,2} + \cdots \phi_{z,N}) \tag{11.3.8}$$

Inserting from Eq. (11.3.4) into Eq. (11.3.7) we get

$$|F| = K_0 \left[\frac{\prod\limits_{i=1}^{M} [\alpha_i^2 + (\omega - \beta_i)^2]}{\prod\limits_{i=1}^{N} [\gamma_i^2 + (\omega - \delta_i)^2]} \right]^{1/2} \tag{11.3.9}$$

This formula gives the *amplitude response*. Similarly for the phases

$$\Phi = \sum_{i=1}^{M} \arctan \frac{\beta_i - \omega}{\alpha_i} - \sum_{i=1}^{N} \arctan \frac{\delta_i - \omega}{\gamma_i} \tag{11.3.10}$$

This expression is the *phase response*.

We will now study the formulas to get additional insight. Figure 11.3.2 shows a pole–zero plot with three poles and two zeros. The poles are to the left of the imaginary axis, the zeros are on the imaginary axis. Suppose that we apply to such network a sinusoidal signal having the amplitude $A = 1$ and frequency ω. This is equivalent to setting $s = j\omega$ and plotting the point in Fig. 11.3.2 (marked by the arrow). Each term of the product in Eq. (11.3.7) is the distance of the pole (or zero) from the point $j\omega$. The formula for the amplitude response says that we must take the product of the distances from the zeros and divide them by the product of the distances from the poles. This gives a value of the amplitude response for the selected frequency. If we wish to do the evaluation for another frequency, all we have to do is shift the dot on the imaginary axis to another position on the imaginary axis, measure the new distances of the poles and zeros to this point and form the ratio again. Doing so for a number of frequencies can result in a quick estimate of the amplitude response of a network described by the given pole–zero plot. To be exact, we should still multiply all such values by K_0, but in such crude plots we are more interested in the shape of the amplitude curve than in exact values and we simply disregard this constant.

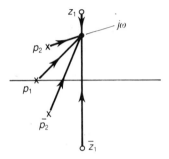

FIGURE 11.3.2   Amplitude response from pole-zero plot.

The phase characteristic, Φ, can be found similarly by adding phase angles from the zeros and subtracting phase angles from the poles, as indicated by Eq. (11.3.10). This is rarely done graphically because of the difficulties with measuring angles.

It is interesting to add the following facts without deriving them. If we want to transfer, without distortion, a signal composed of many sinusoidal components, then we need a network whose amplitude response is constant for all frequencies. It is relatively easy to see how the amplitude response deviates from the desired constant response. Simultaneously, the phase response must be a straight line passing through the origin. It is, unfortunately, much more difficult to interpret the significance of the deviations in the phase response. Because of these difficulties with the phase response, another, much better measure is often used. It is called the *group delay*, τ, and is defined as the derivative of the phase characteristic with respect to ω.

$$\tau(\omega) = -\frac{d\Phi}{d\omega} \tag{11.3.11}$$

The minus sign in front of the expression makes the group delay positive for most pole-zero configurations. Differentiating Eq. (11.3.10) with respect to ω we get

$$\tau(\omega) = \sum_{i=1}^{M} \frac{\alpha_i}{\alpha_i^2 + (\omega - \beta_i)^2} - \sum_{i=1}^{N} \frac{\gamma_i}{\gamma_i^2 + (\omega - \delta_i)^2} \tag{11.3.12}$$

This response is easy to estimate from the pole–zero plot once we realize that the denominators in Eq. (11.3.12) are the squares of the distances we already obtained when we evaluated the amplitude response. The numerators are the real parts of the pole or zero coordinates, with their signs taken into consideration. Thus we can draw the group delay plot simultaneously with the amplitude plot without operating with angles.

What should the group delay response be in order not to distort signals? We stated that the ideal phase response should be a straight line passing through the origin. Since the group delay is the derivative of the phase and the derivative of a straight line is a constant, we conclude that for transfer without distortion the group delay should be constant for all frequencies of interest.

Formulas Eq. (11.3.9), Eq. (11.3.11) and Eq. (11.3.12) are very easy to program, but finding the poles and zeros for a larger network is next to impossible without a computer. The program described in Appendix B plots the responses either from known poles and zeros, or from the network function. If the network function is given, it finds the poles and zeros, provides residues for time domain responses and gives you the choice to study the above three responses and various time responses as well.

Example 1.

Figure 11.3.3 shows a third-order filter called the Butterworth filter. Derive the transfer impedance for this filter. You will get in the denominator a polynomial of third-order which happens to have a real role at $s = -1$. Divide the denominator by the term $s + 1$ to lower the degree and find the remaining two poles. Make an estimate of the amplitude and group delay response. Using nodal formulation we get the system equation

$$\begin{bmatrix} 1 + s + \dfrac{1}{2s} & -\dfrac{1}{2s} \\[2ex] -\dfrac{1}{2s} & 1 + s + \dfrac{1}{2s} \end{bmatrix} \begin{bmatrix} V_1 \\ V_2 \end{bmatrix} = \begin{bmatrix} J \\ 0 \end{bmatrix}$$

The determinant (after a few steps) is

$$D = \frac{s^3 + 2s^2 + 2s + 1}{s}$$

and the numerator is

$$N_2 = \frac{J}{2s}$$

The network function is

$$Z_{TR} = \frac{V_{out}}{J} = \frac{1}{2(s^3 + 2s^2 + 2s + 1)}$$

Since we were given the real pole, we divide the denominator by the term $(s + 1)$. The result is the polynomial $s^2 + s + 1$ which has the roots

$$s_{1,2} = \frac{-1 \pm \sqrt{2}}{2}$$

Use the poles to sketch the amplitude and the group delay for a few frequencies.

FIGURE 11.3.3   Third-order Butterworth filter.

THIRD-ORDER BUTTERWORTH FILTER
AC ANALYSIS
MAGNITUDE OF VOUT

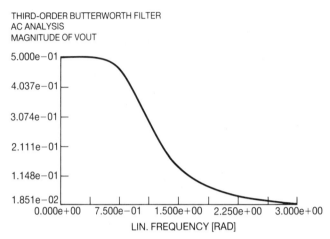

FIGURE 11.3.4   Amplitude response of filter in Fig. 11.3.3.

THIRD-ORDER BUTTERWORTH FILTER
AC ANALYSIS
DELAY [MS] OF VOUT

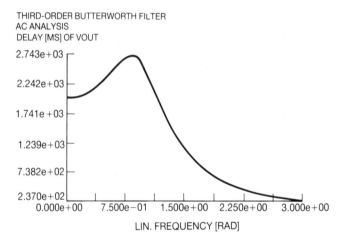

FIGURE 11.3.5   Group delay of the network in Fig. 11.3.3.

We have used NETOPT to find the amplitude response, in Fig. 11.3.4, and the group delay, in Fig. 11.3.5. You can also use the program from Appendix B by giving the network function Z_{TR}.

11.4. BODE PLOTS

Evaluation of the frequency domain responses, explained in the last section, is not very difficult if we use a calculator, and is quite easy when using a computer. In the not very distant past, when these devices were not available, evaluation was quite difficult. About fifty years ago, an American scientist, Bode, came up with a very simple and efficient idea how to plot approximate amplitude and phase responses. This section will explain the theory and its application.

We will use for our explanations a function which has three terms in the denominator,

$$F = K_0 \frac{1}{s(s+c)\left(s^2 + \dfrac{\omega_0}{Q} s + \omega_0^2\right)} \tag{11.4.1}$$

The three terms represent all possibilities, except that similar terms can also appear in the numerator. It is convenient to normalize the terms by factoring out the constant terms

$$F = \frac{K_0}{c\omega_0^2} \frac{1}{s\left(\dfrac{s}{c}+1\right)\left(\dfrac{s^2}{\omega_0^2} + \dfrac{s}{\omega_0 Q} + 1\right)} = \frac{K_0}{c\omega_0^2} F_0 F_1 F_2 \tag{11.4.2}$$

We will substitute $s = j\omega$ and consider separately each of the three functions:

$$F_0 = \frac{1}{j\omega} = \frac{1}{\omega} e^{-j90°} = |F_0| e^{j\phi_0} \tag{11.4.3}$$

$$F_1 = \frac{1}{\dfrac{j\omega}{c}+1} = \left[\dfrac{\omega^2}{c^2} + 1\right]^{-1/2} e^{j\phi_1} = |F_1| e^{j\phi_1} \tag{11.4.4}$$

where

$$\phi_1 = -\arctan \frac{\omega}{c} \tag{11.4.5}$$

$$F_2 = \frac{1}{-\dfrac{\omega^2}{\omega_0^2} + \dfrac{j\omega}{\omega_0 Q} + 1} = \left[\left(1 - \dfrac{\omega^2}{\omega_0^2}\right)^2 + \dfrac{\omega^2}{\omega_0^2 Q^2}\right]^{-1/2} e^{j\phi_2} = |F_2| e^{j\phi_2} \tag{11.4.6}$$

where

$$\phi_2 = \arctan \frac{\dfrac{\omega}{\omega_0 Q}}{1 - \dfrac{\omega^2}{\omega_0^2}} \tag{11.4.7}$$

Using these expressions in Eq. (14.4.2) we can write

$$|F|e^{j\Phi} = \frac{K_0}{c\omega_0^2}|F_0||F_1||F_2|e^{j(\phi_0 + \phi_1 + \phi_2)} \tag{11.4.8}$$

and have for the absolute value

$$|F| = \frac{K_0}{c\omega_0^2}|F_0||F_1||F_2| \tag{11.4.9}$$

and for the phase

$$\Phi = \phi_0 + \phi_1 + \phi_2 \tag{11.4.10}$$

So far all the operations were straightforward, but Eq. (11.4.9) asks for the product of three functions, an operation which is not easy to do graphically. Additions or subtractions are clearly preferable and can be achieved by taking the logarithm of Eq. (11.4.9). It is an accepted practice to use a new measure, decibels, dB, by taking 20 times the decadic logarithm: if we have an absolute value, A, then its value in decibels is

$$|A|_{\mathrm{dB}} = 20 \log |A| \tag{11.4.11}$$

Note that we said absolute value: the number must be positive and must not be zero, because the logarithm is not defined for negative numbers and zero. Expressing (11.4.9) in dB gives

$$|F|_{\mathrm{dB}} = 20 \log \frac{K_0}{c\omega_0^2} + 20 \log |F_0| + 20 \log |F_1| + 20 \log |F_2|$$

For plotting of the function values in dB we use a linear vertical scale and for plotting of the frequency we use a logarithmic horizontal scale. As a consequence of this choice the responses of the three absolute values can be approximated by straight lines, as we will show. Consider first

$$|F_0|_{\mathrm{dB}} = 20 \log \frac{1}{\omega} = -20 \log \omega \tag{11.4.12}$$

and take three values:

$$\text{for } \omega = 0.1 \qquad |F_0|_{\mathrm{dB}} = +20 \text{ dB}$$
$$\text{for } \omega = 1.0 \qquad |F_0|_{\mathrm{dB}} = 0 \text{ dB}$$
$$\text{for } \omega = 10 \qquad |F_0|_{\mathrm{dB}} = -20 \text{ dB}$$

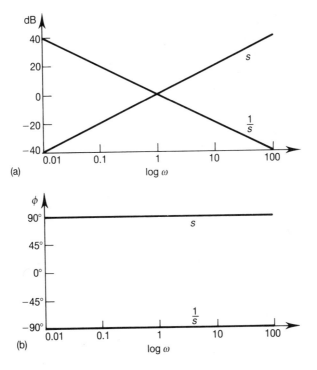

FIGURE 11.4.1   Bode plot for s and $1/s$: (a) Absolute value for $s = j\omega$. (b) Phase for $s = j\omega$.

This is plotted in Fig. 11.4.1a, the straight line marked with the function $1/s$. In the same figure is also the line corresponding to a zero at the origin, marked by the function s. The phase shift is $-90°$ for $1/s$ and $+90°$ for s, both in Fig. 11.4.1b.

The next function is

$$|F_1|_{dB} = 20 \log \left[\frac{\omega^2}{c^2} + 1 \right]^{-1/2} = -10 \log \left[\frac{\omega^2}{c^2} + 1 \right] \qquad (11.4.13)$$

and we take three values similarly as above:

$$\text{for } \frac{\omega}{c} \ll 1 \qquad |F_1|_{dB} = 0 \text{ dB}$$

$$\text{for } \frac{\omega}{c} = 1 \qquad |F_1|_{dB} = -10 \log 2 = -3.01 \text{ dB}$$

$$\text{for } \frac{\omega}{c} \gg 1 \qquad |F_1|_{dB} = -20 \log \frac{\omega}{c} \text{ dB}$$

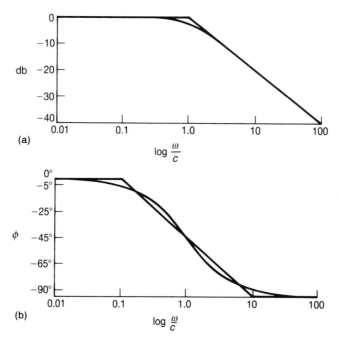

FIGURE 11.4.2   Detailed plot for one real pole: (a) Absolute value. (b) Phase.

Figure 11.4.2a is the plot of this function in the same semilogarithmic scale. The curve can be very nearly approximated by two straight lines, one horizontal at 0 dB, up to the frequency $\omega/c = 1$, and from there by another straight line which drops 20 dB per decade. This approximation commits an error of 3 dB at the frequency $\omega/c = 1$, as derived above, but is still a good approximation. The phase angle is plotted in Fig. 11.4.2b and in the semilogarithmic scale is very nearly replaced by three straight lines: the horizontal line on the left, with $\phi = 0$ up to $\omega/c = 0.1$; another horizontal line $\phi = 90°$ from $\omega/c = 10$; and a straight line connecting the end points. The straight-line approximations are summarized in Fig. 11.4.3a and b for the term in the denominator as well as in the numerator.

For the quadratic term we have

$$|F_2|_{dB} = 20 \log \left[\left(1 - \frac{\omega^2}{\omega_0^2} \right)^2 + \frac{\omega^2}{\omega_0^2 Q^2} \right]^{-1/2}$$

$$= -10 \log \left[\left(1 - \frac{\omega^2}{\omega_0^2} \right)^2 + \frac{\omega^2}{\omega_0^2 Q^2} \right] \qquad (11.4.14)$$

We can plot this equation for various values of Q but first select the same special

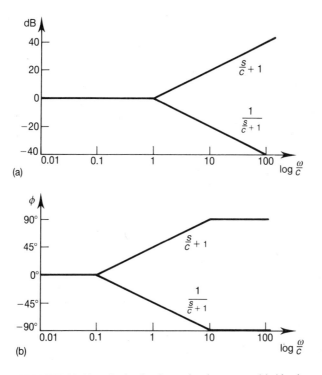

FIGURE 11.4.3   Bode plot for real pole or zero: (a) Absolute value. (b) Phase.

frequencies:

$$\text{for } \frac{\omega}{\omega_0} \ll 1 \qquad |F_2|_{dB} = 0 \text{ dB}$$

$$\text{for } \frac{\omega}{\omega_0} = 1 \qquad |F_2|_{dB} = -10 \log \frac{1}{Q^2} = +20 \log Q \text{ dB}$$

$$\text{for } \frac{\omega}{\omega_0} \gg 1 \qquad |F_2|_{dB} = -10 \log \frac{\omega^4}{\omega_0^4} = -40 \log \frac{\omega}{\omega_0} \text{ dB}$$

The function is plotted in Fig. 11.4.4a for various values of Q. The straight-line approximation is still reasonable for low values of Q. The horizontal line at 0 dB ends at $\omega/\omega_0 = 1$ and then continues as a straight line with a drop of 40 dB per decade. At the break point we are committing an error equal to $20 \log Q$. The phase response can be evaluated and plotted similarly, see Fig. 11.4.4b. For a quick estimate of the responses, we disregard the inaccuracies and use only the straight lines. This is summarized in Fig. 11.4.5a and b for both cases of the function being in the numerator or denominator.

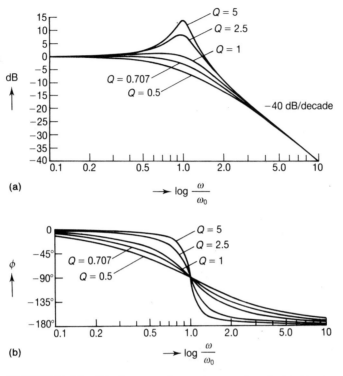

FIGURE 11.4.4   Plot for a complex conjugate pole pair: (a) Absolute value for several Q. (b) Phase for several Q.

In actual plots we do not use normalized forms, but make the breaks at the point c for terms of the form $s + c$ and at the point ω_0 for terms of the form $s^2 + (\omega_0/Q)s + \omega_0^2$. We must also take into account the modification of the accompanying constant by the factored terms. Let us clarify the above points by taking the function

$$F(s) = \frac{s + 0.5}{s(s + 5)} = 0.1 \frac{\dfrac{s}{0.5} + 1}{s\left(\dfrac{s}{5} + 1\right)} \tag{11.4.15}$$

The individual lines and the resulting response are in Fig. 11.4.6. For the pole $1/s$ we have a falling straight line passing through the point $\omega = 1$, with a drop of 20 dB/decade. For the numerator term $s + 0.5$ we have a horizontal line at 0 dB up to $\omega = 0.5$ and then a rising line with 20 dB/decade. For the denominator term $1/(s + 5)$ we have a horizontal line with 0 dB up to $\omega = 5$ and then a falling straight line with 20 dB/decade. Adding these lines we get the Bode plot for Eq.

(a)

$\longrightarrow \log \dfrac{\omega}{\omega_0}$

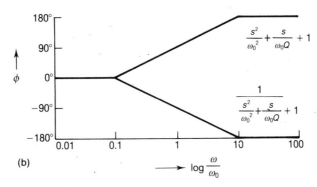

(b)

$\longrightarrow \log \dfrac{\omega}{\omega_0}$

FIGURE 11.4.5   Bode plot for complex conjugate poles or zeros: (a) Amplitude approximation. (b) Phase approximation.

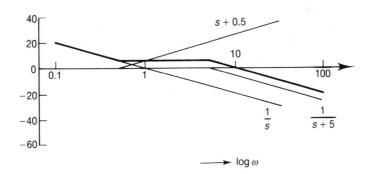

FIGURE 11.4.6   Bode plot for $F(s) = \dfrac{s + 0.5}{s(s + 5)}$.

Fig. 11.4.7   Fig. 11.4.8

FIGURE 11.4.7   Realization of $Z_{in} = \dfrac{s + 0.5}{s(s + 5)}$.

FIGURE 11.4.8   Amplitude response of Z_{in} for Fig. 11.4.7.

(11.4.15), but we must still consider the constant which had to be factored for the normalized functions: $0.5/5 = 0.1$ or -20 dB. Thus the plot should actually start at 0 dB.

It is interesting to compare this plot with an exact plot. It turns out that Eq. (14.4.15) can be realized as input impedance of the network in Fig. 11.4.7. Using a logarithmic horizontal scale in the same frequency range and using dB on the vertical scale, we get the plot in Fig. 11.4.8 (obtained by NETOPT). The plot starts at 0.0427 dB, which is practically the 0 dB obtained by the Bode plot. It drops to -22.97 dB at $\omega = 10$, again not very different from the expected -20 dB. The breaks in the lines do not exist but we can see a flattening portion in the middle of the response. The function in Eq. (11.4.15) can also be plotted using the program from Appendix B.

Example 1.

Draw the Bode plot for the function

$$F(s) = \frac{s(s + 0.5)}{(s + 1)(s + 2)(s + 5)(s + 10)}$$

The individual functions and the resulting plot are in Fig. 11.4.9. This plot starts at 0 dB for $\omega = 0.1$, but we must not forget the constant resulting from normalization, in this case

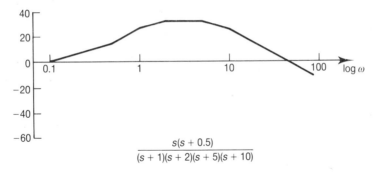

$$\frac{s(s+0.5)}{(s+1)(s+2)(s+5)(s+10)}$$

FIGURE 11.4.9   Bode plot for $F(s) = \dfrac{s(s+0.5)}{(s+1)(s+2)(s+5)(s+10)}$.

$0.5/1 \times 2 \times 5 \times 10 = 0.005$, or $20 \log 0.005 = -46.021\,$dB. The whole response must be pulled down by this amount to reflect a correct result. In many cases we do not need the absolute value, just the shape of the response, and in such case the plot in Fig. 11.4.9 is sufficient. Use the program from Appendix B to see the true response.

PROBLEMS CHAPTER 11

P.11.1 Display the poles and zeros of the following functions in the complex plane. Determine whether the functions are stable or unstable. Using the program from Appendix B, calculate the amplitude response in magnitude and in dB. Also display the phase and group delay responses.

(a) $F(s) - \dfrac{1}{s+1}$

(b) $F(s) = \dfrac{1}{(s+1)(s+2)}$

(c) $F(s) = \dfrac{s+1}{s+2}$

(d) $F(s) = \dfrac{s+3}{(s+2)(s+4)}$

(e) $F(s) = \dfrac{s+3}{s^2 - 3s + 2}$

(f) $S(s) = \dfrac{s-1}{(s+2)(s^2 - 5s + 3)}$

P.11.2 If a complex conjugate pair of poles is close to the imaginary axis, then there is a peak in the

amplitude response. The group delay also has a peak in the vicinity of such a pole. Make a sketch and run the program from Appendix B for the following pole pairs. Note that we are changing the real parts of the poles without changing the imaginary parts.

(a) $F_1(s) = \dfrac{1}{(s+1+j1)(s+1-j1)}$

(b) $F_2(s) = \dfrac{1}{(s+0.75+j1)(s+0.75-j1)}$

(c) $F_3(s) = \dfrac{1}{(s+0.5+j1)(s+0.5-j1)}$

(d) $F_3(s) = \dfrac{1}{(s+0.25+j1)(s+0.25-j1)}$

P.11.3 It is not possible to discover from frequency domain responses whether the function is stable or unstable. Make an estimate on the amplitude response of the following pairs of functions; the second is always an unstable function.

(a) $F_1(s) = \dfrac{1}{s+1}$ and $F_2(s) = \dfrac{1}{s-1}$

(b) $F_1(s) = \dfrac{s-1}{s+2}$ and $F_2(s) = \dfrac{s+1}{s-2}$

(c) $F_1(s) = \dfrac{s-1}{s+1}$ and $F_2(s) = \dfrac{s+1}{s-1}$

P.11.4 Plot the group delay responses of the functions in P.11.3. They will differ but one still cannot discover that some of the functions are unstable.

P.11.5 The so-called Butterworth filters have poles spread equally on a unit circle. The numerator is a constant equal to 1, the coordinates of the poles are

(a) $p_{1,2} = -0.70711 \pm j0.70711$
(b) $p_1 = -1$ and $p_{2,3} = -0.5 \pm j0.86603$
(c) $p_{1,2} = -0.38268 \pm j0.92388$ and
$$p_{3,4} = -0.92388 \pm 0.38268$$

Plot the poles and see how they are placed on the unit circle. Make an estimate on the amplitude and group delay responses. Use the program from Appendix B to plot the responses. Also observe unit step responses of these filters.

P.11.6 Another type of simple filters are the Chebychev filters. They have equal ripples in the magnitude or dB function, in the case given here equal to 0.5 dB. Let the numerator constant be equal to 1 and use the poles

(a) $p_{1,2} = -0.71281 \pm j1.00404$
(b) $p_1 = -0.62646$ and
$$p_{2,3} = -0.31323 \pm j1.02193$$
(c) $p_{1,2} = -1.7535 \pm j1.01625$ and
$$p_{3,4} = -0.42334 \pm j0.42095$$

Plot the poles and make an estimate on the amplitude and group delay responses. Use the program from Appendix B to plot the responses. Also observe unit step responses of these filters.

P.11.7 Make Bode plots for the functions in P.11.1.

P.11.8a–d. Networks which have only resistors and capacitors, or networks which have only resistors and inductors, have network functions with simple poles, all lying on the negative real axis. A pole or zero at the origin is possible. Show that this is true at least for the following networks. Find the input impedance, input admittance, transfer admittance and current transfer functions. Select

any other combinations and run the NETOPT program with its pole–zero option to show that this is true.

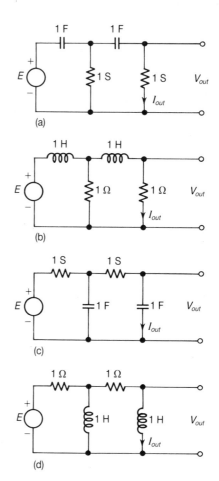

(a)

(b)

(c)

(d)

P.11.9a–b. Networks having resistors, inductors and capacitors can have complex conjugate poles or only real poles, depending on element values. They can never be unstable and all poles will have negative real parts. Select any RLC network without dependent sources and confirm, at least on your examples, that this is true. Use NETOPT and its pole–zero option. Networks with dependent sources can be unstable. Use problems from Chapter 7 and 8, select those with dependent sources and check whether they are stable or unstable. Use your derivations and confirm by NETOPT using its pole–zero option.

P.11.10 Select any pole positions with negative real parts, either real or in complex conjugate pairs. You will get a certain amplitude and group delay responses. If you add to such functions zeros which have exactly the same coordinates but the real parts are positive, you will get functions which are called all-pass. Their amplitude response will be a constant for all frequencies and group delay will be exactly twice as large as for the networks with poles only. Analyze the following functions, sketch their amplitude and group delay responses. Solve using the program from Appendix B.

(a) $F_1(s) = \dfrac{1}{s+1}$ and $F_2(s) = \dfrac{s-1}{s+1}$

(b) $F_1(s) = \dfrac{1}{s^2 + 2s + 2}$ and $F_2 = \dfrac{s^2 - 2s + 2}{s^2 + 2s + 2}$

(c) $F_1(s) = \dfrac{1}{(s+1)(s^2 + 2s + 2)}$ and

$$F_2(s) = \frac{(s-1)(s^2 - 2s + 2)}{(s+1)(s^2 + 2s + 2)}$$

CHAPTER 12

Active Networks

12.0. INTRODUCTION

In the low-frequency regions we can avoid the use of inductors by building RC-active networks. Although this book will not go into the details of the design of such networks, we will present a practical method for their analysis. It is based on nodal formulation and on the fact that all active networks contain voltage sources whose one terminal is grounded.

The amplifiers are almost always operational amplifiers and their idealized equivalent, with infinite gain, is a convenient abstraction. Unfortunately, infinite gain cannot be handled by computers. In order to avoid difficulties, we introduce a modification which makes it possible to deal with ideal or nonideal operational amplifiers in a unified way in hand or computer applications.

Analysis is first explained on networks with finite-gain amplifiers, then on networks with ideal operational amplifiers and finally on networks with nonideal operational amplifiers.

12.1. AMPLIFIERS AND OPERATIONAL AMPLIFIERS

Amplifiers used in active networks are equivalent to voltage-controlled voltage sources with one output terminal grounded. Figure 12.1.1a shows the equivalent

(a)   (b)

FIGURE 12.1.1   Amplifier with gain A: (a) Equivalent VV. (b) Usual symbol.

VV, in b is the usual symbol. If the input voltage is V_{in}, then the output is

$$V_{out} = AV_{in} \qquad\qquad (12.1.1)$$

where A is the gain of the amplifier. If $A > 0$ the amplifier is said to be non-inverting, if $A < 0$, then it is inverting. On the right side of Fig. 12.1.1 is the usual symbol for this type of amplifier. This symbol is used when the input as well as the output voltage are measured with respect to ground. We also note that no current is flowing into the amplifier at the input terminal.

Differential amplifiers are used more often. Figure 12.1.2a shows the equivalent VV, Fig. 12.1.2b the usual symbol. It is understood that V_+ denotes the non-inverting terminal and V_- the inverting terminal. Both voltages are measured with respect to ground. The output of the device is

$$V_o = A(V_+ - V_-) \qquad\qquad (12.1.2)$$

measured with respect to ground. The other terminal of the internal voltage-controlled voltage source is grounded. The symbol for the differential amplifier is on the right of Fig. 12.1.2. The gain of the differential amplifier, A, is always assumed to be positive.

It should be clear that grounding the negative terminal reduces the differential amplifier to the amplifier shown in Fig. 12.1.2. Should we ground the positive terminal, the differential amplifier becomes an inverting amplifier.

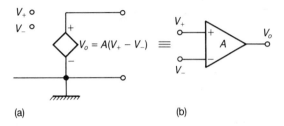

(a)   (b)

FIGURE 12.1.2   Differential amplifier with gain A: (a) Equivalent differential VV. (b) Usual symbol.

For the purposes of analysis, it is convenient to define an idealized element, the *ideal operational amplifier*. Its definition is the same as in Eq. (12.1.2), but the gain of the amplifier is assumed to be infinite. In some cases this is a simplification, in others a complication. If we keep the gain, A, during the analysis and later want to see what the result would be for $A \to \infty$, we must apply mathematical limiting, a process which is more complicated than simple substitution. In addition, infinity cannot be treated by a computer. For these reasons, we modify the operational amplifier and define the inverted gain B

$$B = -\frac{1}{A} \tag{12.1.3}$$

This modifies Eq. (12.1.2) to

$$V_+ - V_- + BV_0 = 0 \tag{12.1.4}$$

Equations (12.1.3) and (12.1.4) will be used extensively in the following sections. Their advantage lies in the fact that at any step of the analysis, even at the very beginning, we can substitute $B = 0$ and convert a nonideal operational amplifier into an ideal one.

12.2. PRINCIPLES OF ACTIVE NETWORK ANALYSIS

As we have seen in the previous section, amplifiers are dependent voltage sources with one output terminal grounded, and no current flows into the amplifier at its input terminals. This fact can be used to simplify analysis. In order to demonstrate the principle of the method to be described, we take two examples.

Consider the simple network in Fig. 12.2.1a; it has one independent voltage source and all passive elements are conductances. If we wish to use nodal analysis, then we must first apply voltage splitting, done in Fig. 12.2.1b. It is then convenient to redraw the network as in Fig. 12.1.1c, with the sources on both sides. Finally we can apply transformation of the voltage sources into current sources and get the final form suitable for nodal formulation, Fig. 12.2.1d. The equations will be

$$(G_1 + G_2 + G_3)V_1 - G_3V_2 = EG_1$$
$$-G_3V_1 + (G_3 + G_4 + G_5)V_2 = EG_5$$

Now suppose that we proceed somewhat differently by arguing that we really do not need to write the KCL equation at the node where the voltage is known. In Fig. 12.2.2 is the same network as in Fig. 12.2.1a, but the node where the voltage source is connected to the network is marked by a large cross; this means that we will not write a KCL equation at this node. Writing them for the remaining

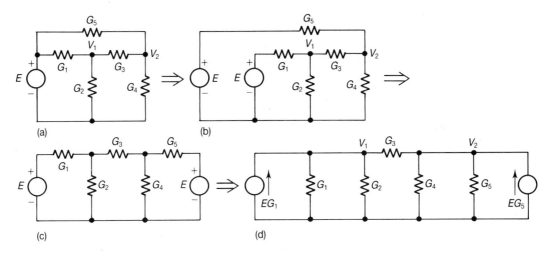

FIGURE 12.2.1   Developing method for nodal analysis with a sequence of transformations (a) to (d).

two nodes results in

$$(G_1 + G_2 + G_3)V_1 - G_3 V_2 - EG_1 = 0$$

$$-G_3 V_1 + (G_3 + G_4 + G_5)V_2 - EG_5 = 0$$

As always we transfer the independent sources (terms with the symbol E) on the right side and we get exactly the same equations as above.

The rule which skips writing the KCL at the nodes with voltage sources can be used in cases when the voltage sources have one node grounded. Fortunately, this happens to be true for practically all active networks, and the rule is valid for dependent sources as well. Consider the network in Fig. 12.2.3. We must mark

Fig. 12.2.2

Fig. 12.2.3

FIGURE 12.2.2   Elimination of a node for KCL equations.

FIGURE 12.2.3   Sallen–Key active network.

voltages at all nodes of the network. Going from the left of the figure we mark the first node with E, a known voltage. The other voltages are unknown; going from left to right we mark them V_1 and V_2. We do not know the voltage of the last node, but we know something about it nevertheless: it will be A times the voltage which appears at the input of the amplifier. We thus mark this last node with the voltage AV_2. The KCL are written at the two nodes which have not been marked. The steps to write KCL are the same as we have used already many times. The equations are

$$(G_1 + G_2 + sC_1)V_1 - (G_2 + sAC_1)V_2 = EG_1$$
$$-V_1G_2 + (G_2 + sC_2)V_2 = 0$$

and in matrix form

$$\begin{bmatrix} (G_1 + G_2 + sC_1) & -(G_2 + sC_1A) \\ -G_2 & (G_2 + sC_2) \end{bmatrix}\begin{bmatrix} V_1 \\ V_2 \end{bmatrix} = \begin{bmatrix} EG_1 \\ 0 \end{bmatrix} \qquad (12.2.1)$$

We postpone solution and discussion to Section 6. At this moment we are interested in the methods how to set up the correct set of equations for various types of amplifiers. We will start with finite-gain amplifiers, because their analysis is somewhat simpler. They will be followed by networks with ideal operational amplifiers and finally with operational amplifiers which are not ideal. At this moment we note that it is advantageous to work with conductances and not resistances.

12.3. NETWORKS WITH FINITE-GAIN AMPLIFIERS

The first network we consider is in Fig. 12.3.1 and we will keep all admittances as symbols. This has certain advantages: any G in the result can be replaced by sC and many configurations can thus be obtained from one analysis. The first and last nodes are marked by crosses because of the voltage sources. The voltages are written at the nodes: E at the first node, then from left to right V_1 and V_2 and, finally, AV_2 at the rightmost node. Only two equations need be written:

$$(G_1 + G_2 + G_3 + G_5)V_1 - G_3V_2 - G_2AV_2 = G_1E$$
$$-G_3V_1 + (G_3 + G_4 + G_6)V_2 - G_6AV_2 = 0$$

In matrix form

$$\begin{bmatrix} (G_1 + G_2 + G_3 + G_5) & -(AG_2 + G_3) \\ -G_3 & (G_3 + G_4 + G_6(1 - A)) \end{bmatrix}\begin{bmatrix} V_1 \\ V_2 \end{bmatrix} = \begin{bmatrix} G_1E \\ 0 \end{bmatrix}$$

FIGURE 12.3.1   Active network with one amplifier.

The transfer function is found by first solving for V_2 and then multiplying the result by A to obtain V_{out}. The voltage transfer function is

$$T_V = \frac{AG_1G_3}{(G_1 + G_2 + G_3 + G_5)[G_3 + G_4 + G_6(1 - A)] - G_3(AG_2 + G_3)}$$

The second network is shown in Fig. 12.3.2. It has one voltage source and two amplifiers. The voltages and crosses are in the figure; we must write the KCL equations for the two uncrossed nodes only:

$$(G_1 + sC_2)V_1 - sC_2A_2V_2 = EG_1$$

$$-G_2A_1V_1 + (G_2 + sC_1)V_2 = EsC_1$$

In matrix form:

$$\begin{bmatrix} (G_1 + sC_2) & -sC_2A_2 \\ -G_2A_1 & (G_2 + sC_1) \end{bmatrix} \begin{bmatrix} V_1 \\ V_2 \end{bmatrix} = \begin{bmatrix} EG_1 \\ EsC_1 \end{bmatrix}$$

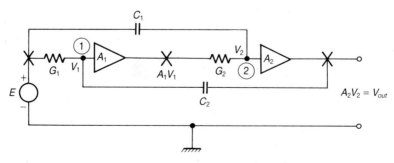

FIGURE 12.3.2   Active network with two amplifiers.

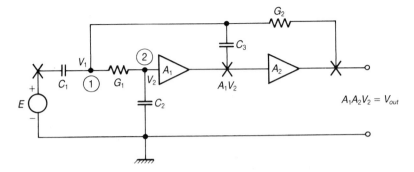

FIGURE 12.3.3   Active network with two amplifiers.

Using Cramer's rule we find the determinant, the numerator N_2 and the output voltage $V_{out} = A_2 V_2$; the voltage transfer function is

$$T_V = \frac{A_2(s^2 C_1 C_2 + s C_1 G_1 + G_1 G_2 A_1)}{s^2 C_1 C_2 + s[C_1 G_1 + C_2 G_2(1 - A_1 A_2)] + G_1 G_2}$$

The third network we consider is shown in Fig. 12.3.3, again with two amplifiers. The process of selecting the nodes for which we write the KCL equations should now be sufficiently clear. The voltages are written into the figure and nodal equations are written only for the two nodes which were not crossed out. They are

$$(G_1 + G_2 + sC_1 + sC_3)V_1 - G_1 V_2 - sC_3 A_1 V_2 - G_2 A_1 A_2 V_2 = EsC_1$$
$$- G_1 V_1 + (G_1 + sC_2)V_2 = 0$$

In matrix form:

$$\begin{bmatrix} (G_1 + G_2 + sC_1 + sC_3) & -(G_1 + sC_3 A_1 + G_2 A_1 A_2) \\ -G_1 & (G_1 + sC_2) \end{bmatrix} \begin{bmatrix} V_1 \\ V_2 \end{bmatrix} = \begin{bmatrix} EsC_1 \\ 0 \end{bmatrix}$$

Using Cramer's rule we find D, N_2 for the voltage V_2 and get the output from $V_{out} = A_1 A_2 V_2$. The transfer function is

$$T = \frac{sC_1 G_1 A_1 A_2}{s^2(C_1 + C_3)C_2 + s[C_1 G_1 + C_2 G_1 + C_2 G_2 + C_3 G_1(1 - A_1)] + G_1 G_2(1 - A_1 A_2)}$$

The next network is in Fig. 12.3.4. All the necessary information regarding the voltages and places where the KCL are written has already been marked. The

FIGURE 12.3.4   Active network with two amplifiers.

equations are

$$(G_1 + G_4 + sC_1)V_1 - sC_1 A_1 V_1 - G_4 A_2 V_2 = EG_1$$

$$-G_3 A_1 V_1 + (G_2 + G_3 + sC_2)V_2 - sC_2 A_2 V_2 = EG_2$$

The system matrix equation is

$$\begin{bmatrix} [G_1 + G_4 + sC_1(1 - A_1)] & -A_2 G_4 \\ -A_1 G_3 & [G_2 + G_3 + sC_2(1 - A_2)] \end{bmatrix} \begin{bmatrix} V_1 \\ V_2 \end{bmatrix} = \begin{bmatrix} EG_1 \\ EG_2 \end{bmatrix}$$

Using standard steps we find

$$D = s^2 C_1 C_2 (1 - A_1)(1 - A_2) + s[C_1(1 - A_1)(G_2 + G_3) + C_2(1 - A_2)(G_1 + G_4)]$$

$$+ (G_1 + G_4)(G_2 + G_3) - A_1 A_2 G_3 G_4$$

$$N = A_2[sC_1 G_2 (1 - A_1) + G_2(G_1 + G_4) + A_1 G_1 G_3]$$

The output is $V_{out} = A_2 V_2$ and the transfer function is

$$T = \frac{V_{out}}{E} = \frac{N}{D}$$

It should be noted here that the above networks were selected for the purpose of practicing the analysis method and not as examples of active networks which could be used for design. Many other considerations must be taken into account before we can be sure that a given network is a good one. In particular, it is necessary to study sensitivity of the network function to changes in element values. This important subject is covered in Chapter 16.

Example 1.

The network in Fig. 12.3.5 has two amplifiers, the first having the gain $A_1 = -1$. Find the output voltage by keeping the gain of the second amplifier, A_2, as a variable. In the result substitute $A_2 = 2$ and determine whether the network is stable. Also find the value of A_2 for which the network will become unstable. The voltages are written at each node; note especially that the voltage at the output of the first amplifier is $-V_1$. The nodal equations for the two nodes without crosses are

$$(5 + s)V_1 + sV_1 - 2A_2V_2 = 3E$$

$$3V_1 + (4 + s)V_2 - A_2V_2 = 0$$

In matrix form

$$\begin{bmatrix} 2s + 5 & -2A_2 \\ 3 & 4 - A_2 + s \end{bmatrix} \begin{bmatrix} V_1 \\ V_2 \end{bmatrix} = \begin{bmatrix} 3E \\ 0 \end{bmatrix}$$

The determinant is, after a few simple steps,

$$D = 2s^2 + s(13 - 2A_2) + 20 + A_2$$

The numerator is $N_2 = -9E$ but we must not forget that the output is $V_{out} = A_2V_2$. The transfer function is

$$T_V = \frac{-9A_2}{2s^2 + s(13 - 2A_2) + 20 + A_2}$$

Substituting $A_2 = 2$ results in the transfer function

$$T_V = \frac{-18}{2s^2 + 9s + 22}$$

FIGURE 12.3.5   Active network with two amplifiers.

The poles are $p_{1,2} = -2.25 \pm 2.437$. Since the real parts of the poles are negative, the network is stable.

The network will become unstable when the poles move on the imaginary axis and will have zero real part. This will happen when the middle term of the denominator becomes zero: $13 - 2A_2 = 0$. The network will be unstable for $A_2 = 13/2 = 6.5$.

Example 2.

Find the voltage transfer function for the network in Fig. 12.3.6. Also determine its poles and zeros and sketch the amplitude and group delay responses. This network leads to somewhat more complicated steps but the style of analysis is the same as for the previous networks. Write the KCL equations for the nodes without crosses yourself. In matrix form this will lead to

$$
\begin{bmatrix}
4s + 2 & 0 & -(2s + 3) \\
0 & 4s + 2 & -1 \\
-2s & -1 & 2s + \dfrac{7}{6}
\end{bmatrix}
\begin{bmatrix}
V_1 \\
V_2 \\
V_3
\end{bmatrix}
=
\begin{bmatrix}
2sE \\
E \\
0
\end{bmatrix}
$$

The determinant is

$$
D = (4s + 2)\left(2s + \frac{7}{6}\right) - 2s(2s + 3)(4s + 2) - (4s + 2)
$$

$$
= (4s + 2)\left[(4s + 2)\left(2s + \frac{7}{6}\right) - (4s^2 + 6s) - 1\right]
$$

The step which took the term $(4s + 2)$ in front of the square bracket is quite crucial for further analysis. After a few additional steps the denominator becomes

$$
D = 4(4s + 2)\left(s^2 + \frac{2}{3}s + \frac{1}{3}\right)
$$

FIGURE 12.3.6   Active network with a Twin-T.

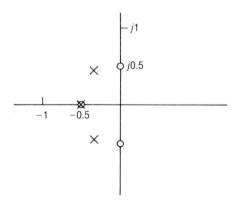

FIGURE 12.3.7   Pole-zero plot for network Fig. 12.3.6.

For output voltage we must first find V_3 by getting the determinant

$$N_3 = \begin{vmatrix} 4s+2 & 0 & 2s \\ 0 & 4s+2 & 1 \\ -2s & -1 & 0 \end{vmatrix} = (4s+2)(4s^2+1)$$

The output is $V_{out} = (3/2)V_3$ and the overall voltage transfer function is

$$T = \frac{3}{2}\frac{(4s+2)(4s^2+1)}{4(4s+2)\left(s^2+\frac{2}{3}s+\frac{1}{3}\right)}$$

This is a third-order network, but we can cancel the term $(4s+2)$ in the numerator against the same term in the denominator and get the final transfer function

$$T = \frac{3}{8}\frac{4s^2+1}{s^2+\frac{2}{3}s+\frac{1}{3}}$$

The network has two purely imaginary zeros at $z_{1,2} = \pm j1/2$ and two complex conjugate poles $p_{1,2} = (-1 \pm j\sqrt{2})/3$. In addition, it has a real zero at the same location as a pole at $s = -1/2$. The pole–zero plot is in Fig. 12.3.7 and cancellation of the pole and zero is indicated by plotting a circle and a cross at the same location. The amplitude response, calculated by NETOPT, is in Fig. 12.3.8, the group delay is in Fig. 12.3.9.

The student should use NETOPT for all the above analyses and confirm his derivations with the symbolic option. Very instructive also is the use of the program described in Appendix B because for any network function or pole–zero plot one can immediately get all frequency domain and time domain responses.

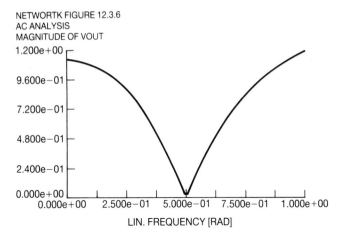

FIGURE 12.3.8   Amplitude response for network Fig. 12.3.6.

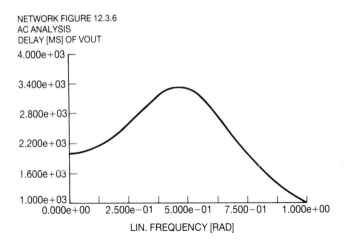

FIGURE 12.3.9   Group delay for network Fig. 12.3.6.

12.4. NETWORKS WITH IDEAL OPAMPS

In this section we will learn how to write equations for networks with ideal operational amplifiers. We have already discussed that it is not convenient to use infinite gain and that it is much more practical to define the inverted gain,

$$B = -\frac{1}{A} \qquad (12.4.1)$$

which simplifies the operational amplifier equation to

$$V_+ - V_- + BV_0 = 0 \qquad (12.4.2)$$

Since all the voltages in Eq. (12.4.2) are unknown, all are placed on the left side of the equal sign.

   Assume now that the gain is increased and becomes infinitely large. This makes $B = 0$ and Eq. (12.4.2) simplifies to

$$V_+ - V_- = 0 \qquad (12.4.3)$$

which is equivalent to

$$V_+ = V_- \qquad (12.4.4)$$

This says that the plus terminal and the minus terminal are *at the same potential*. If both terminals are floating, then whatever voltage, V_i, we write at one of these terminals, the same voltage is written at the other terminal. Sometimes the terminal V_+ is grounded; this means that the other terminal is at ground potential and $V_- = 0$.

   We will apply the above conclusions to several networks, first to Fig. 12.4.1. The voltages are written into the figure and the nodes with voltage sources are marked by crosses. Since the $+$ terminal is grounded, the $-$ terminal of an ideal operational amplifier must be at ground potential and this is indicated by writing the zero voltage at node 1. Only one KCL equation is needed for this network, written for the node which has zero voltage:

$$(G_1 + G_2)0 - G_2 V_{out} = EG_1$$

We have retained the term which is multiplied by zero, but it clearly need not be

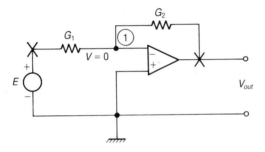

FIGURE 12.4.1   Network with ideal OPAMP.

written at all. The transfer function is

$$T = \frac{V_{out}}{E} = -\frac{G_1}{G_2} = -\frac{R_2}{R_1} \tag{12.4.5}$$

The network is an inverting device with the gain defined by the ratio of the two resistors. However, the network is *not* equivalent to an amplifier with the same gain because a current flows from the voltage source through the conductances G_1 and G_2 to the output. If, for instance, a capacitor is used instead of G_2, the transfer function becomes

$$T = \frac{V_{out}}{E} = -\frac{G_1}{sC_2}$$

and the network becomes an inverting integrator (inverting because of the minus sign and integrator because of the s in the denominator).

The second fundamental network is shown in Fig. 12.4.2. Only one node need be considered for the KCL, the others are marked by crosses:

$$(G_1 + G_2)E - G_2 V_{out} = 0$$

The transfer function is

$$T = \frac{V_{out}}{E} = 1 + \frac{G_1}{G_2} \tag{12.4.6}$$

It is a non-inverting amplifier (no current flows from the voltage source E) and the gain is defined by the ratio of the two conductances. Let $G_1 = 0$ as a special case. This removes G_1 from the figures and in such case the other conductance is meaningless; we can replace it by a short circuit, as shown in Fig. 12.4.3. The network becomes a unity-gain buffer, an amplifier with a gain equal to one.

Fig. 12.4.2   Fig. 12.4.3

FIGURE 12.4.2   Realizing positive gain with an OPAMP.

FIGURE 12.4.3   Unity gain buffer.

FIGURE 12.4.4   Active network with ideal OPAMP.

A more complicated network is shown in Fig. 12.4.4. We have already considered this network in Section 2 because the OPAMP with the two conductances G_4 and G_5 is nothing but an amplifier with finite gain, as we established in the previous case. We proceed as before: cross out the nodes where there are voltage sources and write voltages at all nodes. This has been done in the figure. Since the OPAMP is ideal, the terminals V_+ and V_- must be at the same potential and we take this fact into account by writing *the same voltage* at these two nodes. The KCL are written at the nodes denoted by the numbers in circles:

$$(G_1 + G_2 + G_3 + sC_1)V_1 - G_3 V_2 - sC_1 V_{out} = EG_1$$

$$-G_3 V_1 + (G_3 + sC_2)V_2 = 0$$

$$(G_4 + G_5)V_2 - G_5 V_{out} = 0$$

In matrix form:

$$
\begin{array}{c}
① \\
② \\
③
\end{array}
\begin{bmatrix}
(G_1 + G_2 + G_3 + sC_1) & -G_3 & -sC_1 \\
-G_3 & (G_3 + sC_2) & 0 \\
0 & (G_4 + G_5) & -G_5
\end{bmatrix}
\begin{bmatrix}
V_1 \\
V_2 \\
V_{out}
\end{bmatrix}
=
\begin{bmatrix}
EG_1 \\
0 \\
0
\end{bmatrix}
$$

We will leave the problem of finding the transfer function to the reader.

Example 1.

Find the input impedance of the network in Fig. 12.4.5.

Notice that the OPAMPs force three node voltages of the network to be equal to the input voltage E. Mark with a cross the nodes where the KCL need not be written and mark the remaining nodes with unknown voltages V_1 and V_2. Only two KCL equations

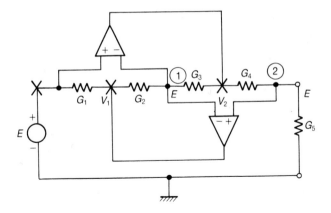

FIGURE 12.4.5   Active network with two ideal OPAMPs.

need be written:

$$E(G_2 + G_3) - V_1 G_2 - V_2 G_3 = 0$$

$$E(G_4 + G_5) - G_4 V_2 = 0$$

In matrix form:

$$\begin{bmatrix} G_2 & G_3 \\ 0 & G_4 \end{bmatrix} \begin{bmatrix} V_1 \\ V_2 \end{bmatrix} = \begin{bmatrix} E(G_2 + G_3) \\ E(G_4 + G_5) \end{bmatrix}$$

The determinant of the system is $D = G_2 G_4$. In order to calculate the input impedance, we must know the current flowing into the network. It is

$$I_{in} = (E - V_1)G_1$$

and we must find V_1 by replacing the first column in the matrix by the right-hand side. The result is

$$V_1 = E - \frac{EG_3 G_5}{G_2 G_4}$$

Inserting into the expression for the current

$$I_{in} = E \frac{G_1 G_3 G_5}{G_2 G_4}$$

and the input impedance is

$$Z_{in} = \frac{E}{I_{in}} = \frac{G_2 G_4}{G_1 G_3 G_5}$$

If we now replace one of the admittances in the numerator, for instance G_2, by the admittance of a capacitor, sC_2, then

$$Z_{in} = \frac{sC_2 G_4}{G_1 G_3 G_5}$$

This expression represents the impedance of an inductor with an equivalent inductance

$$L = \frac{C_2 G_4}{G_1 G_3 G_5}$$

The network acts as an inductor with one terminal grounded.

Example 2.

Find the transfer function for the network in Fig. 12.4.6.

All the necessary voltages have been written into the figure. The nodes where we do not write the KCL are marked by the crosses, the nodes where we must write the equations are indicated by numbers in circles.

The matrix equation is

$$\begin{bmatrix} (G_1 + G_4 + sC_1) & -G_4 & -(sC_1 + G_1) \\ (G_2 + G_3) & 0 & -G_2 \\ (G_5 + sC_2) & -sC_2 & 0 \end{bmatrix} \begin{bmatrix} V_1 \\ V_2 \\ V_{out} \end{bmatrix} = \begin{bmatrix} 0 \\ 0 \\ EG_5 \end{bmatrix}$$

and the transfer function is

$$T = \frac{V_{out}}{E} = \frac{G_4 G_5 (G_2 + G_3)}{s^2 C_1 C_2 G_3 + sC_2 G_1 G_3 + G_2 G_4 G_5}$$

FIGURE 12.4.6   Active network with two ideal OPAMPs.

FIGURE 12.4.7   Active network with three ideal OPAMPs.

Example 3.

Figure 12.4.7 shows a practical active network with three ideal operational amplifiers. Write the nodal equations and put them into matrix form. Try to solve using Cramer's rule and determinant expansion, as explained in Chapter 3. Also find the numerators N_2, N_3, N_4.

All the preliminary steps needed for analysis are marked in the figure: nodes where we do not write KCL are marked by crosses, nodes where we must write the KCL are marked by numbers in circles. We directly write the matrix form

$$
\begin{bmatrix}
(G_3 + G_4) & -G_4 & 0 & -G_3 \\
(G_1 + G_2) & 0 & -G_2 & 0 \\
0 & -G_5 & -sC_1 & 0 \\
0 & 0 & -G_6 & -sC_2
\end{bmatrix}
\begin{bmatrix}
V_1 \\
V_2 \\
V_3 \\
V_4
\end{bmatrix}
=
\begin{bmatrix}
0 \\
EG_1 \\
0 \\
0
\end{bmatrix}
$$

The solutions are

$$
D = s^2 C_1 C_2 G_4 (G_1 + G_2) + s C_2 G_2 G_5 (G_3 + G_4) + G_3 G_5 G_6 (G_1 + G_2)
$$
$$
N_2 = s^2 C_1 C_2 G_1 (G_3 + G_4) E
$$
$$
N_3 = -s C_2 G_1 G_5 (G_3 + G_4) E
$$
$$
N_4 = G_1 G_5 G_6 (G_3 + G_4) E
$$

NETOPT gives the possibility of checking transfer functions with up to five elements kept as variables. Instructions on how to use the program and how to interpret the results are in Chapter 18, Section 5.

12.5. NETWORKS WITH NONIDEAL OPAMPS

In this section we are coming to situations where we can no longer assume that the OPAMPs are ideal. In order to make an easy comparison with the method of the previous section, we will consider the same networks.

Once the amplifiers are nonideal we cannot make any assumptions about the various voltages in the network. All are unknown except where we apply the independent voltage sources. We assign consecutive subscripts to the voltages of all nodes, cross out the nodes which are connected to voltage sources and write the KCL for the remaining nodes. This does not differ from all previous steps. Finally, we use, for each operational amplifier, Eq. (12.1.4), in which we substitute actual nodal voltages for the symbols V_+, V_- and V_0.

Let us take the network in Fig. 12.5.1. The input terminals of the OPAMP are no longer at the same potential and we assign voltage V_1 to the internal node. The KCL for this node is

$$(G_1 + G_2)V_1 - G_2 V_{out} = EG_1$$

For the OPAMP we substitute V_1 for V_-, 0 for V_+, and V_{out} for V_0. The second equation is

$$-V_1 + BV_{out} = 0$$

Calculate V_{out} and obtain the transfer function

$$T = \frac{V_{out}}{E} = \frac{G_1}{-G_2 + B(G_1 + G_2)}$$

If we set $B = 0$ then the expression simplifies to the result we obtained in the previous section.

FIGURE 12.5.1   Network Fig. 12.4.1 with nonideal OPAMP.

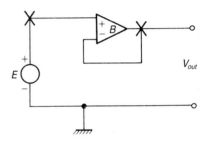

Fig. 12.5.2   Fig. 12.5.3

FIGURE 12.5.2   Network Fig. 12.4.2 with nonideal OPAMP.

FIGURE 12.5.3   Unity gain buffer with nonideal OPAMP.

The second network is in Fig. 12.5.2; only one KCL need be written, followed by the equation expressing the properties of the OPAMP:

$$(G_1 + G_2)V_1 - G_2 V_{out} = 0$$
$$E - V_1 + B V_{out} = 0$$

The solution is

$$T = \frac{V_{out}}{E} = \frac{G_1 + G_2}{G_2 - (G_1 + G_2)B}$$

If we set $B = 0$, the expression reduces to Eq. (12.4.6).

The unity-gain buffer, now with a nonideal OPAMP, is shown in Fig. 12.5.3. In this case no KCL need be written (all nodes are crossed out). Only the OPAMP equation remains:

$$E - V_{out} + B V_{out} = 0$$

The property of this network with a nonideal OPAMP is described by the transfer function

$$T = \frac{V_{out}}{E} = \frac{1}{1 - B}$$

The next case is the network in Fig. 12.5.4. We must write three KCL equations at the nodes marked by the voltages V_1, V_2 and V_3 and attach the equation for

FIGURE 12.5.4   Network Fig. 12.4.4 with nonideal OPAMP.

the OPAMP:

$$(G_1 + G_2 + G_3 + sC_1)V_1 - G_3V_2 - sC_1V_{out} = EG_1$$
$$-G_3V_1 + (G_3 + sC_2)V_2 = 0$$
$$(G_4 + G_5)V_3 - G_5V_{out} = 0$$
$$V_2 - V_3 + BV_{out} = 0$$

In matrix form:

$$
\begin{bmatrix}
(G_1 + G_2 + G_3 + sC_1) & -G_3 & 0 & -sC_1 \\
-G_3 & (G_3 + sC_2) & 0 & 0 \\
0 & 0 & (G_4 + G_5) & -G_5 \\
0 & 1 & -1 & B
\end{bmatrix}
\begin{bmatrix}
V_1 \\
V_2 \\
V_3 \\
V_{out}
\end{bmatrix}
=
\begin{bmatrix}
EG_1 \\
0 \\
0 \\
0
\end{bmatrix}
$$

Example 1.

Write the nodal equations and the necessary matrix equation for the network in Fig. 12.5.5, now with nonideal OPAMPs.

Only two nodal equations need be written, followed by the equations of the two OPAMPs:

$$-G_2V_1 + (G_2 + G_3)V_2 - G_3V_3 = 0$$
$$-G_4V_3 + (G_4 + G_5)V_4 = 0$$
$$E - V_2 + B_1V_3 = 0$$
$$-V_2 + V_4 + B_2V_1 = 0$$

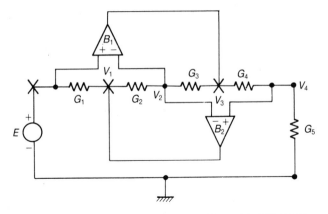

FIGURE 12.5.5   Network Fig. 12.4.5 with nonideal OPAMPs.

In matrix form:

$$\begin{bmatrix} -G_2 & (G_2 + G_3) & -G_3 & 0 \\ 0 & 0 & -G_4 & (G_4 + G_5) \\ 0 & -1 & B_1 & 0 \\ B_2 & -1 & 0 & 1 \end{bmatrix} \begin{bmatrix} V_1 \\ V_2 \\ V_3 \\ V_4 \end{bmatrix} = \begin{bmatrix} 0 \\ 0 \\ -E \\ 0 \end{bmatrix}$$

Example 2.

The network in Fig. 12.5.6 has two nonideal OPAMPs. Write the nodal equations and put them into matrix form.

All the preliminary operations have been written into the figure. We give only the matrix form:

$$\begin{bmatrix} (G_1 + G_4 + sC_1) & 0 & -G_4 & 0 & -(G_1 + sC_1) \\ 0 & (G_2 + G_3) & 0 & 0 & -G_2 \\ 0 & 0 & -sC_2 & (G_5 + sC_2) & 0 \\ -1 & 1 & B_1 & 0 & 0 \\ -1 & 0 & 0 & 1 & B_2 \end{bmatrix} \begin{bmatrix} V_1 \\ V_2 \\ V_3 \\ V_4 \\ V_{out} \end{bmatrix} = \begin{bmatrix} 0 \\ 0 \\ EG_5 \\ 0 \\ 0 \end{bmatrix}$$

Example 3.

Write the nodal equations for the network in Fig. 12.5.7 and put them in matrix form. The OPAMPs are nonideal.

FIGURE 12.5.6   Network Fig. 12.4.6 with nonideal OPAMPs.

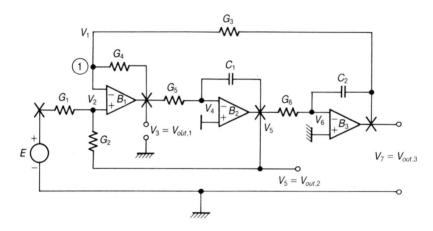

FIGURE 12.5.7   Network Fig. 12.4.7 with nonideal OPAMPs.

All the preliminary steps have been applied to the figure. The nodal equations are

$$(G_3 + G_4)V_1 - G_4V_3 - G_3V_7 = 0$$

$$(G_1 + G_2)V_2 - G_2V_5 = EG_1$$

$$-G_5V_3 + (G_5 + sC_1)V_4 - sC_1V_5 = 0$$

$$-G_6V_5 + (G_6 + sC_2)V_6 - sC_2V_7 = 0$$

$$-V_1 + V_2 + B_1V_3 = 0$$

$$-V_4 + B_2V_5 = 0$$

$$-V_6 + B_3V_7 = 0$$

The matrix equation is

$$
\begin{bmatrix}
(G_3 + G_4) & 0 & -G_4 & 0 & 0 & 0 & -G_3 \\
0 & (G_1 + G_2) & 0 & 0 & -G_2 & 0 & 0 \\
0 & 0 & -G_5 & (G_5 + sC_1) & -sC_1 & 0 & 0 \\
0 & 0 & 0 & 0 & -G_6 & (G_6 + sC_2) & -sC_2 \\
-1 & +1 & B_1 & 0 & 0 & 0 & 0 \\
0 & 0 & 0 & -1 & B_2 & 0 & 0 \\
0 & 0 & 0 & 0 & 1 & -1 & B_3
\end{bmatrix}
\begin{bmatrix}
V_1 \\ V_2 \\ V_3 \\ V_4 \\ V_5 \\ V_6 \\ V_7
\end{bmatrix}
=
\begin{bmatrix}
0 \\ EG_1 \\ 0 \\ 0 \\ 0 \\ 0 \\ 0
\end{bmatrix}
$$

The last example shows that hand calculations are really difficult if we must consider several nonideal OPAMPs. For this reason hand analysis is usually done by assuming that the OPAMPs are ideal and conclusions are taken for such results. For networks with nonideal OPAMPs use NETOPT. It has a built-in model for the inverted gain,

$$
B = -\frac{s}{GB}
$$

where GB is the gain–bandwidth product in Hertz, Hz. This nonideal operational amplifier is given the symbol N. Details about modeling amplifiers are in Chapter 15 and instructions on how to use NETOPT are in Chapter 18.

12.6. PRACTICAL USE OF RC ACTIVE NETWORKS

All previous sections of this chapter were devoted to methods of writing equations of active networks. In several cases we also solved the matrix equations. Some of the networks we considered are practical, some were invented solely for the purpose of teaching the method of writing the minimum set of equations directly. In this section we take a few practical networks and discuss typical ways how the results of the analysis are used.

As an introduction consider Fig. 12.6.1a where we have the symbol for an amplifier with $A \geq 1$. Its practical realization by means of an ideal OPAMP and two resistors is in Fig. 12.6.1b. In active networks this is always used whenever we need $A > 1$.

The first network with some practical importance was the network in Fig. 12.2.3, system matrix equation Eq. (12.2.1). We now solve it. The determinant is

$$
D = s^2 C_1 C_2 + s[C_2(G_1 + G_2) + C_1 G_2(1 - A)] + G_1 G_2
$$

Using Cramer's rule we obtain the numerator for V_2 and $V_{out} = AV_2$. The transfer

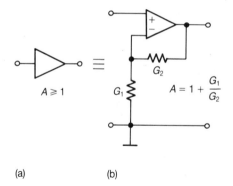

(a)   (b)

FIGURE 12.6.1   Amplifier with positive gain: (a) Symbol. (b) Realization with ideal OPAMP.

function is

$$T = \frac{V_{out}}{E} = \frac{AG_1G_2}{s^2C_1C_2 + s[C_2(G_1 + G_2) + C_1G_2(1 - A)] + G_1G_2}$$

In applications, we are interested in Q and ω_0 and we now derive practical formulas to get both these quantities directly from the second-order transfer function. The denominator polynomial is of the type

$$D = Us^2 + Vs + W$$

where U, V and W are functions of network elements. Rewrite as

$$D = U\left(s^2 + \frac{V}{U}s + \frac{W}{U}\right)$$

and compare with the general form introduced earlier [Eq. (8.4.5)]:

$$D = s^2 + \frac{\omega_0}{Q}s + \omega_0^2$$

We can identify

$$\omega_0^2 = \frac{W}{U} \tag{12.6.1}$$

and

$$\frac{\omega_0}{Q} = \frac{V}{U} \tag{12.6.2}$$

Substituting ω_0 from Eq. (12.6.1) we obtain

$$Q = \frac{\sqrt{UW}}{V} \qquad (12.6.3)$$

Equations (12.6.1)–(12.6.3) simplify our work. For the above example we immediately have

$$\omega_0^2 = \frac{G_1 G_2}{C_1 C_2}$$

and

$$Q = \frac{\sqrt{C_1 C_2 G_1 G_2}}{C_2 G_1 + C_2 G_2 + C_1 G_2 (1 - A)}$$

These expressions can be used for the design of the network. Suppose that we wish to realize $\omega_0 = 1$ and control Q by the gain of the amplifier only. We can select, more or less arbitrarily, $G_1 = G_2 = 1$ and $C_1 = C_2 = 1$. Then

$$Q = \frac{1}{3 - A}$$

or

$$A = \frac{3Q - 1}{Q}$$

Should we wish to realize the network for $Q = 2$, then we need $A = 2.5$.

The next practical network is in Fig. 12.6.2. Writing the nodal equations:

$$(G_1 + sC_1)V_1 - sC_1 V_2 - G_1 V_{out} = 0$$

$$(G_2 + G_3)V_1 - G_2 V_{out} = EG_3$$

$$(G_4 + G_5 + sC_2)V_1 - G_4 V_2 = EsC_2$$

In matrix form

$$\begin{bmatrix} (G_1 + sC_1) & -sC_1 & -G_1 \\ (G_2 + G_3) & 0 & -G_2 \\ (G_4 + G_5 + sC_2) & -G_4 & 0 \end{bmatrix} \begin{bmatrix} V_1 \\ V_2 \\ V_{out} \end{bmatrix} = \begin{bmatrix} 0 \\ G_3 E \\ sC_2 E \end{bmatrix}$$

The determinant of the matrix is the denominator of the transfer function:

$$D = s^2 C_1 C_2 G_2 + sC_1 G_2 G_5 + G_1 G_3 G_4$$

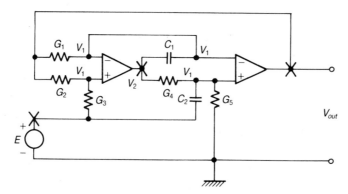

FIGURE 12.6.2   All-pass with OPAMPs.

Using Cramer's rule for the output voltage we obtain the numerator:

$$N = E(s^2 C_1 C_2 G_2 - s C_1 G_3 G_5 + G_1 G_3 G_4)$$

Let us now select

$$G_2 = G_3 = G$$

The transfer function becomes

$$T = \frac{V_{out}}{E} = \frac{s^2 C_1 C_2 G - s C_1 G_5 G + G_1 G G_4}{s^2 C_1 C_2 G + s C_1 G_5 G + G_1 G G_4}$$

We note that the expressions in the numerator and denominator differ only by the sign of the middle terms. Networks which have this property are called *all-pass* networks because they do not influence the amplitude of the signal, only the phase (and group delay). Such networks are used in the design of filters. Applying our formulas we get

$$\omega_0^2 = \frac{G_1 G_4}{C_1 C_2}$$

and

$$Q = \frac{1}{G_5} \left[\frac{C_2 G_1 G_4}{C_1} \right]^{1/2}$$

Suppose that we select $C_1 = C_2 = 1$, $G_5 = 0.5$ and all other conductances $G_i = 1$; this results in $Q = 2$. Because the network is an all-pass, the amplitude response is equal to 1 for all frequencies, but the group delay is a function of frequency. Figure 12.6.3 is the plot of this group delay, obtained by NETOPT.

FIGURE 12.6.3   Group delay of network Fig. 12.6.2.

We will consider one more practical network, this time with three operational amplifiers. It is shown in Fig. 12.6.4. The voltages at the input terminals of the OPAMPs are equal to zero; we write the nodal equations at these points. In matrix form:

$$
\begin{bmatrix}
-(G_3 + sC_1) & 0 & -G_2 \\
-G_4 & -sC_2 & 0 \\
0 & -G_5 & -G_6
\end{bmatrix}
\begin{bmatrix}
V_1 \\
V_2 \\
V_3
\end{bmatrix}
=
\begin{bmatrix}
EG_1 \\
0 \\
0
\end{bmatrix}
$$

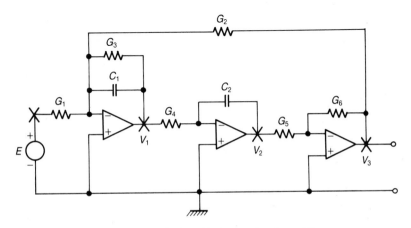

FIGURE 12.6.4   Network with three ideal operational amplifiers.

The denominator of all network functions is

$$D = -(s^2 C_1 C_2 G_6 + s C_2 G_3 G_6 + G_2 G_4 G_5)$$

Solving for all three voltages at the outputs of the OPAMPs, we get the transfer functions

$$T_1 = \frac{V_1}{E} = \frac{s C_2 G_1 G_6}{D}$$

$$T_2 = \frac{V_2}{E} = \frac{-G_1 G_4 G_6}{D}$$

$$T_3 = \frac{V_3}{E} = \frac{G_1 G_4 G_5}{D}$$

Finally, using Eq. (12.6.1) and Eq. (12.6.3) we obtain

$$\omega_0^2 = \frac{G_2 G_4 G_5}{C_1 C_2 G_6}$$

and

$$Q = \frac{1}{G_3} \left[\frac{C_1 G_2 G_4 G_5}{C_2 G_6} \right]^{1/2}$$

If we select, for instance, $C_1 = C_2 = 1$ and $G_2 = G_4 = G_5 = G_6 = 1$, then Q is controlled only by the conductance G_3. We thus have a very simple way to construct networks with various Q.

The considerations presented above are not sufficient for reliable design of active networks. It is also necessary to check how changes in element values influence the properties of the network. If small changes in the values result in large changes of the network responses, then such networks are not suitable for practical applications. These topics are generally lumped under the name of network sensitivity. This important subject is covered in Chapter 16.

PROBLEMS CHAPTER 12

You can use the symbolic option in NETOPT to confirm your solutions. Keep the symbolic results for sensitivity studies in Chapter 16.

P.12.1a. The amplifier has unity gain, the other elements are variable. Derive the transfer function in terms of variable elements. Make $C_1 = C_2 = C$ and $G_1 = G_2 = G$ and find the expression for ω_0 and Q. What is the meaning of $Q = 0.5$ in terms of the poles of the network?

(a)

P.12.1b. Do the same as in a. What is the difference?

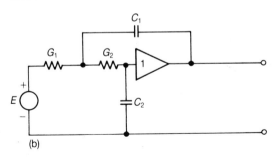

(b)

P.12.1c. Do the same as in a.

(c)

P.12.2a–e. The operational amplifiers are ideal. Derive expressions for the output in terms of variable elements.

(a)

(b)

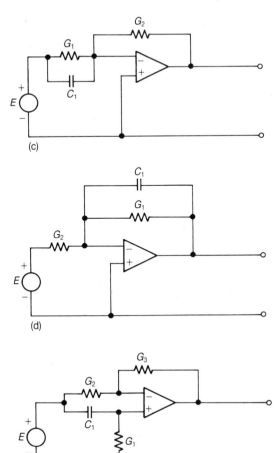

(c)

(d)

(e)

P.12.3 The operational amplifier in this configuration acts as a unity gain buffer. Derive the expression for the output in terms of variable elements. Afterwards make $G_1 = G_2 = G$ and $C_1 = C_2 = C$ and plot the poles. Next remove the operational amplifier by short-circuiting its input and output and derive the transfer function for equal G and C. Note that in this case you do not get a double pole.

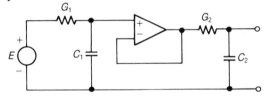

P.12.4 Analyze the network by keeping all elements as variables.

P.12.5 Derive the transfer function in terms of variable elements.

P.12.6 Use numerical values of elements and derive the transfer function. What are the poles of the network?

P.12.7 Use $C_1 = C_2 = 1$ but keep the conductances and the gain of the VV as variables. Derive the transfer function. Set $G_1 = G_2 = G$ and get the expression for ω_0 and Q. If we wish to make $Q = 2$, what gain of the amplifier must we select?

P.12.8 Derive the transfer function using given numerical values. Keep the gain, A, as a variable.

P.12.9 Use given element values but keep the amplifier gain, A, as a variable. Derive the transfer function.

P.12.10 Find the input impedance in terms of variable elements.

P.12.11 Find the transfer function by keeping the elements as variables. Also give expression for ω_0 and Q. Note that some conductances and the two capacitors have equal values.

P.12.12 Derive the transfer function by keeping G_1 and G_2 as variables. Notice that the term $(1 + G_2)$ can be factored out of the denominator. Select $G_2 = 4G_1$ and notice the type of the numerator and denominator polynomial. They have numerically equal coefficients but the sign at the odd power of s differs. Such networks are called all-pass because they have constant frequency response (pass all frequencies equally) but do influence the group delay. Find the poles and zeros for $G_1 = 1$, show that the amplitude response is independent of frequency and sketch the group delay. You can use either the program from Appendix B or NETOPT to plot the responses.

P.12.13 Use given numerical values but keep G_1 and G_2 as variables. Derive the transfer function.

P.12.14 Keep the capacitors as variables and set $G_1 = G_2 = 1$. Find the transfer function.

P.12.15 Set $G_1 = G_2 = G_3 = 1$ and keep the capacitors as variables. Find the transfer function.

P.12.16 Let all conductances be equal to 1 and keep the capacitors as variables. Find the transfer function.

P.12.17 Find the transfer function by keeping all elements as variables. The operational amplifiers are ideal.

P.12.18 This network is an example how small active filters are built. The dotted line separates two networks, in this case with identical structures. Because the second network is getting its input from the operational amplifier, the networks do not influence each other. We can realize one of them for a given pair of complex conjugate poles and the other one for another pair. The overall responses will be equal to the product of both transfer functions. In both networks set $G_2 = 1$ and $C_1 = 1$ but keep G_1 and C_2 and the gain A as variables. Derive the transfer function. Design the first network to realize the denominator $D_1 = s^2 + 2s + 2$, and the second network to realize $D_2 = s^2 + 2s + 5$. Plot the poles. Evaluate the frequency domain response of each of the networks separately and then when they are connected together. Use either the program from Appendix B or NETOPT to display the results.

CHAPTER 13

Two-Ports

13.0. INTRODUCTION

The theory of two-ports was invented as a method to analyze complicated networks with limited computational resources: a larger network is broken into smaller networks with precisely defined interconnections and special two-port rules are applied. Since the arrival of computers, two-port theory has become less important for computations, but has retained its importance as a theoretical tool.

13.1. DEFINITION OF TWO-PORTS

In order to introduce two-port theory, consider an arbitrary network in which we select only four terminals and make them the only places where connections to other networks or elements can be made. The situation is sketched in Fig. 13.1.1. Instead of taking each terminal separately and measuring the voltages at these terminals with respect to ground, we group them into two pairs. Each pair is called a *port* and we thus have a *two-port* network. Anything can be inside the box but to make the theory practical we usually consider only a few components. It is a common practice to denote by V_1 the voltage at the left port and by V_2 the voltage at the right port, with $+$ and $-$ as indicated. There will also be currents and the directions are *always* assumed to be as shown in the figure. *Note the currents again!* At the $+$ terminals they flow *into* the network, at the $-$ terminals they flow *out of* the network.

FIGURE 13.1.1   Two-port orientations of voltages and currents.

In our earlier chapters we always considered the current as positive when it was *leaving* the node, but here the currents go away from the top nodes and go to the nodes at the bottom of the figure. We were also considering the voltages with respect to ground, but here we are taking the voltage between two nodes. Why are there two such ways of considering the voltages and currents? The reasons are historical. Two-port theory was developed many years ago, and at the beginning even the definition shown in Fig. 13.1.1 was not unified; in German literature the currents on the right was originally reversed. The directions shown in Fig. 13.1.1 are now accepted as standard in the two-port theory. In all other applications, voltages are measured with respect to ground and currents are considered as positive when leaving a node.

Since we have four variables, V_1, V_2, I_1 and I_2, we can consider any two of them as independent variables and the other two as dependent variables. Each dependent variable will, in general, depend on the two independent variables.

We can have several choices. For instance, we can select the currents I_1 and I_2 as independent variables; then the dependent variables will be V_1 and V_2. Other choices exist as well and the following pages will describe some of them.

Before we go into the details, let us give some *practical* advice. The various two-port definitions are related and they can be transformed from one to another by means of the Fig. 13.4.1, derived later. There are only a few simple exceptions and we will cover them. The student should learn how to derive any of the parameters directly, from their definition. It is, however, worth remembering that the first two, the impedance and admittance parameters, are the simplest ones and their use, in connection with the transformations, can save a considerable amount of work.

13.1.1. Impedance Parameters

Let the currents be the independent variables. Then we can write two general equations

$$V_1 = z_{11}I_1 + z_{12}I_2$$
$$V_2 = z_{21}I_1 + z_{22}I_2$$

(13.1.1)

which couple the independent and dependent variables in a most general form. The subscripts are chosen to take advantage of standard matrix notation:

$$\begin{bmatrix} V_1 \\ V_2 \end{bmatrix} = \begin{bmatrix} z_{11} & z_{12} \\ z_{21} & z_{22} \end{bmatrix} \begin{bmatrix} I_1 \\ I_2 \end{bmatrix}$$

(13.1.2)

The network inside the rectangle in Fig. 13.1.1 is given and we cannot go inside it but we can attach sources and do measurements (or calculations). The steps will become clear from the following.

For the system Eq. (13.1.1) we can set $I_2 = 0$, which means that we *open circuit* the second port and attach a source at the first port. This simplifies Eq. (13.1.1) to

$$V_1 = z_{11}I_1$$
$$V_2 = z_{21}I_1$$

Dividing by I_1 we obtain

$$z_{11} = \left.\frac{V_1}{I_1}\right|_{I_2=0} \tag{13.1.3}$$

$$z_{21} = \left.\frac{V_2}{I_1}\right|_{I_2=0} \tag{13.1.4}$$

Similarly, setting $I_1 = 0$ (open-circuiting the first port) and placing the source to the second port we get

$$V_1 = z_{12}I_2$$
$$V_2 = z_{22}I_2$$

Dividing by I_2

$$z_{12} = \left.\frac{V_1}{I_2}\right|_{I_1=0} \tag{13.1.5}$$

$$z_{22} = \left.\frac{V_2}{I_2}\right|_{I_1=0} \tag{13.1.6}$$

The parameters z_{ij} have the dimensions of an impedance and are sometimes referred to as *two-port impedance parameters*. Since they were obtained by open-circuiting the ports, they are also known as *open-circuit parameters*.

We will demonstrate the derivation of these parameters on the example of a T network, shown in Fig. 13.1.2. First we attach a source at the first port. We can use either a voltage or a current source; in this case we take the voltage source as shown in Fig. 13.1.2a. Since $I_2 = 0$, we have

$$I_1 = \frac{V_1}{Z_1 + Z_3}$$

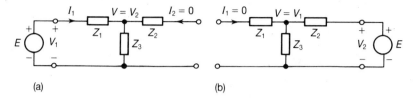

FIGURE 13.1.2 Z parameters for a T network: (a) For z_{11} and z_{21}. (b) For z_{12} and z_{22}.

from which we get immediately

$$z_{11} = \frac{V_1}{I_1} = Z_1 + Z_3$$

The voltage at the middle node must be equal to V_2 since no current flows through Z_2 and there can be no voltage drop across it,

$$V_2 = I_1 Z_3$$

and

$$z_{21} = \frac{V_2}{I_1} = Z_3$$

In the next step we place the voltage source at the second port and open-circuit the first port, as shown in Fig. 13.1.2b. Proceeding the same way, we obtain

$$z_{12} = \frac{V_1}{I_2} = Z_3$$

$$z_{22} = \frac{V_2}{I_2} = Z_2 + Z_3$$

Collecting the terms in the form of a matrix

$$\mathbf{Z} = \begin{bmatrix} Z_1 + Z_3 & Z_3 \\ Z_3 & Z_2 + Z_3 \end{bmatrix} \tag{13.1.7}$$

By setting Z_1 or Z_2 or both equal to zero, we get the open-circuit parameters for three additional networks, shown in Fig. 13.1.3 with their matrices. Note that in the simplest case, when only Z_3 is left, the open-circuit matrix becomes singular (the determinant of the matrix is zero). Another point to remember: if the network does not have dependent sources and is composed of only R, L and C elements, then z_{12} is *always* equal to z_{21}. Such networks are called *reciprocal*.

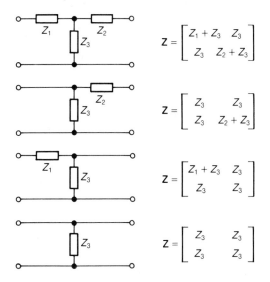

$$Z = \begin{bmatrix} Z_1 + Z_3 & Z_3 \\ Z_3 & Z_2 + Z_3 \end{bmatrix}$$

$$Z = \begin{bmatrix} Z_3 & Z_3 \\ Z_3 & Z_2 + Z_3 \end{bmatrix}$$

$$Z = \begin{bmatrix} Z_1 + Z_3 & Z_3 \\ Z_3 & Z_3 \end{bmatrix}$$

$$Z = \begin{bmatrix} Z_3 & Z_3 \\ Z_3 & Z_3 \end{bmatrix}$$

FIGURE 13.1.3 T network and its simplifications.

Example 1.

Find the open-circuit parameters for the network in Fig. 13.1.4.

It is advantageous to use current sources when deriving the z parameters. We must proceed with two solutions, as given in Fig. 13.1.4a and b.

(a) Note how we numbered the nodes, to have the subscripts 1 and 2 for the required results. We must calculate V_1 and V_2. Using standard nodal formulation (write the KCL

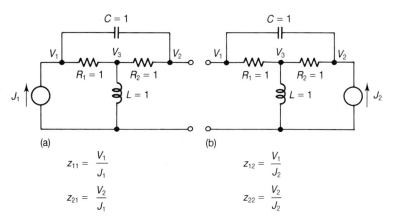

$$z_{11} = \frac{V_1}{J_1}$$

$$z_{21} = \frac{V_2}{J_1}$$

$$z_{12} = \frac{V_1}{J_2}$$

$$z_{22} = \frac{V_2}{J_2}$$

FIGURE 13.1.4 Z parameters for a bridged T network: (a) Connection to calculate z_{11} and z_{21}. (b) Connection to calculate z_{12} and z_{22}.

equations) we arrive at the system matrix

$$\begin{bmatrix} s+1 & -s & -1 \\ -s & s+1 & -1 \\ -1 & -1 & \dfrac{2s+1}{s} \end{bmatrix} \begin{bmatrix} V_1 \\ V_2 \\ V_3 \end{bmatrix} = \begin{bmatrix} J_1 \\ 0 \\ 0 \end{bmatrix}$$

The determinant is

$$D = \frac{2s+1}{s}$$

Replacing the first column by the right-hand side we get

$$N_1 = \frac{2s^2 + 2s + 1}{s} J_1$$

and

$$z_{11} = \frac{V_1}{J_1} = \frac{2s^2 + 2s + 1}{2s + 1}$$

Replacing the second column by the right-hand side we get

$$N_2 = (2s + 2) J_1$$

and

$$z_{21} = \frac{V_2}{J_1} = \frac{2s^2 + 2s}{2s + 1}$$

(b) The second solution differs from the first one in the right-hand-side vector only; J_2 will be in the second position and zeros in the first and third. The determinant is already known. Since the network is reciprocal, it must be true that $z_{12} = z_{21}$. In addition, since the network is symmetrical, it will also be true that $z_{22} = z_{11}$ (check!). The open circuit impedance matrix is

$$\mathbf{Z} = \begin{bmatrix} \dfrac{2s^2 + 2s + 1}{2s + 1} & \dfrac{2s^2 + 2s}{2s + 1} \\ \dfrac{2s^2 + 2s}{2s + 1} & \dfrac{2s^2 + 2s + 1}{2s + 1} \end{bmatrix}$$

13.1.2. Admittance Parameters

The other obvious selection is to take the voltages as independent variables. Then the general two-port equations will be

$$I_1 = y_{11}V_1 + y_{12}V_2$$
$$I_2 = y_{21}V_1 + y_{22}V_2$$

(13.1.8)

or in matrix form,

$$\begin{bmatrix} I_1 \\ I_2 \end{bmatrix} = \begin{bmatrix} y_{11} & y_{12} \\ y_{21} & y_{22} \end{bmatrix} \begin{bmatrix} V_1 \\ V_2 \end{bmatrix}$$

(13.1.9)

The equations can be clearly simplified if we select $V_2 = 0$, which is achieved by short-circuiting the second port. Attaching a source at the first port we get

$$I_1 = y_{11}V_1$$
$$I_2 = y_{21}V_1$$

Dividing by V_1

$$y_{11} = \frac{I_1}{V_1}\bigg|_{V_2=0}$$

(13.1.10)

$$y_{21} = \frac{I_2}{V_1}\bigg|_{V_2=0}$$

(13.1.11)

Transferring the source to the second port and short-circuiting the first port

$$I_1 = y_{12}V_2$$
$$I_2 = y_{22}V_2$$

and dividing by V_2

$$y_{12} = \frac{I_1}{V_2}\bigg|_{V_1=0}$$

(13.1.12)

$$y_{22} = \frac{I_2}{V_2}\bigg|_{V_1=0}$$

(13.1.13)

The parameters y_{ij} have the dimension of an admittance and are sometimes referred to as *two-port admittance parameters*. Since they were obtained by short-circuiting the ports, they are also known as *short-circuit parameters*.

We will demonstrate derivation of the parameters on the example of a Π network, shown in Fig. 13.1.5. Short-circuit the second port, attach a current source at the first port and note the direction of I_2 in Fig. 13.1.5a. Because of the short-circuit, we have

$$I_1 = V_1(Y_1 + Y_3)$$

from which we get

$$y_{11} = \frac{I_1}{V_1} = Y_1 + Y_3$$

In determining I_2 we realize that this current is exactly equal to the current flowing through Y_3 but *in opposite direction*:

$$-I_2 = V_1 Y_3$$

and

$$y_{21} = \frac{I_2}{V_1} = -Y_3$$

In the next step we place the current source at the second port and short-circuit the first port, as shown in Fig. 13.1.5b. Proceeding similarly as above, we obtain

$$y_{12} = \frac{I_1}{V_2} = -Y_3$$

$$y_{22} = \frac{I_2}{V_2} = Y_2 + Y_3$$

Collecting the terms

$$\mathbf{Y} = \begin{bmatrix} Y_1 + Y_3 & -Y_3 \\ -Y_3 & Y_2 + Y_3 \end{bmatrix} \tag{13.1.4}$$

By setting Y_1 or Y_2 or both equal to zero we get the short-circuit parameters for

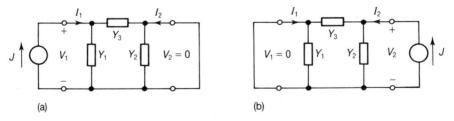

(a)   (b)

FIGURE 13.1.5 Y parameters for a Π network. (a) For y_{11} and y_{21}. (b) For y_{12} and y_{22}.

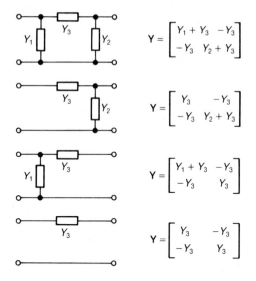

FIGURE 13.1.6   Π network and its simplifications.

three additional networks. The networks and their matrices are shown in Fig. 13.1.6. We note again that the simplest network, consisting of only Y_3, has a singular admittance matrix. If the network is without dependent sources, then y_{12} is always equal to y_{21} (reciprocal network).

Example 2.

Find the short-circuit parameters using the network in Fig. 13.1.7a and b. The network is the same as in Fig. 13.1.4, but is already redrawn for this purpose.

We are free to select the independent sources (current or voltage), but in order to go via nodal formulation we select current sources. They are in the figure.

(a) Nodal equations for the network are

$$\begin{bmatrix} s + 1 & -1 \\ -1 & \dfrac{2s + 1}{s} \end{bmatrix} \begin{bmatrix} V_1 \\ V_2 \end{bmatrix} = \begin{bmatrix} J_1 \\ 0 \end{bmatrix}$$

The determinant is

$$D = \frac{2s^2 + 2s + 1}{s}$$

$$V_1 = \frac{2s + 1}{2s^2 + 2s + 1} J_1$$

$$V_2 = \frac{s}{2s^2 + 2s + 1} J_1$$

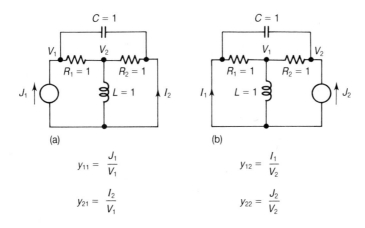

$$y_{11} = \frac{J_1}{V_1}$$

$$y_{21} = \frac{I_2}{V_1}$$

$$y_{12} = \frac{I_1}{V_2}$$

$$y_{22} = \frac{J_2}{V_2}$$

FIGURE 13.1.7 Y parameters for a bridged T network: (a) Connection to calculate y_{11} and y_{21}. (b) Connection to calculate y_{12} and y_{22}.

and we have immediately

$$y_{11} = \frac{J_1}{V_1} = \frac{2s^2 + 2s + 1}{2s + 1}$$

We also need the current I_2. It flows in the opposite direction and therefore

$$-I_2 = sV_1 + V_2$$

with the result

$$I_2 = -\frac{2s^2 + 2s}{2s^2 + 2s + 1} J_1$$

Forming the ratio

$$y_{21} = \frac{I_2}{V_1} = -\frac{2s^2 + 2s}{2s + 1}$$

The network is reciprocal and symmetric and thus $y_{12} = y_{21}$ and also $y_{11} = y_{22}$ (check!). The short-circuit admittance matrix is

$$\mathbf{Y} = \begin{bmatrix} \dfrac{2s^2 + 2s + 1}{2s + 1} & -\dfrac{2s^2 + 2s}{2s + 1} \\[2ex] -\dfrac{2s^2 + 2s}{2s + 1} & \dfrac{2s^2 + 2s + 1}{2s + 1} \end{bmatrix}$$

13.1.3. Cascade Parameters

One useful selection of independent and dependent variables leads to the *cascade* or *transmission* or *ABCD* parameters. All three names apply to the same situation and the choice

$$V_1 = t_{11}V_2 - t_{12}I_2$$
$$I_1 = t_{21}V_2 - t_{22}I_2$$

(13.1.15)

In matrix form,

$$\begin{bmatrix} V_1 \\ I_1 \end{bmatrix} = \begin{bmatrix} t_{11} & t_{12} \\ t_{21} & t_{22} \end{bmatrix} \begin{bmatrix} V_2 \\ -I_2 \end{bmatrix}$$

(13.1.16)

Note especially the minus sign in the right-hand-side vector. Since in some books these parameters are also referred to as *ABCD* parameters, remember that

$$\begin{bmatrix} A & B \\ C & D \end{bmatrix} = \begin{bmatrix} t_{11} & t_{12} \\ t_{21} & t_{22} \end{bmatrix}$$

(13.1.17)

We will use the t_{ij} notation.

The equations can be simplified if we select $I_2 = 0$, which is achieved by *open-circuiting* the second port and attaching a source at the first port. This leads to

$$V_1 = t_{11}V_2$$
$$I_1 = t_{21}V_2$$

and dividing by V_2

$$t_{11} = \left. \frac{V_1}{V_2} \right|_{I_2=0}$$

(13.1.18)

$$t_{21} = \left. \frac{I_1}{V_2} \right|_{I_2=0}$$

(13.1.19)

The other set of parameters is obtained by setting $V_2 = 0$. This means that we must short-circuit the second port and leave the source at the first port. Doing so we get

$$V_1 = -t_{12}I_2$$
$$I_1 = -t_{22}I_2$$

and dividing by I_2

$$t_{12} = \left. -\frac{V_1}{I_2} \right|_{V_2=0}$$

(13.1.20)

$$t_{22} = \left. -\frac{I_1}{I_2} \right|_{V_2=0}$$

(13.1.21)

FIGURE 13.1.8   Transmission parameters for an admittance: (a) Connection to calculate t_{11} and t_{21}. (b) Connection to calculate t_{12} and t_{22}.

FIGURE 13.1.9   Transmission parameters for an impedance: (a) Connection to calculate t_{11} and t_{21}. (b) Connection to calculate t_{12} and t_{22}.

The parameters t_{ij} have mixed dimensions: t_{11} and t_{22} are dimensionless while t_{21} has the dimension of an admittance and t_{12} the dimension of an impedance. They can be measured relatively easily because we do not have to move the source. All we must do is open-circuit or short-circuit the second port.

We will demonstrate derivation of the parameters on two special cases. The first one is an admittance connected as shown in Fig. 13.1.8. Attach a current source at the first port and leave the second port open as shown in Fig. 13.1.8a. Clearly, $V_1 = V_2$ and, as follows from Eq. (13.1.18), $t_{11} = 1$. From Eq. (13.1.19), we have $I_1/V_2 = I_1/V_1 = Y$ and $t_{21} = Y$. In Fig. 13.1.8b is the next situation, with the second port short-circuited. Since now $I_1 = -I_2$, it follows that $t_{22} = 1$. Finally, since in this case $V_2 = 0$, the last element of the transmission matrix is $t_{12} = 0$. The **T** matrix is in Fig. 13.1.8.

The second example is an impedance connecting the input and output ports, Fig. 13.1.9. Let us now use a voltage source and open the second port, as in a. Since there is no current, $V_1 = V_2$, $t_{11} = 1$ and $t_{21} = 0$. In the next step, short-circuit the output, as shown in b. It is immediately seen that $I_1 = -I_2$ and $t_{22} = 1$. We also have $V_1 = ZI_1 = -ZI_2$ and $t_{12} = Z$. The **T** matrix is given in the same figure.

Example 3.

Find the cascade parameters for the network in Fig. 13.1.10a and b. It is the same network as in Fig. 13.1.4 but already redrawn for this purpose.

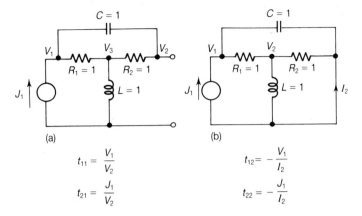

$$t_{11} = \frac{V_1}{V_2}$$

$$t_{21} = \frac{J_1}{V_2}$$

$$t_{12} = -\frac{V_1}{I_2}$$

$$t_{22} = -\frac{J_1}{I_2}$$

FIGURE 13.1.10   Transmission parameters for a bridged T: (a) For t_{11} and t_{21}. (b) For t_{12} and t_{22}.

(a) This case was already solved in Example 1, Fig. 13.1.4a; we obtained there

$$V_1 = \frac{2s^2 + 2s + 1}{2s + 1} J_1$$

$$V_2 = \frac{2s^2 + 2s}{2s + 1} J_1$$

Their ratio gives

$$t_{11} = \frac{V_1}{V_2} = \frac{2s^2 + 2s + 1}{2s^2 + 2s}$$

and we also get directly

$$t_{21} = \frac{J_1}{V_1} = \frac{2s + 1}{2s^2 + 2s}$$

(b) This case was already solved in Example 2, Fig. 13.1.7a. We obtained

$$V_1 = \frac{2s + 1}{2s^2 + 2s + 1} J_1$$

$$-I_2 = \frac{2s^2 + 2s}{2s^2 + 2s + 1} J_1$$

Their ratio provides

$$t_{21} = -\frac{V_1}{I_2} = \frac{2s + 1}{2s^2 + 2s}$$

and also directly

$$t_{22} = -\frac{J_1}{I_2} = \frac{2s^2 + 2s + 1}{2s^2 + 2s}$$

The cascade matrix is

$$\mathbf{T} = \begin{bmatrix} \dfrac{2s^2 + 2s + 1}{2s^2 + 2s} & \dfrac{2s + 1}{2s^2 + 2s} \\[3mm] \dfrac{2s + 1}{2s^2 + 2s} & \dfrac{2s^2 + 2s + 1}{2s^2 + 2s} \end{bmatrix}$$

13.1.4. Hybrid Parameters

Other choices of independent and dependent variables are possible and two more will be discussed. One of them is the system called g parameters,

$$\begin{bmatrix} I_1 \\ V_2 \end{bmatrix} = \begin{bmatrix} g_{11} & g_{12} \\ g_{21} & g_{22} \end{bmatrix} \begin{bmatrix} V_1 \\ I_2 \end{bmatrix} \tag{13.1.22}$$

We now proceed routinely: The choice $I_2 = 0$ simplifies the equations to

$$I_1 = g_{11} V_1$$
$$V_2 = g_{21} V_1$$

and

$$g_{11} = \left. \frac{I_1}{V_1} \right|_{I_2 = 0} \tag{13.1.23}$$

$$g_{21} = \left. \frac{V_2}{V_1} \right|_{I_2 = 0} \tag{13.1.24}$$

To determine the other two parameters, set $V_1 = 0$. This simplifies the equations to

$$I_1 = g_{12} I_2$$
$$V_2 = g_{22} I_2$$

and

$$g_{12} = \left. \frac{I_1}{I_2} \right|_{V_1 = 0} \tag{13.1.25}$$

$$g_{22} = \left. \frac{V_2}{I_2} \right|_{V_1 = 0} \tag{13.1.26}$$

The other set of hybrid parameters are h parameters,

$$\begin{bmatrix} V_1 \\ I_2 \end{bmatrix} = \begin{bmatrix} h_{11} & h_{12} \\ h_{21} & h_{22} \end{bmatrix} \begin{bmatrix} I_1 \\ V_2 \end{bmatrix}$$

(13.1.27)

Following the above procedures we get

$$h_{11} = \frac{V_1}{I_1}\bigg|_{V_2 = 0}$$

$$h_{21} = \frac{I_2}{I_1}\bigg|_{V_2 = 0}$$

(13.1.28)

$$h_{12} = \frac{V_1}{V_2}\bigg|_{I_1 = 0}$$

$$h_{22} = \frac{I_2}{V_2}\bigg|_{I_1 = 0}$$

These parameters are sometimes used to define properties of transistors.

13.1.5. Two-Port Equivalent Networks

It is advantageous to find equivalent networks to the various two-port definitions; we will do so for the impedance and admittance parameters. The transformations will be used in Chapter 15. Consider the first equation of Eq. (13.1.1),

$$V_1 = z_{11}I_1 + z_{12}I_2$$

The two terms on the right describe voltages which are added; this means that they are in series. z_{11} is multiplied by I_1 and thus it must be an impedance. The second term expresses a voltage which depends on the current at the other port; it is a current-controlled voltage source. Similar argument applies for the second equation of Eq. (13.1.1)

$$V_2 = z_{21}I_1 + z_{22}I_2$$

We thus can drawn an equivalent network in terms of impedances and dependent sources, as shown in Fig. 13.1.11.

Small modifications of the equations can get us another equivalent network. Suppose that we simultaneously add and subtract the term $z_{12}I_1$ in the second equation and rearrange the terms:

$$V_1 = z_{11}I_1 + z_{12}I_2$$
$$V_2 = z_{12}I_1 + z_{22}I_2 + (z_{21} - z_{12})I_1$$

Fig. 13.1.11   Fig. 13.1.12

FIGURE 13.1.11   Equivalent network for Z parameters.

FIGURE 13.1.12   Another equivalent network for Z parameters.

Disregard for a moment the term in the brackets and look at the equations; they express a reciprocal two-port. The term in the brackets is a current-controlled voltage source, appearing at the second port. Thus we can draw an equivalent network, as shown in Fig. 13.1.12.

Similar steps can be applied to the admittance parameters

$$I_1 = y_{11}V_1 + y_{12}V_2$$
$$I_2 = y_{21}V_1 + y_{22}V_2$$

The sums of terms on the right represent elements which are connected in parallel, y_{11} is an admittance, $y_{12}V_2$ represents a voltage-controlled current source. Similar conclusions can be obtained for the second equation. The equivalent network, in terms of admittances and voltage-controlled current sources, is shown in Fig. 13.1.13.

If we simultaneously add and subtract the term $y_{12}V_1$ to the second equation and rearrange terms, we get

$$I_1 = y_{11}V_1 + y_{12}V_2$$
$$I_2 = y_{12}V_1 + y_{22}V_2 + (y_{21} - y_{12})V_1$$

Disregard for a moment the term in the brackets; without it the equations describe a reciprocal two-port. The term which we did not consider is a voltage-controlled

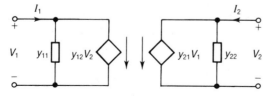

FIGURE 13.1.13   Equivalent network for Y parameters.

FIGURE 13.1.14   Another equivalent network for Y parameters.

current source, appearing at the second port only. We can thus draw the reciprocal network, as shown in Fig. 13.1.14, and add a current source at the second port; it is controlled by the voltage of the first port (V_1) and has a transconductance which is equal to $g = y_{21} - y_{12}$. Thus we can draw an equivalent network as shown in Fig. 13.1.14.

13.2. INTERCONNECTIONS OF TWO-PORTS

In this section we explain the main application of two-port parameters and the interconnections of two ports.

13.2.1. Series Connection of Two-Ports

Take first the open-circuit parameters and two networks, denoted by superscripts (1) and (2). Their equations are

$$\begin{bmatrix} V_1^{(1)} \\ V_2^{(1)} \end{bmatrix} = \begin{bmatrix} z_{11}^{(1)} & z_{12}^{(1)} \\ z_{21}^{(1)} & z_{22}^{(1)} \end{bmatrix} \begin{bmatrix} I_1^{(1)} \\ I_2^{(1)} \end{bmatrix} \tag{13.2.1}$$

$$\begin{bmatrix} V_1^{(2)} \\ V_2^{(2)} \end{bmatrix} = \begin{bmatrix} z_{11}^{(2)} & z_{12}^{(2)} \\ z_{21}^{(2)} & z_{22}^{(2)} \end{bmatrix} \begin{bmatrix} I_1^{(2)} \\ I_2^{(2)} \end{bmatrix} \tag{13.2.2}$$

Connect these networks as shown in Fig. 13.2.1 (series connection of two-ports). After the connection, we have

$$V_1 = V_1^{(1)} + V_1^{(2)}$$
$$V_2 = V_2^{(1)} + V_2^{(2)} \tag{13.2.3}$$

and we also note that

$$I_1 = I_1^{(1)} = I_1^{(2)}$$
$$I_2 = I_2^{(1)} = I_2^{(2)} \tag{13.2.4}$$

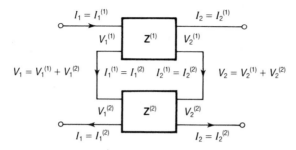

FIGURE 13.2.1   Series connection of two-ports.

Substitute Eq. (13.2.4) into the right-hand-side vectors of Eq. (13.2.1) and Eq. (13.2.2) and then add these two equations

$$\begin{bmatrix} V_1^{(1)} \\ V_2^{(1)} \end{bmatrix} + \begin{bmatrix} V_1^{(2)} \\ V_2^{(2)} \end{bmatrix} = \begin{bmatrix} z_{11}^{(1)} & z_{12}^{(1)} \\ z_{21}^{(1)} & z_{22}^{(1)} \end{bmatrix} \begin{bmatrix} I_1 \\ I_2 \end{bmatrix} + \begin{bmatrix} z_{11}^{(2)} & z_{12}^{(2)} \\ z_{21}^{(2)} & z_{22}^{(2)} \end{bmatrix} \begin{bmatrix} I_1 \\ I_2 \end{bmatrix}$$

On the left side we add the two vectors and use Eq. (13.2.3). On the right side the two vectors are now identical and we can write

$$\begin{bmatrix} V_1 \\ V_2 \end{bmatrix} = [\mathbf{Z}^{(1)} + \mathbf{Z}^{(2)}] \begin{bmatrix} I_1 \\ I_2 \end{bmatrix} \qquad (13.2.5)$$

For this type of connection, all we have to do is add the two matrices to get the overall matrix of the combined network. To be entirely general, we should also consider certain restrictions; they will be discussed at the end of this section.

The impedance interconnection is not particularly useful and we took it as an introduction. We will discuss the other cases more briefly because they follow the same type of reasoning.

13.2.2. Parallel Connection of Two-Ports

Parallel connection of two-ports has a number of practical applications. Consider again two networks, this time described by their admittance two-port matrices

$$\begin{bmatrix} I_1^{(1)} \\ I_2^{(1)} \end{bmatrix} = \begin{bmatrix} y_{11}^{(1)} & y_{12}^{(1)} \\ y_{21}^{(1)} & y_{22}^{(1)} \end{bmatrix} \begin{bmatrix} V_1^{(1)} \\ V_2^{(1)} \end{bmatrix} \qquad (13.2.6)$$

$$\begin{bmatrix} I_1^{(2)} \\ I_2^{(2)} \end{bmatrix} = \begin{bmatrix} y_{11}^{(2)} & y_{12}^{(2)} \\ y_{21}^{(2)} & y_{22}^{(2)} \end{bmatrix} \begin{bmatrix} V_1^{(2)} \\ V_2^{(2)} \end{bmatrix} \qquad (13.2.7)$$

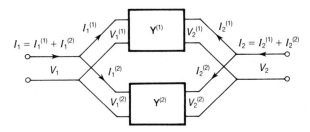

FIGURE 13.2.2   Parallel connection of two-ports.

and connect them as shown in Fig. 13.2.2. The connection clearly makes

$$V_1 = V_1^{(1)} = V_1^{(2)}$$
$$V_2 = V_2^{(1)} = V_2^{(2)}$$

(13.2.8)

and also

$$I_1 = I_1^{(1)} + I_1^{(2)}$$
$$I_2 = I_2^{(1)} + I_2^{(2)}$$

(13.2.9)

If we substitute V_1 and V_2 into Eq. (13.2.6) and Eq. (13.2.7), the right-hand-side vectors will be the same ones and we can add both equations. As a result, we get

$$\begin{bmatrix} I_1 \\ I_2 \end{bmatrix} = [\mathbf{Y}^{(1)} + \mathbf{Y}^{(2)}] \begin{bmatrix} V_1 \\ V_2 \end{bmatrix}$$

(13.2.10)

and we conclude that if we connect two-ports as indicated in Fig. 13.2.2, then the admittance matrix of the resulting network will be the sum of the matrices of the original networks. Again, some restrictions apply and will be mentioned at the end of the section.

13.2.3. Cascade Connection of Two-Ports

In this quite important case we consider two networks whose two-port descriptions have been obtained in the form

$$\begin{bmatrix} V_1^{(1)} \\ I_1^{(1)} \end{bmatrix} = \begin{bmatrix} t_{11}^{(1)} & t_{12}^{(1)} \\ t_{21}^{(1)} & t_{22}^{(1)} \end{bmatrix} \begin{bmatrix} V_2^{(1)} \\ -I_2^{(1)} \end{bmatrix}$$

(13.2.11)

$$\begin{bmatrix} V_1^{(2)} \\ I_1^{(2)} \end{bmatrix} = \begin{bmatrix} t_{11}^{(2)} & t_{12}^{(2)} \\ t_{21}^{(2)} & t_{22}^{(2)} \end{bmatrix} \begin{bmatrix} V_2^{(2)} \\ -I_2^{(2)} \end{bmatrix}$$

(13.2.12)

FIGURE 13.2.3   Cascade connection of two-ports.

Connect the networks as shown in Fig. 13.2.3; this will enforce

$$\begin{bmatrix} V_2^{(1)} \\ -I_2^{(1)} \end{bmatrix} = \begin{bmatrix} V_1^{(2)} \\ I_1^{(2)} \end{bmatrix}$$

(13.2.13)

and we can substitute the right-hand-side of Eq. (13.2.12) for the right-hand-side vector in Eq. (13.2.11). The result is

$$\begin{bmatrix} V_1^{(1)} \\ I_1^{(1)} \end{bmatrix} = [\mathbf{T}^{(1)}\mathbf{T}^{(2)}] \begin{bmatrix} V_2^{(2)} \\ -I_2^{(2)} \end{bmatrix}$$

(13.2.14)

The cascade matrix of the overall network is equal to the *product* of the two cascade matrices. This is the only case for which the matrices are multiplied. In all other cases they are added.

13.2.4. Hybrid Connections of Two-Ports

The hybrid connections have little practical application but using the same steps as above, we can find that the overall matrix will be $\mathbf{G} = \mathbf{G}^{(1)} + \mathbf{G}^{(2)}$ if the networks are connected as shown in Fig. 13.2.4. This connection is called *parallel-series* (parallel on the left and series on the right).

If the two-ports are connected as shown in Fig. 13.2.5, the H matrices are added, $\mathbf{H} = \mathbf{H}^{(1)} + \mathbf{H}^{(2)}$. This connection is called *series-parallel* (series on the left and parallel on the right).

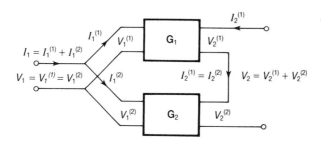

FIGURE 13.2.4   Parallel-series connection of two-ports.

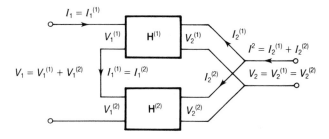

FIGURE 13.2.5   Series-parallel connection of two-ports.

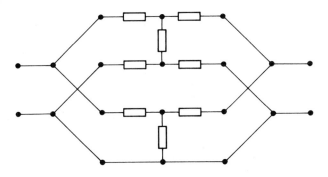

FIGURE 13.2.6   Addition of **Y** matrices not possible.

We will conclude this section by showing, in Fig. 13.2.6, parallel connection of two two-ports for which the **Y** matrices *cannot* be added. In this combination, the lower network short-circuits some elements of the upper circuit. A theoretical remedy is available in the use of an ideal 1:1 transformer. Unfortunately, ideal transformers in reality do not exist, they are only an abstraction introduced into network theory. Our suggestion is that whenever the reader encounters a similar suspicious situation, then he should apply standard analysis methods and avoid the two-port theory and its interconnections.

13.3. SIMPLE TWO-PORTS

We will start with dependent sources, because they are already known; all we have to do is put their equations into proper two-port form. The other elements we consider in this section are the ideal transformer, the gyrator and the impedance converter. All two-port equations in this section will be in the cascade form.

13.3.1. Dependent Sources

The controlled sources are collected in Fig. 13.3.1. The voltage-controlled current source, VC, is defined by the equations

$$I_1 = 0$$
$$I_2 = gV_1$$

To give the properties in cascade form, we rewrite them

$$V_1 = \frac{I_2}{g}$$

$$I_1 = 0$$

(13.3.1)

and in matrix form

$$\begin{bmatrix} V_1 \\ I_1 \end{bmatrix} = \begin{bmatrix} 0 & -1/g \\ 0 & 0 \end{bmatrix} \begin{bmatrix} V_2 \\ -I_2 \end{bmatrix}$$

(13.3.2)

The voltage-controlled voltage source, VV, is described by the equations

$$I_1 = 0$$
$$V_2 = \mu V_1$$

FIGURE 13.3.1   Dependent sources as two-ports.

and the cascade matrix takes the form

$$\begin{bmatrix} V_1 \\ I_1 \end{bmatrix} = \begin{bmatrix} 1/\mu & 0 \\ 0 & 0 \end{bmatrix} \begin{bmatrix} V_2 \\ -I_2 \end{bmatrix}$$

((13.3.3))

The current-controlled voltage source, CV, is described by

$$V_1 = 0$$
$$V_2 = rI_1$$

and the cascade form is

$$\begin{bmatrix} V_1 \\ I_1 \end{bmatrix} = \begin{bmatrix} 0 & 0 \\ 1/r & 0 \end{bmatrix} \begin{bmatrix} V_2 \\ -I_2 \end{bmatrix}$$

(13.3.4)

Finally, the current-controlled current source, CC, is defined by the equations

$$V_1 = 0$$
$$I_2 = \alpha I_1$$

and in matrix form

$$\begin{bmatrix} V_1 \\ I_1 \end{bmatrix} = \begin{bmatrix} 0 & 0 \\ 0 & -1/\alpha \end{bmatrix} \begin{bmatrix} V_2 \\ -I_2 \end{bmatrix}$$

(13.3.5)

13.3.2. Ideal Transformer

The ideal transformer is a theoretical abstraction, but actual transformers can sometimes have properties which resemble those of the ideal transformer. Details are postponed to the next chapter. Here we give the two-port definition of an ideal transformer

$$V_1 = nV_2$$
$$I_1 = -\frac{1}{n}I_2$$

(13.3.6)

and its symbol in Fig. 13.3.2. The minus sign can also be in the first equation instead of the second. Properties of this idealized device do not depend on frequency and apply equally for dc and ac.

FIGURE 13.3.2   Ideal transformer.

Placing the equations into the two-port cascade form

$$\begin{bmatrix} V_1 \\ I_1 \end{bmatrix} = \begin{bmatrix} n & 0 \\ 0 & 1/n \end{bmatrix} \begin{bmatrix} V_2 \\ -I_2 \end{bmatrix}$$

(13.3.7)

13.3.3. Gyrator

The gyrator is a device which has been extensively used in the design of low-frequency active filters. It has the property that when loaded by a capacitor, its input impedance becomes that of an inductor. A gyrator can be realized by means of two operational amplifiers and we have already analyzed one network having similar properties; see Fig. 13.4.5 and the discussion relating to this network. Additional application details are in Section 13.4.5.

The gyrator is often denoted by the symbol in Fig. 13.3.3a and is defined by the equations

$$I_1 = -g_2 V_2$$
$$I_2 = g_1 V_1$$

(13.3.8)

The minus sign can be placed into the second equation instead of the first one. Looking carefully at these equatons, we see that they describe two voltage-controlled current sources, connected as shown in Fig. 13.3.3b. The equations can also be modified into the impedance form by rewriting them as

$$V_1 = \frac{I_2}{g_1}$$
$$V_2 = -\frac{I_1}{g_2}$$

(13.3.9)

If we set $r_1 = 1/g_2$ and $r_2 = 1/g_1$, then the equations describe two current-controlled voltage sources connected as shown in Fig. 13.3.3c. The equations have

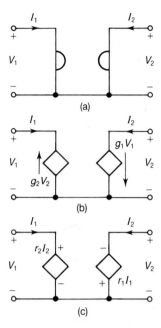

FIGURE 13.3.3   Gyrator: (a) Symbol. (b) Realization with two *VC*. (c) Realization with two *CV*.

the cascade form

$$\begin{bmatrix} V_1 \\ I_1 \end{bmatrix} = \begin{bmatrix} 0 & -1/g_1 \\ -g_2 & 0 \end{bmatrix} \begin{bmatrix} V_2 \\ -I_2 \end{bmatrix} \qquad (13.3.10)$$

The gyrator is an important device which has been used extensively in the construction of active filters and we will give its application later in this chapter.

13.3.4. Impedance Convertor

The impedance convertor is a device described by equations

$$V_1 = k_1 V_2$$
$$I_1 = -k_2 I_2 \qquad (13.3.11)$$

The minus sign can be moved to the upper equation. As long as there is one minus sign in Eq. (13.3.11), the device is called the *positive impedance convertor*. Its cascade matrix equation has the form

$$\begin{bmatrix} V_1 \\ I_1 \end{bmatrix} = \begin{bmatrix} k_1 & 0 \\ 0 & k_2 \end{bmatrix} \begin{bmatrix} V_2 \\ -I_2 \end{bmatrix} \qquad (13.3.12)$$

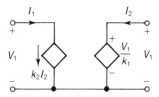

FIGURE 13.3.4   Modeling a *NIC* by one *CC* and one *VV*.

If the signs in both equations in Eq. (13.3.11) are the same, either $+$ or $-$, then the device is called the *negative impedance converter*. Reasons for these names will become clear in Section 13.4.6. At low frequencies, this device can be realized with operational amplifiers or transistors. If we compare the cascade matrix in Eq. (13.3.12) with the cascade matrix of the ideal transformer, Eq. (13.3.7), then we see that the positive impedance convertor will become an ideal transformer if $k_1 = n$ and $k_2 = 1/n$. The impedance convertor is a more general device than the ideal transformer. No standard figure is used for it in the literature; it can be drawn as a two-port box with the letters PIC (for positive) or NIC (for negative) impedance convertor written into it. It can also be modeled by means of dependent sources. Rewriting the NIC equations in the form

$$V_2 = \frac{V_1}{k_1}$$

$$I_1 = k_2 I_2$$

we see that the element can be modeled by one *CC* and one *VV* as shown in Fig. 13.3.4.

13.4. APPLICATION OF TWO-PORT THEORY

In this section we will show a number of applications, some quite practical, some given more for historical reasons.

13.4.1. Transformations of Two-Port Parameters

First we establish that, in the case of a general network, any two-port description can be transformed into another one.

We will start by transforming the impedance parameters into cascade parameters. The impedance parameters were defined in Eq. (13.1.1), here written as two equations

$$V_1 = z_{11}I_1 + z_{12}I_2 \tag{13.4.1}$$

$$V_2 = z_{21}I_1 + z_{22}I_2 \tag{13.4.2}$$

The cascade parameters require V_1 and I_1 on the left side of the equations, see Eq. (13.1.15). In order to achieve this modification, separate the current I_1 in Eq. (13.4.2)

$$I_1 = \frac{1}{z_{21}} V_2 - \frac{z_{22}}{z_{21}} I_2 \qquad (13.4.3)$$

and insert into Eq. (13.4.1)

$$V_1 = \frac{z_{11}}{z_{21}} V_2 - \frac{z_{11}z_{22} - z_{12}z_{21}}{z_{21}} I_2 \qquad (13.4.4)$$

These last two equations already have the necessary form, but we note that

$$\Delta_z = z_{11}z_{22} - z_{12}z_{21} \qquad (13.4.5)$$

is the determinant of the $\mathbf{Z}$ matrix. Inserting into Eq. (13.4.4) and writing in matrix form

$$\begin{bmatrix} V_1 \\ I_1 \end{bmatrix} = \begin{bmatrix} \dfrac{z_{11}}{z_{21}} & \dfrac{\Delta_z}{z_{21}} \\ \dfrac{1}{z_{21}} & \dfrac{z_{22}}{z_{21}} \end{bmatrix} \begin{bmatrix} V_2 \\ -I_2 \end{bmatrix} \qquad (13.4.6)$$

We have expressed the cascade matrix in terms of the impedance parameters. Similar conversions are possible between any pair of two-port parameters and the expressions are summarized in Fig 13.4.1. At the bottom of the figure are the necessary equations and definitions of the independent and dependent variables. The symbol Δ denotes the determinant of the matrix. It is interesting to note that all conversions require evaluation of the determinant of the two-port matrix and for some networks the determinant may be zero. No harm is done if the transformation puts zero determinant into the numerator of the expression, but the transformation is impossible if a zero determinant appears in the denominator. We have already encountered such networks, see the bottom of Fig. 13.1.3 and Fig. 13.1.5. The matrices given there are singular and cannot be inverted. This means that the single impedance Z_3 in Fig. 13.1.3 cannot be expressed in terms of the admittance parameters. Similarly, the single admittance Y_3 in Fig. 13.1.6 cannot be expressed in terms of impedance parameters.

Figure 13.4.1 can save us a lot of work. The parameters which are easiest to derive are the open-circuit (or impedance) parameters. All we have to do is to attach sources, first at port one and next at the port two. It does not matter which type of independent sources we chose, but since nodal formulation is generally preferred, we suggest the use of independent current sources. The use of the conversion table in Fig. 13.4.1 then provides any other set of two-port parameters.

	Z	Y	T	H	G
Z	$\begin{matrix} z_{11} & z_{12} \\ z_{21} & z_{22} \end{matrix}$	$\begin{matrix} \dfrac{y_{22}}{\Delta_y} & -\dfrac{y_{12}}{\Delta_y} \\[2mm] -\dfrac{y_{21}}{\Delta_y} & \dfrac{y_{11}}{\Delta_y} \end{matrix}$	$\begin{matrix} \dfrac{t_{11}}{t_{21}} & \dfrac{\Delta_t}{t_{21}} \\[2mm] \dfrac{1}{t_{21}} & \dfrac{t_{22}}{t_{21}} \end{matrix}$	$\begin{matrix} \dfrac{\Delta_h}{h_{22}} & \dfrac{h_{12}}{h_{22}} \\[2mm] -\dfrac{h_{21}}{h_{22}} & \dfrac{1}{h_{22}} \end{matrix}$	$\begin{matrix} \dfrac{1}{g_{11}} & -\dfrac{g_{12}}{g_{11}} \\[2mm] \dfrac{g_{21}}{g_{11}} & \dfrac{\Delta_g}{g_{11}} \end{matrix}$
Y	$\begin{matrix} \dfrac{z_{22}}{\Delta_z} & -\dfrac{z_{12}}{\Delta_z} \\[2mm] -\dfrac{z_{21}}{\Delta_z} & \dfrac{z_{11}}{\Delta_z} \end{matrix}$	$\begin{matrix} y_{11} & y_{12} \\ y_{21} & y_{22} \end{matrix}$	$\begin{matrix} \dfrac{t_{22}}{t_{12}} & -\dfrac{\Delta_t}{t_{12}} \\[2mm] \dfrac{-1}{t_{12}} & \dfrac{t_{11}}{t_{12}} \end{matrix}$	$\begin{matrix} \dfrac{1}{h_{11}} & -\dfrac{h_{12}}{h_{11}} \\[2mm] \dfrac{h_{21}}{h_{11}} & \dfrac{\Delta_h}{h_{11}} \end{matrix}$	$\begin{matrix} \dfrac{\Delta_g}{g_{22}} & \dfrac{g_{12}}{g_{22}} \\[2mm] -\dfrac{g_{21}}{g_{22}} & \dfrac{1}{g_{22}} \end{matrix}$
T	$\begin{matrix} \dfrac{z_{11}}{z_{21}} & \dfrac{\Delta_z}{z_{21}} \\[2mm] \dfrac{1}{z_{21}} & \dfrac{z_{22}}{z_{21}} \end{matrix}$	$\begin{matrix} -\dfrac{y_{22}}{y_{21}} & -\dfrac{1}{y_{21}} \\[2mm] -\dfrac{\Delta_y}{y_{21}} & -\dfrac{y_{11}}{y_{21}} \end{matrix}$	$\begin{matrix} t_{11} & t_{12} \\ t_{21} & t_{22} \end{matrix}$	$\begin{matrix} -\dfrac{\Delta_h}{h_{21}} & -\dfrac{h_{11}}{h_{21}} \\[2mm] -\dfrac{h_{22}}{h_{21}} & -\dfrac{1}{h_{21}} \end{matrix}$	$\begin{matrix} \dfrac{1}{g_{21}} & \dfrac{g_{22}}{g_{21}} \\[2mm] \dfrac{g_{11}}{g_{21}} & \dfrac{\Delta_g}{g_{21}} \end{matrix}$
H	$\begin{matrix} \dfrac{\Delta_z}{z_{22}} & \dfrac{z_{12}}{z_{22}} \\[2mm] -\dfrac{z_{21}}{z_{22}} & \dfrac{1}{z_{22}} \end{matrix}$	$\begin{matrix} \dfrac{1}{y_{11}} & -\dfrac{y_{12}}{y_{11}} \\[2mm] \dfrac{y_{21}}{y_{11}} & \dfrac{\Delta_y}{y_{11}} \end{matrix}$	$\begin{matrix} \dfrac{t_{12}}{t_{22}} & \dfrac{\Delta_t}{t_{22}} \\[2mm] \dfrac{-1}{t_{22}} & \dfrac{t_{21}}{t_{22}} \end{matrix}$	$\begin{matrix} h_{11} & h_{12} \\ h_{21} & h_{22} \end{matrix}$	$\begin{matrix} \dfrac{g_{22}}{\Delta_g} & -\dfrac{g_{12}}{\Delta_g} \\[2mm] -\dfrac{g_{21}}{\Delta_g} & \dfrac{g_{11}}{\Delta_g} \end{matrix}$
G	$\begin{matrix} \dfrac{1}{z_{11}} & -\dfrac{z_{12}}{z_{11}} \\[2mm] \dfrac{z_{21}}{z_{11}} & \dfrac{\Delta_z}{z_{11}} \end{matrix}$	$\begin{matrix} \dfrac{\Delta_y}{y_{22}} & \dfrac{y_{12}}{y_{22}} \\[2mm] -\dfrac{y_{21}}{y_{22}} & \dfrac{1}{y_{22}} \end{matrix}$	$\begin{matrix} \dfrac{t_{21}}{t_{11}} & -\dfrac{\Delta_t}{t_{11}} \\[2mm] \dfrac{1}{t_{11}} & \dfrac{t_{12}}{t_{11}} \end{matrix}$	$\begin{matrix} \dfrac{h_{22}}{\Delta_h} & -\dfrac{h_{12}}{\Delta_h} \\[2mm] -\dfrac{h_{21}}{\Delta_h} & \dfrac{h_{11}}{\Delta_h} \end{matrix}$	$\begin{matrix} g_{11} & g_{12} \\ g_{21} & g_{22} \end{matrix}$
	$\begin{bmatrix} V_1 \\ V_2 \end{bmatrix} = \mathbf{Z} \begin{bmatrix} I_1 \\ I_2 \end{bmatrix}$	$\begin{bmatrix} I_1 \\ I_2 \end{bmatrix} = \mathbf{Y} \begin{bmatrix} V_1 \\ V_2 \end{bmatrix}$	$\begin{bmatrix} V_1 \\ I_1 \end{bmatrix} = \mathbf{T} \begin{bmatrix} V_2 \\ -I_2 \end{bmatrix}$	$\begin{bmatrix} V_1 \\ I_2 \end{bmatrix} = \mathbf{H} \begin{bmatrix} I_1 \\ V_2 \end{bmatrix}$	$\begin{bmatrix} I_1 \\ V_2 \end{bmatrix} = \mathbf{G} \begin{bmatrix} V_1 \\ I_2 \end{bmatrix}$

FIGURE 13.4.1   Conversion table for two-ports.

Example 1.

Use the open-circuit impedance parameters derived in Section 1 for the network in Fig. 13.1.4 and get the short-circuit admittance and cascade parameters using Fig. 13.4.1. Compare the results with the results of Example 2 and Example 3 in Section 1.

13.4.2. Star–Mesh Transformation

The *star–mesh* (or star-delta, or Y/Δ) transformation changes the *star* or *T* network into a *mesh* or Δ network. Both networks have exactly the same properties as far as their port currents and voltages are concerned. The transformation is indicated in Fig. 13.4.2a and b. Consider the *star* network in Fig. 13.4.2a and its impedance

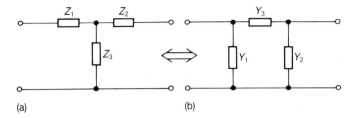

FIGURE 13.4.2   Star–mesh transformation: (a) Star network. (b) Mesh network.

parameters

$$\begin{bmatrix} V_1 \\ V_2 \end{bmatrix} = \begin{bmatrix} Z_1 + Z_3 & Z_3 \\ Z_3 & Z_2 + Z_3 \end{bmatrix} \begin{bmatrix} I_1 \\ I_2 \end{bmatrix}$$

(13.4.7)

Separate the currents by inverting the **Z** matrix

$$\begin{bmatrix} I_1 \\ I_2 \end{bmatrix} = \frac{1}{\Delta_z} \begin{bmatrix} Z_2 + Z_3 & -Z_3 \\ -Z_3 & Z_1 + Z_3 \end{bmatrix} \begin{bmatrix} V_1 \\ V_2 \end{bmatrix}$$

(13.4.8)

and compare with the short-circuit admittance parameters of the mesh network in Fig. 13.4.2b

$$\begin{bmatrix} I_1 \\ I_2 \end{bmatrix} = \begin{bmatrix} Y_1 + Y_3 & -Y_3 \\ -Y_3 & Y_2 + Y_3 \end{bmatrix} \begin{bmatrix} V_1 \\ V_2 \end{bmatrix}$$

(13.4.9)

If the networks have to be exactly equivalent, the matrices in Eq. (13.4.8) and Eq. (13.4.9) must be equal. Comparing term by term we arrive at

$$Y_1 = \frac{Z_2}{\Delta_z}$$

$$Y_2 = \frac{Z_1}{\Delta_z}$$

(13.4.10)

$$Y_3 = \frac{Z_3}{\Delta_z}$$

where

$$\Delta_z = Z_1 Z_2 + Z_1 Z_3 + Z_2 Z_3$$

(13.4.11)

is the determinant of the **Z** matrix. Equations (13.4.10) and (13.4.11) are used for transformation of the star network into the mesh network.

In order to go in the reverse direction, it is easiest to invert the system Eq. (13.4.9) to get

$$\begin{bmatrix} V_1 \\ V_2 \end{bmatrix} = \frac{1}{\Delta_y} \begin{bmatrix} Y_2 + Y_3 & Y_3 \\ Y_3 & Y_1 + Y_3 \end{bmatrix} \begin{bmatrix} I_1 \\ I_2 \end{bmatrix}$$
(13.4.12)

and compare the matrix with the matrix in (13.4.7). The result is

$$Z_1 = \frac{Y_2}{\Delta_y}$$

$$Z_2 = \frac{Y_1}{\Delta_y}$$
(13.4.13)

$$Z_3 = \frac{Y_3}{\Delta_y}$$

where Δ_y is the determinant of the **Y** matrix,

$$\Delta_y = Y_1 Y_2 + Y_1 Y_3 + Y_2 Y_3$$
(13.4.14)

Equations (13.4.13) and (13.4.14) are used when transforming the mesh network into the star network.

Example 2.

The network in Fig. 13.4.3a has a star-portion marked by the dashed lines. Transform the Y portion in the dashed box into a Δ network in Fig. 13.4.3b.

(a)

$Z_1 = Z_2 = 1$
$Z_3 = s$

(b)

FIGURE 13.4.3   Applying star–mesh transformation: (a) Original network. (b) Transformed network.

FIGURE 13.4.4   Equivalent to network in Fig. 13.4.3(a).

Using Eq. (13.4.11) we calculate $\Delta = 2s + 1$ and using Eq. (13.4.1) we get

$$Y_1 = \frac{1}{2s + 1}$$

$$Y_2 = Y_1$$

$$Y_3 = \frac{s}{2s + 1}$$

It turns out that these admittances can be realized as combinations of elements; we give them as a matter of interest in Fig. 13.4.4.

13.4.3. Cascade Connection

One of the situations where two-port parameters can be practical is analysis of ladder networks consisting of impedances in series and admittances in parallel. We will use the network in Fig. 13.4.5 to indicate this type of application, but it should be understood that direct analysis based on nodal or loop formulation is more efficient. Partition the network into subnetworks as indicated by the dashed lines and use the cascade parameters from Fig. 13.1.8 and Fig. 13.1.9. The overall

FIGURE 13.4.5   Multiplication of cascade matrices.

matrix is the product of the individual matrices,

$$\mathbf{T} = \begin{bmatrix} 1 & Z_1 \\ 0 & 1 \end{bmatrix} \begin{bmatrix} 1 & 0 \\ Y_2 & 1 \end{bmatrix} \begin{bmatrix} 1 & Z_3 \\ 0 & 1 \end{bmatrix} \begin{bmatrix} 1 & 0 \\ Y_4 & 1 \end{bmatrix}$$ (13.4.15)

Multiplication of the matrices results in

$$\begin{bmatrix} (1 + Z_1 Y_2 + Z_1 Y_4 + Z_3 Y_4 + Z_1 Y_2 Z_3 Y_4) & (Z_1 + Z_3 + Z_1 Y_2 Z_3) \\ (Y_2 + Y_4 + Y_2 Z_3 Y_4) & (1 + Y_2 Z_3) \end{bmatrix}$$ (13.4.16)

13.4.4. Mixed Connection

When the types of connections are mixed, it is necessary to use transformations given in Fig. 13.4.1 and much of the usefulness of the two-port definitions is lost. We will nevertheless indicate how one could proceed.

Consider the network in Fig. 13.4.6 with subnetworks indicated by the boxes and with two-port matrices as indicated.

In the first step, multiply $\mathbf{T}^{(1)} \cdot \mathbf{T}^{(2)}$ to get the cascade matrix $\mathbf{T}^{(A)}$ of the first two networks. In order to connect this resulting subnetwork in parallel to network 3, it is first necessary to convert $\mathbf{T}^{(A)}$ into $\mathbf{Y}^{(A)}$ by using Fig. 13.4.1; after this step we can add the matrices and get $\mathbf{Y}^{(B)} = \mathbf{Y}^{(A)} + \mathbf{Y}^{(3)}$. In order to continue to the last subnetwork, we must again convert $\mathbf{Y}^{(B)}$ into $\mathbf{T}^{(B)}$ and only then can we get the overall $\mathbf{T}$ as the product of $\mathbf{T}^{(B)} \cdot \mathbf{T}^{(4)}$. We stress again that this is an obsolete method of analysis and should be avoided.

13.4.5. Gyrator

In the low-frequency region gyrators have been used extensively to simulate inductors by means of capacitors. We will show this type of gyrator application.

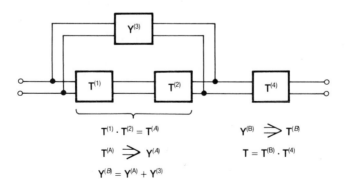

FIGURE 13.4.6   Steps to obtain overall **T** matrix.

FIGURE 13.4.7   Gyrator loaded by an admittance.

Consider first the gyrator loaded by an admittance, Fig. 13.4.7. We are looking for the input impedance, $Z_{in} = V_1/I_1$. The gyrator equations (13.3.8) are repeated here

$$I_1 = -g_2 V_2$$
$$I_2 = g_1 V_1$$

The loading impedance couples V_2 and I_2 by the equation

$$Y V_2 = -I_2$$

Here we note the negative sign because, by definition, the current I_2 flows into the two-port. Applying an independent current source $J = I_1$ and calculating V_1, we obtain the input impedance

$$Z_{in} = \frac{Y}{g_1 g_2}$$

If the loading admittance is a capacitor, we substitute $Y = sC$ and

$$Z_{in} = s \frac{C}{g_1 g_2} = sL$$

where

$$L = \frac{C}{g_1 g_2} \tag{13.4.17}$$

We conclude that input impedance of a gyrator loaded by a capacitor is equal to the impedance of an inductor.

We can use yet another way to obtain this result. Consider the cascade matrix of the gyrator Eq. (13.3.10) and the cascade matrix of an admittance. Fig. 13.1.8. The product of these two cascade matrices is equal to

$$\begin{bmatrix} V_1 \\ I_1 \end{bmatrix} = \begin{bmatrix} -Y/g_1 & -1/g_1 \\ -g_2 & 0 \end{bmatrix} \begin{bmatrix} V_2 \\ -I_2 \end{bmatrix}$$

If we do not attach any more networks on the right, then $I_2 = 0$ and the equations simplify to

$$V_1 = -\frac{Y}{g_1}$$

$$I_1 = -g_2$$

Their ratio is

$$Z_{in} = \frac{Y}{g_1 g_2} \tag{13.4.18}$$

the same as above.

Another example is in Fig. 13.4.8. The network consists of two gyrators with a capacitor in the middle, all grounded. Since the three elements are connected in cascade, we take the product of their cascade matrices to obtain the cascade matrix of the overall circuit. The matrices are in the figure. We discover that if we make the two gyrators equal, the result is equivalent to the cascade matrix of an inductor, both of whose terminals are floating. The resulting inductance is equal to

$$L = \frac{sC}{g^2} \tag{13.4.19}$$

This is very interesting, because three *grounded* devices simulate a *floating* inductor. It would be quite difficult to discover this property if we did not have two-port theory.

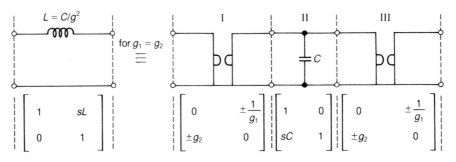

FIGURE 13.4.8   Realization of a floating inductor.

FIGURE 13.4.9   Negative impedance convertor with a load.

13.4.6. Negative Impedance Convertor

The negative impedance convertor was defined as a device with the following properties:

$$V_1 = k_1 V_2$$
$$I_1 = k_2 I_2 \tag{13.4.20}$$

If we load this network with an impedance Z, then

$$V_2 = -I_2 Z \tag{13.4.21}$$

the minus sign coming from the definition of the I_2 direction. The situation is sketched in Fig. 13.4.9. Substitute Eq. (13.4.21) into Eq. (13.4.20) and find the input impedance. The result is

$$Z_{in} = -\frac{k_1}{k_2} Z \tag{13.4.22}$$

and we conclude that a negative impedance convertor loaded by an impedance has an input impedance equal to the negative of its load. Scaling is possible by the choice of k_1 and k_2.

PROBLEMS CHAPTER 13

P.13.1 Find the z parameters for the bridged network which we used in the text, Fig. 13.1.4, but now keep all elements as variables.

P.13.2 Find the z parameters.

P.13.3 Find the cascade parameters for the network in P.13.2.

P.13.4 Find the z parameters. Keep elements as variables.

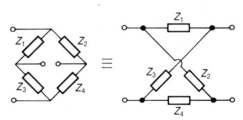

P.13.5 Find the cascade parameters for the network in P.13.4. Use $Z_1 = Z_4$ and $Z_2 = Z_3$.

P.13.6 Find the cascade parameters by keeping the elements as variables.

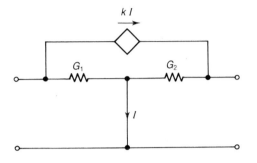

P.13.7 Find the cascade parameters by using the result in P.13.6 and Fig. 13.1.8.

P.13.8 Find the H parameters for the network in P.13.6.

P.13.9 Find the cascade parameters. The network is used to model a field-effect transistor.

P.13.10 Find the impedance and cascade parameters.

P.13.11 Find the cascade matrix of the network by using the cascade matrix of problem P.13.10. Set all element values equal to one.

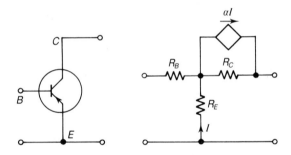

P.13.12 Find the impedance parameters.

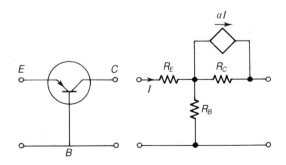

P.13.13 Find the h parameters for P.13.12.

P.13.14 Find the cascade parameters for P.13.12.

P.13.15 Find the z and h parameters.

P.13.16 Find the cascade parameters by using the matrices in Fig. 13.1.8 and Fig. 13.1.9 and multiplying them.

CHAPTER 14

Transformers

14.0. INTRODUCTION

Transformers are the last linear elements which we will consider. For application in low-frequency regions, they have two or more coils on a magnetic core. In high-frequency applications, transformers may be constructed with or without magnetic cores. This chapter will cover all the important aspects of transformers and will explain how to analyze networks which use them.

14.1. PRINCIPLE OF TRANSFORMERS

A current flowing through a coil forms around it a magnetic field. If we place another coil sufficiently close that the magnetic field reaches this second coil, then a voltage is induced in it. Such an arrangement is the simplest form of a transformer.

The transformer usually has a magnetic core which concentrates the magnetic field for better coupling of the coils. Although technical construction may differ from case to case, two simple examples of transformers are shown in Fig. 14.1.1. One of the coils, for instance the left one, may be called the primary winding, the other is then the secondary winding. Assume that we let a current flow through the left coil as indicated, for instance by connecting to it an independent current source. The magnetic flux which is created by the current is determined by the right-hand rule:

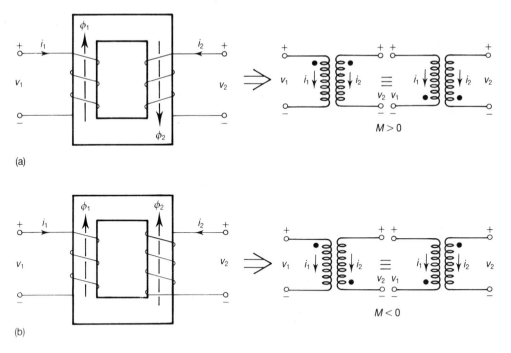

FIGURE 14.1.1   Transformer and symbols: (a) Fluxes in the same direction. (b) Fluxes in opposite directions.

Open the palm of your *right* hand, with the thumb pointing sidewise. Place the palm on the coil so that the fingers point in the direction of the current; your thumb indicates the direction of the magnetic flux through the core.

In both cases drawn in Fig. 14.1.1, the flux of the left coil, Φ_1, points upwards and goes clockwise around the core. If we move the current source to the right side so that the current flows into the upper terminal, then the right-hand rule tells us that the magnetic flux of the right coil, Φ_2, will also point clockwise in Fig. 14.1.1a, in *the same* direction as the flux of the first coil. A different situation is in Fig. 14.1.1b, where the right-hand rule indicates that the two fluxes act in opposite directions. We will return to these examples later.

In the case of a single inductor, the voltage across it is given by the derivative of the flux with respect to time. In the case of a transformer, the situation is more complicated; the fluxes are

$$\Phi_1 = \Phi_{11}(i_1) + \Phi_{12}(i_2)$$
$$\Phi_2 = \Phi_{21}(i_1) + \Phi_{22}(i_2)$$

(14.1.1)

Differentiation with respect to time gives

$$v_1 = \frac{d\Phi_1}{dt} = \frac{\partial\Phi_{11}}{\partial i_1}\frac{di_1}{dt} + \frac{\partial\Phi_{12}}{\partial i_2}\frac{di_2}{dt}$$

$$v_2 = \frac{d\Phi_2}{dt} = \frac{\partial\Phi_{21}}{\partial i_1}\frac{di_1}{dt} + \frac{\partial\Phi_{22}}{\partial i_2}\frac{di_2}{dt}$$

These expressions are still complicated but if we make the assumption that the partial derivatives are constants (the transformer is a linear device), then we can write

$$v_1 = L_{11}\frac{di_1}{dt} + L_{12}\frac{di_2}{dt}$$

$$v_2' = L_{21}\frac{di_1}{dt} + L_{22}\frac{di_2}{dt}$$

(14.1.2)

It is always true in all transformers that $L_{12} = L_{21} = M$ and we can simplify our notation by dropping the second subscript:

$$v_1 = L_1\frac{di_1}{dt} + M\frac{di_2}{dt}$$

$$v_2 = M\frac{di_1}{dt} + L_2\frac{di_2}{dt}$$

(14.1.3)

We could now take advantage of our knowledge of the Laplace transform, use

$$\mathscr{L}\left[\frac{di_1}{dt}\right] = sI_1 - J_{10}$$

$$\mathscr{L}\left[\frac{di_2}{dt}\right] = sI_2 - J_{20}$$

(14.1.4)

and take the Laplace transform of Eq. (14.1.3):

$$V_1 = sL_1I_1 + sMI_2 - L_1J_{10} - MJ_{20}$$
$$V_2 = sMI_1 + sL_2I_2 - MJ_{10} - L_2J_{20}$$

This would prepare our expressions for a time domain solution with initial conditions.

If the initial currents are zero, then

$$V_1 = sL_1I_1 + sMI_2$$
$$V_2 = sMI_1 + sL_2I_2$$

(14.1.5)

Here L_1 is the *primary inductance*, L_2 the *secondary inductance* and M is the *mutual inductance*. We can also write a matrix equation

$$\begin{bmatrix} V_1 \\ V_2 \end{bmatrix} = \begin{bmatrix} sL_1 & sM \\ sM & sL_2 \end{bmatrix} \begin{bmatrix} I_1 \\ I_2 \end{bmatrix}$$

(14.1.6)

and realize that this is nothing but the impedance form of a two-port.

We now return to Fig. 14.1.1. If we reverse the direction of the current flowing through the secondary coils, the direction of the secondary flux will change. We clearly need some simple method to draw a transformer, simultaneously clearly indicating how the winding of the coil influences the flux. This is done with the dot convention. We draw the coils as simple inductors, as shown on the right of Fig. 14.1.1, and indicate the direction of the fluxes by two dots and by two currents flowing through the coils. If the directions of currents through the coils are such that they *both* point either away from the dots or to the dots, then the fluxes in the core flow in the same direction and mutual inductance, M, has a positive sign. If one of the currents points to the dot and the other away from the dot, as shown in Fig. 14.1.1b, then the mutual inductance, M, is negative.

The dot convention is useful only for two coupled coils but transformers may have several coils. An example with four coils is given in Fig. 14.1.2. Using the right-hand rule we easily find the signs of the mutual inductances; they are written in the figure. The equations will be

$$V_1 = sL_{11}I_1 + sL_{12}I_2 - sL_{13}I_3 - sL_{14}I_4$$
$$V_2 = sL_{21}I_1 + sL_{22}I_3 - sL_{23}I_3 - sL_{24}I_4$$
$$V_3 = -sL_{31}I_1 - sL_{32}I_2 + sL_{33}I_3 + sL_{34}I_4$$
$$V_4 = -sL_{41}I_1 - sL_{42}I_2 + sL_{43}I_3 + sL_{44}I_4$$

(14.1.7)

or in matrix form

$$\begin{bmatrix} V_1 \\ V_2 \\ V_3 \\ V_4 \end{bmatrix} = \begin{bmatrix} +sL_{11} & +sL_{12} & -sL_{13} & -sL_{14} \\ +sL_{21} & +sL_{22} & -sL_{23} & -sL_{24} \\ -sL_{31} & -sL_{32} & +sL_{33} & +sL_{34} \\ -sL_{41} & -sL_{42} & +sL_{43} & +L_{44} \end{bmatrix} \begin{bmatrix} I_1 \\ I_2 \\ I_3 \\ I_4 \end{bmatrix}$$

(14.1.8)

We must always keep in mind that $L_{ij} = L_{ji}$. Another mathematical way to

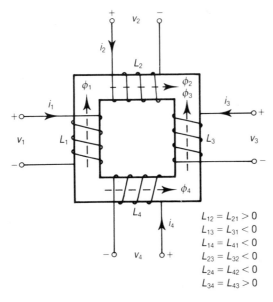

FIGURE 14.1.2   Transformer with four coils.

indicate that we have a matrix equation is to use bold letters

$$\mathbf{V} = \mathbf{LI} \tag{14.1.9}$$

If we invert the matrix,

$$\mathbf{M} = \mathbf{L}^{-1} \tag{14.1.10}$$

we get the admittance form of the equations

$$\mathbf{I} = \mathbf{MV} \tag{14.1.11}$$

The above examples have shown that it is easy to write the equations if we know the directions of the windings and the directions of the currents. In the following we will restrict our explanations to transformers having only two coils.

There still remains the question of how to find the value of the mutual inductance. This can be done experimentally. For the transformer in Fig. 14.1.1, attach a current source to the primary winding and open-circuit the secondary winding, thus making $I_2 = 0$. If the source frequency is $s = j\omega$, then Eq. (14.1.5) reduce to

$$V_1 = j\omega L_1 I_1$$
$$V_2 = j\omega M I_1$$

Measurement of V_2 gives us the possibility of calculating M. The same steps can be applied to Eq. (14.1.7): attach a similar source to one of the windings and open-circuit all the others. The equations simplify and from the measurement we calculate some of the mutual inductances. In the next step, transfer the source to another winding and do the same. Repeat until all mutual inductances have been found. We can reduce the number of such measurements by realizing that $L_{ij} = L_{ji}$.

14.2. COUPLING COEFFICIENT

When we apply a current to the first coil only, a voltage is induced in the other coil; if we do not attach any load, the current I_2 will be zero. Equations (14.1.5) simplify to

$$V_1 = sL_1I_1$$
$$V_2 = \pm sMI_1$$

and their ratio is

$$A = \left.\frac{V_1}{V_2}\right|_{I_2=0} = \frac{L_1}{\pm M} \tag{14.2.1}$$

Doing the same from the other side of the transformer without loading the first coil we get

$$V_1 = \pm sMI_2$$
$$V_2 = sL_2I_2$$

and

$$B = \left.\frac{V_1}{V_2}\right|_{I_1=0} = \pm\frac{M}{L_2} \tag{14.2.2}$$

The ratio of Eq. (14.2.2) and Eq. (14.2.1) is

$$\frac{B}{A} = \frac{M^2}{L_1L_2} = k^2$$

The constant k is called the *coupling coefficient* and is always positive, regardless of which sign we took with the mutual inductance.

Using the concept of stored energy we will now establish that the coefficient of coupling always satisfies the inequality

$$0 \le k \le 1 \tag{14.2.3}$$

The power delivered by a source connected to the first coil is

$$p_1 = v_1 i_1 = L_1 \frac{di_1}{dt} i_1 \pm M \frac{di_2}{dt} i_1$$

Power delivered by the source at the other coil is

$$p_2 = v_2 i_2 = \pm M \frac{di_1}{dt} i_2 + L_2 \frac{di_2}{dt} i_2$$

and the total power is

$$p = p_1 + p_2 = L_1 i_1 \frac{di_1}{dt} \pm M \left(i_1 \frac{di_2}{dt} + i_2 \frac{di_1}{dt} \right) + L_2 i_2 \frac{di_2}{dt}$$

This can be rewritten in the form

$$p = \frac{d}{dt} \left(\frac{L_1 i_1^2}{2} \pm M i_1 i_2 + \frac{L_2 i_2^2}{2} \right)$$

An integral over certain time will provide the energy stored in the transformer. The energy cannot be negative,

$$w = \frac{L_1 I_1^2}{2} + \frac{L_2 I_2^2}{2} \pm M I_1 I_2 \geq 0 \qquad (14.2.4)$$

We can modify Eq. (14.2.4) by simultaneously adding and subtracting the value $\sqrt{L_1 L_2 I_1 I_2}$:

$$w = \frac{1}{2} (\sqrt{L_1} I_1 - \sqrt{L_2} I_2)^2 + (\sqrt{L_1 L_2} \pm M) I_1 I_2 \geq 0$$

The first term is always positive, but can be made zero, for instance by selecting $L_1 = L_2$ and $I_1 = I_2$. If we select a positive sign of M, the second term will be always positive. However, if we select a negative sign for M, then the limiting case will be

$$\sqrt{L_1 L_2} - M \geq 0$$

which is the same as writing

$$k = \frac{M}{\sqrt{L_1 L_2}} \qquad 0 \leq k \leq 1 \qquad (14.2.5)$$

14.3. PERFECT AND IDEAL TRANSFORMERS

The name *perfect transformer* is given to a transformer with a coupling coefficient $k = 1$. In such case

$$L_1 L_2 = M^2 \tag{14.3.1}$$

Figure 14.3.1 shows a transformer with a loading resistor. If the resistor is *not* connected, then $I_2 = 0$,

$$V_1 = sL_1 I_1$$
$$V_2 = sM I_1$$

and the voltage transfer is

$$\frac{V_2}{V_1} = \frac{M}{L_1} = \left[\frac{L_2}{L_1}\right]^{1/2} = \frac{1}{n} \tag{14.3.2}$$

For a perfect (or nearly perfect) transformer, the ratio $1/n$ is equal to the ratio of the turns of the primary and secondary coils. In some cases, this is even written as n_1/n_2. The choice of the dots indicates that the transformer equations are

$$V_1 = sL_1 I_1 + sM I_2$$
$$V_2 = sM I_1 + sL_2 I_2 \tag{14.3.3}$$

If we attach the loading resistor and select the loop currents, called now I_a and I_b, to circulate in clockwise directions, then

$$I_a = I_1$$
$$I_b = -I_2 \tag{14.3.4}$$

Substituting into the transformer equations Eq. (14.3.3)

$$V_1 = sL_1 I_a - sM I_b$$
$$V_2 = sM I_a - sL_2 I_b \tag{14.3.5}$$

FIGURE 14.3.1   Transformer with a loading resistor.

We see from the figure that

$$V_1 = E$$
$$-V_2 + RI_b = 0$$

(14.3.6)

Combining with Eq. (14.3.5)

$$\begin{bmatrix} sL_1 & -sM \\ -sM & sL_2 + R \end{bmatrix} \begin{bmatrix} I_a \\ I_b \end{bmatrix} = \begin{bmatrix} E \\ 0 \end{bmatrix}$$

The determinant is

$$D = s^2(L_1 L_2 - M^2) + sL_1 R$$

and substituting condition Eq. (14.3.1)

$$D = sL_1 R$$

(14.3.7)

The currents are

$$I_a = \frac{sL_2 + R}{sL_1 R} E$$

$$I_b = \frac{sM}{sL_1 R} E = \left[\frac{L_2}{L_1}\right]^{1/2} \frac{1}{R} E = \frac{1}{n} \frac{1}{R} E$$

(14.3.8)

The output voltage is

$$V_{out} = I_b R = \frac{1}{n} E$$

(14.3.9)

the input impedance is

$$\frac{E}{I_a} = Z_{in} = \frac{sL_1 R}{sL_2 + R} = \frac{sL_1 R}{sL_2\left(1 + \frac{R}{sL_2}\right)} = n^2 \frac{1}{\frac{1}{R} + \frac{1}{sL_2}}$$

(14.3.10)

and the current ratio is

$$\frac{I_b}{I_a} = \frac{sM}{sL_2\left(1 + \frac{R}{sL_2}\right)} = \left[\frac{L_1}{L_2}\right]^{1/2} \frac{1}{1 + \frac{R}{sL_2}} = n \frac{1}{1 + \frac{R}{sL_2}}$$

(14.3.11)

The voltage ratio is independent of the load and the input impedance is represented by parallel connection of the resistor and the secondary inductor, multiplied by n^2. If both inductances are large, we can simplify Eq. (14.3.10) and Eq. (14.3.11) to

$$Z_{in} = n^2 R \qquad (14.3.12)$$

$$I_b = nI_a \qquad (14.3.13)$$

and still have

$$V_2 = \frac{1}{n} V_1 \qquad (14.3.14)$$

The product $V_2 I_b = V_1 I_a$ indicates that this simplified device does not consume any power. If we return to currents I_1 and I_2 by using Eq. (14.3.4), we get

$$V_1 = nV_2$$

$$I_1 = -\frac{1}{n} I_2 \qquad (14.3.15)$$

These are two-port equations which define an *ideal transformer*. Its symbol is in Fig. 14.3.2.

Ideal transformers have been used extensively in network theory but they cannot be realized technically. Their equations indicate that they transform the dc current, something we cannot achieve with coils.

Technical transformers which are used in power transfers are always realized with as high a coefficient of coupling as possible. Properties of such transformers are very similar to those of a perfect transformer, Eq. (14.3.9)–(14.3.11). If we need only approximate answers, we could use even Eq. (14.3.12)–(14.3.14), but we must keep in mind that they do not apply for dc.

FIGURE 14.3.2   Ideal transformer.

14.4. NETWORKS WITH TRANSFORMERS

In this section we give a method to analyze networks with transformers. We assume that the current directions and winding directions have already been determined and that the networks have the necessary dots at the proper positions.

The first network is in Fig. 14.4.1; I_1, I_2, V_1 and V_2 refer to Eq. (14.1.5), repeated here for convenience:

$$V_1 = sL_1I_1 + sMI_2$$
$$V_2 = sMI_1 + sL_2I_2$$

(14.4.1)

Since we have chosen both currents I_1 and I_2 to point in the direction away from the dots, the mutual inductance in Eq. (14.4.1) has positive sign.

The loop currents are denoted by I_a and I_b and we always select clockwise direction for them. The two sets of currents are related:

$$I_1 = I_a$$
$$I_2 = I_a - I_b$$

(14.4.2)

Substituting Eq. (14.4.2) into Eq. (14.4.1) we obtain

$$V_1 = sL_1I_a + sM(I_a - I_b)$$
$$V_2 = sMI_a + sL_2(I_a - I_b)$$

(14.4.3)

We are now ready to write the loop equations:

$$V_1 + V_2 = E$$
$$RI_b - V_2 = 0$$

(14.4.4)

Substituting for V_1 and V_2 from Eq. (14.4.3) we get

$$(sL_1 + 2sM + sL_2)I_a - (sL_2 + sM)I_b = E$$
$$-(sL_2 + sM)I_a + (sL_2 + R)I_b = 0$$

FIGURE 14.4.1   Network with a transformer.

As always, it is convenient to put these equations into matrix form

$$\begin{bmatrix} s(L_1 + L_2 + 2M) & -s(L_2 + M) \\ -s(L_2 + M) & sL_2 + R \end{bmatrix} \begin{bmatrix} I_a \\ I_b \end{bmatrix} = \begin{bmatrix} E \\ 0 \end{bmatrix} \qquad (14.4.5)$$

and solution can proceed along established lines.

Example 1.

Analyze the network in Fig. 14.4.2.
    The currents are

$$I_1 = I_a$$
$$I_2 = -I_b \qquad\qquad (14.4.6)$$

Inserting into Eq. (14.4.1) results in

$$V_1 = sL_1 I_a - sM I_b$$
$$V_2 = sM I_a - sL_2 I_b \qquad\qquad (14.4.7)$$

The loop equations are

$$R_1 I_a + V_1 + R_2(I_a - I_b) = E$$
$$R_2(I_b - I_a) - V_2 + R_3 I_b = 0 \qquad (14.4.8)$$

Substitution of V_1 and V_2 from Eq. (14.4.7) gives the final set, written here directly in matrix form:

$$\begin{bmatrix} R_1 + R_2 + sL_1 & -R_2 - sM \\ -R_2 - sM & R_2 + R_3 + sL_2 \end{bmatrix} \begin{bmatrix} I_a \\ I_b \end{bmatrix} = \begin{bmatrix} E \\ 0 \end{bmatrix}$$

FIGURE 14.4.2   Network with a transformer.

FIGURE 14.4.3   Network with a transformer.

Example 2.

Analyze the network in Fig. 14.4.3.

The currents are

$$I_1 = I_a - I_b$$
$$I_2 = I_b$$

When substituted into the transformer equations, we get

$$V_1 = sL_1(I_a - I_b) + sMI_b$$
$$V_2 = sM(I_a - I_b) + sL_2I_b \qquad (14.4.9)$$

The loop equations for the network are

$$R_1I_a + V_1 = E$$
$$-V_1 + V_2 + R_2I_b = 0 \qquad (14.4.10)$$

When we now substitute V_1 and V_2 from Eq. (14.4.9), we get a system of equations which leads to the following matrix form

$$\begin{bmatrix} R_1 + sL_1 & s(M - L_1) \\ s(M - L_1) & s(L_1 + L_2 - 2M) + R_2 \end{bmatrix} \begin{bmatrix} I_a \\ I_b \end{bmatrix} = \begin{bmatrix} E \\ 0 \end{bmatrix}$$

Example 3.

Analyze the network in Fig. 14.4.4.

The currents are

$$I_1 = I_c - I_a$$
$$I_2 = I_b - I_c \qquad (14.4.11)$$

When substituted into the transformer equations, we get

$$V_1 = sL_1(I_c - I_a) + sM(I_b - I_c)$$
$$V_2 = sM(I_c - I_a) + sL_2(I_b - I_c) \qquad (14.4.12)$$

FIGURE 14.4.4   Network with three meshes.

The loop equations for this network are

$$-V_1 + R_1(I_a - I_c) + R_2(I_a - I_b) = E$$

$$R_2(I_b - I_a) + R_3(I_b - I_c) + V_2 + R_5 I_b = 0 \qquad (14.4.13)$$

$$R_4 I_c + R_3(I_c - I_b) + R_1(I_c - I_a) + V_1 - V_2 = 0$$

When we substitute from Eq. (14.4.12), we get fairly lengthy expressions; they can be simplified and written in matrix form:

$$\begin{bmatrix} R_1 + R_2 + sL_1 & -R_2 - sM & sM - sL_1 - R_1 \\ -R_2 - sM & R_2 + R_3 + R_5 + sL_2 & sM - sL_2 - R_3 \\ sM - sL_1 - R_1 & sM - sL_2 - R_3 & R_1 + R_3 + R_4 + s(L_1 + L_2 - 2M) \end{bmatrix} \begin{bmatrix} I_a \\ I_b \\ I_c \end{bmatrix} = \begin{bmatrix} E \\ 0 \\ 0 \end{bmatrix}$$

The examples have given a reliable method for how to handle networks with transformers. We write a bit more than absolutely necessary, but we avoid mistakes which usually happen when we try to do too many steps simultaneously.

14.5. EQUIVALENT TRANSFORMER NETWORKS

The transformer matrix equation

$$\begin{bmatrix} V_1 \\ V_2 \end{bmatrix} = \begin{bmatrix} sL_1 & sM \\ sM & sL_2 \end{bmatrix} \begin{bmatrix} I_1 \\ I_2 \end{bmatrix} \qquad (14.5.1)$$

is actually an open-circuit two-port description [compare with Eq. (13.1.2)]. If we succeed in drawing another network which will be described by the same matrix equation, then the two networks will behave in the same way. If any of them is placed into a box which cannot be opened, no amount of measurements can determine which of the networks is inside.

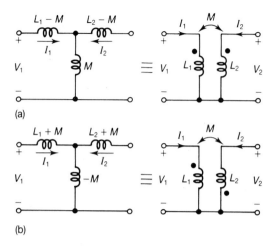

FIGURE 14.5.1   Equivalent T network for a transformer: (a) With positive M. (b) With negative M.

Consider the network on the left of Fig. 14.5.1a and derive its two-port impedance equations. We get exactly the expression Eq. (14.5.1) and thus the network is equivalent to the transformer on the right. Another network, in Fig. 14.5.1b is equivalent to the transformer with negative M. Note that the two windings of the transformer are connected at the bottom of the drawings.

We can also invert the matrix equation above and get

$$\begin{bmatrix} I_1 \\ I_2 \end{bmatrix} = \begin{bmatrix} y_{11} & y_{12} \\ y_{21} & y_{22} \end{bmatrix} \begin{bmatrix} V_1 \\ V_2 \end{bmatrix}$$

(14.5.2)

where

$$y_{11} = \frac{L_2}{s(L_1 L_2 - M^2)}$$

$$y_{12} = y_{21} = \frac{-M}{s(L_1 L_2 - M^2)}$$

(14.5.3)

$$y_{22} = \frac{L_1}{s(L_1 L_2 - M^2)}$$

We try the network in Fig. 14.5.2 as an equivalent. Its admittance two-port

FIGURE 14.5.2   Π network and its two-port **Y** matrix.

parameters are

$$y_{11} = \frac{1}{sL_a} + \frac{1}{sL_c}$$

$$y_{12} = y_{21} = \frac{-1}{sL_c} \tag{14.5.4}$$

$$y_{22} = \frac{1}{sL_b} + \frac{1}{sL_c}$$

Comparing Eq. (14.5.4) with Eq. (14.5.3) we get

$$\frac{1}{sL_a} + \frac{1}{sL_c} = \frac{L_2}{s(L_1 L_2 - M^2)}$$

$$\frac{1}{sL_c} = \frac{M}{s(L_1 L_2 - M^2)}$$

$$\frac{1}{sL_b} + \frac{1}{sL_c} = \frac{L_1}{s(L_1 L_2 - M^2)}$$

and finally

$$L_a = \frac{L_1 L_2 - M^2}{L_2 - M}$$

$$L_b = \frac{L_1 L_2 - M^2}{L_1 - M} \tag{14.5.5}$$

$$L_c = \frac{L_1 L_2 - M^2}{M}$$

The equivalent network is drawn in Fig. 14.5.3a for positive M and in b for negative M. Again, note that the bottoms of the coils are connected.

Inversion of the matrix, used in the above steps, is not possible if the determinant $L_1 L_2 - M_2$ is zero. We showed in Section 14.3 that this condition describes a

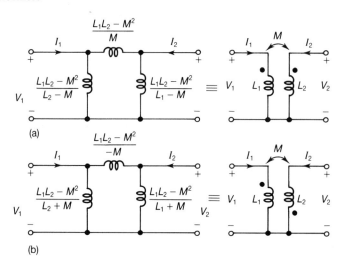

FIGURE 14.5.3   Equivalent Π network for a transformer: (a) With positive M. (b) With negative M.

perfect transformer and we conclude that it is not possible to get the admittance form of two-port equations for a perfect transformer.

The above derivations can be applied to solve the following two examples. Figure 14.5.4 shows a transformer with a series connection of its coils. We would like to know what the value of the resulting equivalent inductance is. We start with the transformer equations

$$V_1 = sL_1 I_1 + sM I_2$$
$$V_2 = sM I_1 + sL_2 I_2$$

FIGURE 14.5.4   Transformer with series connection of coils.

and realize that $I_1 = I_2 = I$. Inserting into the equations:

$$V_1 = (sL_1 + sM)I$$
$$V_2 = (sM + sL_2)I$$

The overall voltage is equal to $V = V_1 + V_2$. Adding the two equations we obtain

$$V = s(L_1 + L_2 + 2M)I$$

from which we conclude that the equivalent inductance is

$$L_{EQ} = L_1 + L_2 + 2M$$

If we connect the two coils in parallel, as shown in Fig. 14.5.5, then we must start with the admittance equations

$$I_1 = y_{11}V_1 + y_{12}V_2$$
$$I_2 = y_{21}V_1 + y_{22}V_2$$

Since both coils have the same voltage, $V = V_1 = V_2$, we substitute and get

$$I_1 = (y_{11} + y_{12})V$$
$$I_2 = (y_{21} + y_{22})V$$

The overall current is $I = I_1 + I_2$, as can be seen from the figure. We add the equations and obtain

$$I = (y_{11} + y_{12} + y_{21} + y_{22})V$$

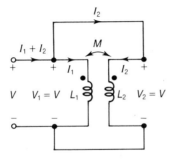

FIGURE 14.5.5   Transformer with parallel connection of coils.

The values were derived in Eq. (14.5.3). Inserting we get

$$I = \frac{L_1 + L_2 - 2M}{s(L_1 L_2 - M^2)} V$$

The equivalent input impedance is

$$Z_{in} = \frac{V}{I} = s \frac{L_1 L_2 - M^2}{L_1 + L_2 - 2M}$$

and the equivalent inductance is

$$L_{EQ} = \frac{L_1 L_2 - M^2}{L_1 + L_2 - 2M}$$

14.6. DOUBLE TUNED CIRCUITS

Double tuned circuits are formed by two RLC circuits with a loose magnetic coupling of the coils. They are used in radios for filtering purposes and have quite interesting properties. We give the derivations and results as a matter of interest.

Consider the network in Fig. 14.6.1a. If we use the equivalent circuit of the transformer from the previous section, we get the network shown in Fig. 14.6.1b. To simplify our work we select $C_1 = C_2 = C$, $L_1 = L_2 = L$ and $G_1 = G_2 = G$. We also substitute $L_e = L - M$ for easier writing. The nodal matrix equation is

$$\begin{bmatrix} sC + G + \dfrac{1}{sL_e} & -\dfrac{1}{sL_e} & 0 \\[2ex] -\dfrac{1}{sL_e} & \dfrac{2M + L_e}{sL_e M} & -\dfrac{1}{sL_e} \\[2ex] 0 & -\dfrac{1}{sL_e} & sC + G + \dfrac{1}{sL_e} \end{bmatrix} \begin{bmatrix} V_1 \\ V_2 \\ V_3 \end{bmatrix} = \begin{bmatrix} J(s) \\ 0 \\ 0 \end{bmatrix}$$

Because we have equal values for the elements, the determinant has terms which can be factored. After a few simple steps we first get

$$D = \frac{1}{s^3 L_e^3} (s^2 L_e C + s L_e G + 1) \left[(s^2 L_e C + s L_e G + 1) \frac{2M + L_e}{M} - 2 \right]$$

and then

$$D = \frac{C^2}{s^3 L_e} \frac{2M + L_e}{M} \left(s^2 + s \frac{G}{C} + \frac{1}{L_e C} \right) \left(s^2 + s \frac{G}{C} + \frac{1}{C(2M + L_e)} \right)$$

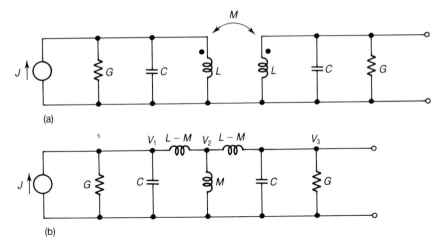

FIGURE 14.6.1   Double tuned circuit with magnetic coupling: (a) Original network. (b) Equivalent for analysis.

It is now convenient to substitute back $L - M$ for L_e:

$$D = \frac{(L + M)C^2}{s^3 M(L - M)}\left(s^2 + s\frac{G}{C} + \frac{1}{C(L - M)}\right)\left(s^2 + s\frac{G}{C} + \frac{1}{C(L + M)}\right)$$

Since the numerator for V_3 is

$$N_3 = \frac{J}{s^2 L_e^2} = \frac{J}{s^2(L - M)^2}$$

the transfer impedance becomes

$$\frac{V_3}{J} = \frac{M}{(L^2 - M^2)C^2}\frac{s}{\left(s^2 + s\frac{G}{C} + \frac{1}{C(L - M)}\right)\left(s^2 + s\frac{G}{C} + \frac{1}{C(L + M)}\right)}$$

For the first quadratic term of the denominator we get

$$Q_1 = R\left[\frac{C}{L - M}\right]^{1/2}$$

$$\omega_1 = \frac{1}{[C(L - M)]^{1/2}}$$

and for the second term

$$Q_2 = R \left[\frac{C}{L+M} \right]^{1/2}$$

$$\omega_2 = \frac{1}{[C(L+M)]^{1/2}}$$

Linear networks can be scaled without losing any information. For our scaling we choose $C = 1\,F$ and $L = 1\,H$; this means each of the tuned circuits would separately resonate at $\omega = 1$. The transfer impedance simplifies to

$$\frac{V_3}{J} = \frac{M}{(1-M^2)} \frac{2}{\left(s^2 + sG + \dfrac{1}{1-M}\right)\left(s^2 + sG + \dfrac{1}{1+M}\right)}$$

where $0 < M < 1$. The transfer function has one zero at the origin and four poles. Resonant frequencies of the poles are above and below $\omega = 1$. If M is small, the two Q are approximately equal. If M grows, the two resonant frequencies pull apart and the Q will start to differ.

Amplitude response of this network is plotted in Fig. 14.6.2a and b for element values $C_1 = C_2 = 1\,F$, $L_1 = L_2 = 1\,H$ and $G_1 = G_2 = 0.05\,S$. Curve a is for $M = 0.05\,H$, curve b for $M = 0.075\,H$. If we select M between these two values, the peaks will get closer to each other. The plot was produced by NETOPT.

It is interesting to note that had we used the other equivalent circuit of the

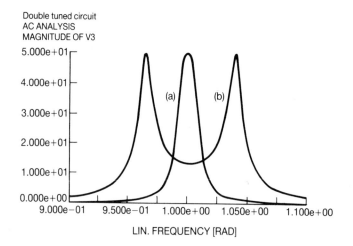

Double tuned circuit
AC ANALYSIS
MAGNITUDE OF V3

LIN. FREQUENCY [RAD]

FIGURE 14.6.2   Amplitude responses for the network in Fig. 14.6.1. Element values are $C_1 = C_2 = 1\,F$, $L_1 = L_2 = 1\,H$, $G_1 = G_2 = 0.05$. Curve (a) is for $M = 0.05\,H$. Curve (b) is for $M = 0.075\,H$.

transformer in our derivations, the resulting matrix equation would be only 2 × 2, but it would not be possible to decompose the denominator into its two quadratic terms in symbolic form.

PROBLEMS CHAPTER 14

The NETOPT program does not have transformer models but it is possible to analyze networks with transformers by using the equivalents from Fig. 14.5.1 and Fig. 14.5.3.

P.14.1 Find the voltage transfer function using (a) the method of this chapter, (b) the equivalents in Fig. 14.5.1, (c) the equivalents in Fig. 14.5.3.

P.14.2 Do the same as in P.14.1.

P.14.3 Do the same as in P.14.1.

P.14.4 Use the transformation in Fig. 14.5.1 and the mesh method to find the voltage transfer function.

P.14.5 Find the voltage transfer function using the given element values.

P.14.6 Do the same as in P.14.5.

P.14.7 Do the same as in P.14.5.

P.14.8 Use numerical values and the transformation in Fig. 14.5.1 to write the mesh equations. Find the transfer impedance.

P.14.9 Calculate the expression for the output current. Use numerical values.

P.14.10 Find the voltage transfer function.

P.14.11 Find the transfer function.

P.14.12 Find the transfer function.

P.14.13 Find the transfer function.

CHAPTER 15

Modeling

15.0. INTRODUCTION

Modeling of systems and devices, especially nonlinear devices like transistors, is an important area which requires considerable knowledge of physics, mathematics and other disciplines. A very small part of the modeling problem, modeling of linear devices, is suitable for this text and we will first derive linear models for amplifiers and transistors.

Afterwards we turn our attention to the consequences resulting from the presence of nonlinearities. Since the purpose of this chapter is to introduce concepts without going into details, we will only use two nonlinear devices, the bipolar diode and the field-effect transistor. With these two elements we will show that writing equations for networks with nonlinear devices is not different from what we have learned so far. Solutions of such equations must be done, in general, by iteration and we will explain the Newton–Raphson method for systems of nonlinear equations. As a result we will see that solutions of nonlinear systems are reduced to repeated solutions of linear systems written in matrix form. Linear elements appear in the matrix exactly the same way as they did in all our studies so far. Nonlinear elements also appear in the same positions as if the elements were linear, only their values are replaced by the values of their derivatives. In other words, everything we learned in linear network theory is reused in nonlinear systems.

The linear part should be covered in any introductory course. The nonlinear part is recommended for understanding but can be considered as optional.

455

15.1. MODELING OF SOURCES

An ideal voltage source (independent or dependent) should, theoretically, deliver an arbitrary current and maintain the specified voltage, even if it is short-circuited. An ideal current source (independent or dependent), should deliver a given amount of current anywhere, theoretically even into an open circuit. Such sources do not exist in nature and the current or voltage are influenced by some internal property. We can model a nonideal voltage source as a combination of an ideal voltage source and an *internal resistance*, R_S, see Fig. 15.1.1. If we do not draw current by connecting some load at the output terminals, the voltage at these terminals is E. If we short-circuit this model, then the current will be limited, $I = E/R_S$ and the power $W = E^2/R_S$ will be dissipated in the internal resistance.

There is also a limitation on the maximum power which a nonideal source can deliver into a load. To find the condition for maximum power transfer consider Fig. 15.1.2 with a variable load resistor R_L. The current flowing from the ideal voltage source is

$$I = \frac{E}{R_S + R_L}$$

the output voltage is

$$V = \frac{ER_L}{R_S + R_L}$$

and the power dissipated in the load is

$$W = E^2 \frac{R_L}{(R_S + R_L)^2}$$

Fig. 15.1.1

Fig. 15.1.2

FIGURE 15.1.1   Nonideal voltage source.

FIGURE 15.1.2   Nonideal voltage source loaded by R_L.

We wish to find the condition under which a source with an internal resistance R_S delivers the maximum amount of power into the load resistor.

As is well known from mathematics, the maximum is found by first obtaining the derivative with respect to the variable and then setting the derivative equal to zero. Differentiating W with respect to R_L gets us the condition

$$\frac{dW}{dR_L} = E^2 \frac{R_S - R_L}{(R_S + R_L)^3} = 0$$

which is satisfied for

$$R_S = R_L \tag{15.1.1}$$

In this case the output voltage is

$$V = \frac{E}{2} \tag{15.1.2}$$

and the power dissipated in the load (as well as in the internal resistance) is

$$W_{max} = \frac{E^2}{4R_L} = \frac{E^2}{4R_S} \tag{15.1.3}$$

If the load is not a pure resistor but is, at some fequency, equal to

$$Z_L = R_L + jX_L$$

then it can be shown that the maximum power transfer occurs for complex source impedance

$$Z_S = R_S + jX_S$$

such that

$$R_S = R_L \tag{15.1.4}$$

$$X_S = -X_L \tag{15.1.5}$$

This situation applies when we are dealing with networks in steady state using phasors.

Ideal current sources are very convenient for theoretical considerations, especially when we are using nodal formulation, but there are certain peculiarities which we must be aware of. First of all, an ideal current source cannot be left on its own, unconnected to anything else, because, by definition, it must deliver a given current

Fig. 15.1.3

Fig. 15.1.4

FIGURE 15.1.3   Nonideal current source.

FIGURE 15.1.4   Load for maximum power transfer.

somewhere. For this reason, there should always be at least one element in parallel with the current source. A nonideal current source can be modeled by the network in Fig. 15.1.3. The Thévenin and Norton theorems show that this source can be equivalent to the voltage source in Fig. 15.1.1 *as far as the loading network is concerned.* Maximum power delivered into a load is subject to the same conditions as derived above. However, it is important to realize that the overall behaviour of a nonideal current source model with a resistance in parallel differs from the model of the nonideal voltage source with a resistance in series. If the model in Fig. 15.1.3 is not connected to any other network, then the internal resistance is consuming power, which is not the case of the nonideal voltage source.

Example 1.

In the network Fig. 15.1.4 find such values R_2 and C to get maximum power transfer at the angular frequency $\omega = 2$.

The source impedance is $Z_S = R_1 + j\omega L = 1 + j2$. Impedance of the load is $Z_L = R_2 - j/2C$. Condition for maximum power transfer will be $R_1 = R_2$ and $C = 1/4$.

Example 2.

Find the values of R_2 and L to get maximum power transfer at $\omega = 2$ for the network in Fig. 15.1.5.

The source impedance is $Z_S = 1/(1 + j2) = (1 - j2)/5$. The load impedance is $Z_L = R_2 + j2L$. Comparison of the real and imaginary parts gives $R_2 = 1/5$, $L = 1/5$.

FIGURE 15.1.5   Load for maximum power transfer.

15.2. LINEAR AMPLIFIERS

Ideal amplifiers with gain A are nothing but voltage-controlled voltage sources. The usual symbol for a *differential* amplifer, already discussed in Chapter 12, is in Fig. 15.2.1.

In the case of an ideal amplifier the gain A is considered to be independent of frequency. This is not true in actual life. Due to various internal capacitances which are inherently present in any actual realization, the gain drops as we go to higher frequencies. Such a frequency dependence is usually simulated by the formula

$$A = \frac{A_0 \omega_0}{s + \omega_0} \tag{15.2.1}$$

Inserting $s = 0$ we see that A_0 is the gain of the amplifier for dc. Suppose that we now select signal frequency $s = j\omega_0$. In such case

$$A = \frac{A_0 \omega_0}{j\omega_0 + \omega_0} = \frac{A_0}{j + 1}$$

The absolute value becomes

$$|A| = 0.7071 A_0$$

and in decibels

$$\left| \frac{A}{A_0} \right|_{dB} = 20 \log 0.7071 \approx -3 \text{ dB}$$

Thus ω_0 is a signal frequency (in radians per second) at which the gain of the amplifier drops 3 dB as compared to the gain at dc. The product in the numerator of Eq. (15.2.1) was given the name *gain–bandwidth product*,

$$GB_\omega = A_0 \omega_0 \tag{15.2.2}$$

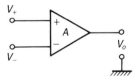

FIGURE 15.2.1   Differential amplifier with gain A.

Specification sheets of manufacturers usually give the gain–bandwidth product in Hz,

$$GB_f = \frac{GB_\omega}{2\pi} \tag{15.2.3}$$

The gain–bandwidth product is sometimes called the unity-gain frequency. The name becomes clear when we substitute $A_0 = 1$ into the definition Eq. (15.2.2). As a rule of thumb we will remember that commercial operational amplifiers may have a dc gain A_0 approximately equal to 100,000 and f_0 may be a low as 10 Hz.

Work with operational amplifiers is simplier if we define the inverted gain

$$B = -\frac{1}{A} \tag{15.2.4}$$

which modifies the mutual relationship of the voltages to

$$V_+ - V_- + BV_o = 0 \tag{15.2.5}$$

If the OPAMP is ideal, with infinitely large gain, then $B = 0$ is substituted into this equation, making $V_+ = V_-$. If we must consider the frequency-dependent properties of the amplifier, we substitute Eq. (15.2.1) into Eq. (15.2.4) and get

$$B = -\frac{s + \omega_0}{GB_\omega} \tag{15.2.6}$$

The expression can be simplified if we realize that just about all practical networks work at frequencies which are considerably higher than ω_0. In such cases $s = j\omega$ is, in absolute value, much larger than ω_0 and we simplify the above expression to

$$B = -\frac{s}{GB_\omega} \tag{15.2.7}$$

This is a practical approximation of a linear nonideal operational amplifier. It is available in NETOPT as an option, with GB defined in Hz.

If we wish to simulate the property Eq. (15.2.1) by a linear network, possibly taking into account an output resistance, we can use Fig. 15.2.2. It has two voltage-controlled voltage sources. The input voltage, $V_{in} = V_+ - V_-$, controls the first VV with the gain μ_1. The voltage V_1 is obtained from the RC voltage divider,

$$V_1 = \frac{\mu_1 V_{in}}{CR\left(s + \dfrac{1}{CR}\right)} = \frac{\omega_0 \mu_1 V_{in}}{s + \omega_0}$$

FIGURE 15.2.2   Equivalent network for amplifier with one pole.

where we defined

$$\omega_0 = \frac{1}{RC}$$

The overall output voltage without loading is

$$V_{out} = \frac{\mu_1 \mu_2 \omega_0 V_{in}}{s + \omega_0}$$

If we now select

$$\mu_1 = 1$$
$$\mu_2 = A_0$$

then

$$\frac{V_{out}}{V_{in}} = \frac{A_0 \omega_0}{s + \omega_0},$$

exactly the expression above. The network in Fig. 15.2.2 is suitable for mesh formulation. For nodal formulation we can transform the controlled voltage sources into controlled current sources by the Thévenin–Norton transformation. The resulting equivalent network is in Fig. 15.2.3.

FIGURE 15.2.3   Equivalent network for amplifier with one pole.

15.3. LINEAR TRANSISTOR MODELS

Transistors are widely used elements and a huge amount of literature deals with their modeling. They are nonlinear devices and we will consider their nonlinear models later in this chapter; here we derive their linear models from their most basic properties.

15.3.1. Field-Effect Transistor

The symbol for a field-effect transistor (FET) is in Fig. 15.3.1. It is a three-terminal device with a current flowing between terminals D, drain, and S, source. The amount of the current is controlled by the voltage between the terminals G (gate) and S but no current is drawn at G. To simplify notation, it is common to denote $v_{GS} = v_G - v_S$ and $v_{DS} = v_D - v_S$. Since no current flows from the gate into the transistor, the properties are described by two general equations

$$i_G = 0$$
$$i_D = f(v_{GS}, v_{DS}) \tag{15.3.1}$$

If we keep the voltages constant, a constant current

$$i_{D,0} = f(v_{GS,0}, v_{DS,0})$$

will flow through the device. This is the operating point of the transistor. Increments about the operating point can be obtained by differentiation. Since i_D is a function of two voltages, we must use the mathematical formula for total differential

$$di_D = \frac{\partial f(v_{GS}, v_{DS})}{\partial v_{GS}} dv_{GS} + \frac{\partial f(v_{GS}, v_{DS})}{\partial v_{DS}} dv_{DS} \tag{15.3.2}$$

The first partial derivative is the derivative of the current flowing from D to S with respect to the voltage between G and S. This indicates that the term is a transconductance of a voltage-controlled current source; we denote it by g_m. The second partial derivative is with respect to the voltage between D and S and is

FIGURE 15.3.1   Field-effect transistor (FET) symbol.

Fig. 15.3.2

Fig. 15.3.3

FIGURE 15.3.2   Small-signal equivalent for a FET.

FIGURE 15.3.3   Small-signal equivalent for a FET with capacitances.

thus a simple conductance, G_{DS}. We can rewrite Eq. (15.3.2) in terms of increments

$$I_D = g_m V_{GS} + G_{DS} V_{DS} \tag{15.3.3}$$

This equation corresponds to the network in Fig. 15.3.2. Since there are invariably some parasitic capacitances between the three terminals, a more complete linear model of the field effect transistor is in Fig. 15.3.3.

15.3.2. Bipolar Transistor

A bipolar transistor symbol is shown in Fig. 15.3.4. The terminals are given the names collector (C), base (B), and emitter (E). To simplify notation, we use $v_{BE} = v_B - v_E$ and $v_{CE} = v_C - v_E$. The device is more complicated than the field-effect transistor since currents flow at all terminals; this can be expressed by two general equations

$$i_B = f_1(v_{BE}, v_{CE})$$
$$i_C = f_2(v_{BE}, v_{CE}) \tag{15.3.4}$$

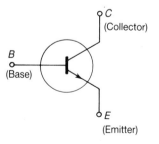

FIGURE 15.3.4   Bipolar transistor symbol.

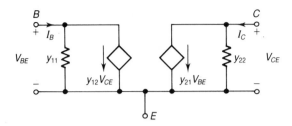

FIGURE 15.3.5   Two-port equivalent for a bipolar transistor.

Following the same steps as above we find total differentials

$$di_B = \frac{\partial f_1(v_{BE}, v_{CE})}{\partial v_{BE}} dv_{BE} + \frac{\partial f_1(v_{BE}, v_{CE})}{\partial v_{CE}} dv_{CE}$$

$$di_C = \frac{\partial f_2(v_{BE}, v_{CE})}{\partial v_{BE}} dv_{BE} + \frac{\partial f_2(v_{BE}, v_{CE})}{\partial v_{CE}} dv_{CE}$$

(15.3.5)

Looking at Eq. (15.3.5) we recognize that this is a two-port admittance equation. Using capital letters for the increments

$$I_B = y_{11} V_{BE} + y_{12} V_{CE}$$

$$I_C = y_{21} V_{BE} + y_{22} V_{CE}$$

(15.3.6)

These equations have their equivalent network in Fig. 15.3.5, but certain modifications, already applied in Chapter 13, can remove one of the dependent sources. Rewrite Eq. (15.3.6) by simultaneously adding and subtracting $y_{12} V_{BE}$ in the second equation,

$$I_B = y_{11} V_{BE} + y_{12} V_{CE}$$

$$I_C = y_{12} V_{BE} + y_{22} V_{CE} + (y_{21} - y_{12}) V_{BE}$$

(15.3.7)

The term in the brackets represents a VC at the right-hand port and the rest describes a reciprocal network with $y_{12} = y_{21}$. We can thus use the equivalent network shown in Fig. 15.3.6. Capacitances, which are invariably present, were also drawn into the equivalent network. More complicated models exist as well but we will not need them.

Figure 15.3.7a shows a simple FET amplifier. On the left the voltage source delivers the signal, the output can be taken either at the point D or S, with respect to ground. In order to make the steps similar to our previous developments, we have marked the voltages by V_1 and V_2 and the resistors by G_1 and G_2. V_{DD} is commonly used to denote the dc voltage supply; it is not normally drawn in the figures, but we have done so on the right for clarification. Every dc source is always

FIGURE 15.3.6   Small-signal network for a bipolar transistor.

put in parallel with a capacitor large enough to effectively ground the point V_{DD} for all nonzero signal frequencies. When we redraw the network for analysis, Fig. 15.3.7b, the resistor G_1 is connected to ground. We also copy the transistor model between the nodes marked by G, D and S, in this case taken from Fig. 15.3.3. The model is inside the dashed box. The network is prepared for nodal analysis and we can use all the steps we learned earlier. First, we take advantage of the fact that the signal voltage source is grounded, mark the node with a cross and use the method we learned in the study of active networks. We do not write the KCL equations at this node. Nodal equations for the other nodes are

$$(G_1 + G_3 + sC_2 + sC_3)V_1 - (G_3 + sC_3)V_2 + g_m(E - V_2) - sC_2E = 0$$

$$-(G_3 + sC_3)V_1 + (G_2 + G_3 + sC_1 + sC_3)V_2 - g_m(E - V_2) - sC_1E = 0$$

Expressions multiplied by E are transferred to the right side of the equations, with the result

$$\begin{bmatrix} (G_1 + G_3 + sC_2 + sC_3) & -(G_3 + sC_3 + g_m) \\ -(G_3 + sC_3) & (G_2 + G_3 + sC_1 + sC_3 + g_m) \end{bmatrix} \begin{bmatrix} V_1 \\ V_2 \end{bmatrix} = \begin{bmatrix} (sC_2 - g_m)E \\ (sC_1 + g_m)E \end{bmatrix}$$

In this example we used all the components of the FET model. At low frequencies we may be able to neglect the capacitors; this is achieved by setting their values equal to zero. The same applies for any missing resistors whose conductance values are also set equal to zero.

Example 1.

Draw the small-signal equivalent circuit for the bipolar amplifier in Fig. 15.3.8a. The only function of the capacitor is to separate the dc source from the signal source; it is large and acts as a short circuit for signal frequencies. For simplicity, neglect all capacitances in the transistor model.

In Fig. 15.3.8b, the voltage source with G_1 in series is transformed into a current source by the Thévenin transformation. Conductances G_2 and G_3 are grounded for signal

FIGURE 15.3.7   FET amplifier: (a) Original network with a dc source. (b) Small-signal equivalent.

frequencies. The model of the transistor, taken from Fig. 15.3.6, is in the dashed box and voltages are indicated at all nodes. Nodal formulation will lead to a 3×3 matrix equation.

15.4. NONLINEAR ELEMENTS

Until now we assumed that the properties of all network elements did not change with changes in voltage, current or environment. Real life devices do not behave as nicely. Even a simple resistor changes its value with time: The current flowing through it influences its temperature and consequently also influences its resistance. Some kind of nonlinear behavior will exist in any element we can think of.

Nonlinear properties are sometimes undesirable but sometimes we need them. To name some useful applications of nonlinearities let us mention modulation, frequency multiplication or frequency division, and class C or E amplifiers. Applications where we need linearity are high-fidelity amplifiers and all kinds of filters.

Although every element is to some extent non-linear, proper choice of conditions

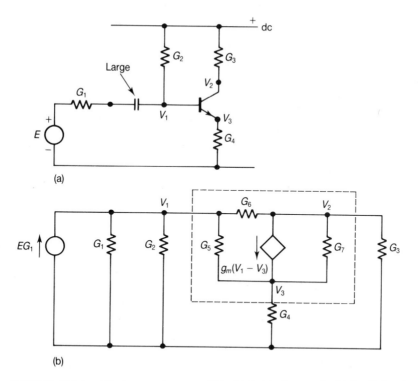

FIGURE 15.3.8   Bipolar transistor amplifier: (a) Original network. (b) Small-signal equivalent.

can make it behave linearly. If the current through the resistor is small, the temperature and thus the resistance will not change noticeably. If a small signal is applied to a nonlinear device, it operates over a small portion of the device characteristic and we can approximate the nonlinearity by a tangent straight line. This is called *linearization* of the characteristic and linear models are sometimes called *small-signal* models. The transistor models in Section 3 are such small-signal models.

The moment we can consider the network as a linear one, all the methods explained in this book can be applied: time domain analysis, frequency domain analysis, poles and zeros, symbolic functions, the superposition principle and frequency domain sensitivities, a subject of the next chapter. In linearized networks we do not even consider the source of dc power. All we need are the signal sources.

The most widely used nonlinear elements, transistors, need a source of dc power (a battery) to make them work. If this dc source has been applied, the network first adjusts the voltages at its various nodes according to the nonlinear properties of the devices. We say that the network reaches its *operating point*. Afterwards we can apply small signals and consider the network as linear, or apply large signals, and then take the nonlinear properties fully into account.

Nonlinearities of elements may be quite complicated but fundamental principles can be explained with the help of simple concepts. Let the current flowing through a nonlinear two-terminal element be a continuous and infinitely differentiable function of voltage,

$$i = f(v) \tag{15.4.1}$$

Such a function can be expanded into a Taylor series around $v = 0$

$$f(v) = f(0) + \frac{f^{(1)}(0)}{1!} v + \frac{f^{(2)}(0)}{2!} v^2 + \frac{f^{(3)}(0)}{3!} v^3 + \cdots \tag{15.4.2}$$

We could consider any number of such terms but for simplicity take only the quadratic term and let the multiplying constant be one,

$$i = v^2 \tag{15.4.3}$$

If the voltage is a composed signal

$$v = 1 + \alpha_1 \cos \omega_1 t + \alpha_2 \cos \omega_2 t \tag{15.4.4}$$

then the current flowing through such element will be the square of this expression,

$$i = 1 + \alpha_1^2 \cos^2 \omega_1 t + \alpha_2^2 \cos^2 \omega_2 t + 2\alpha_1 \cos \omega_1 t$$
$$+ 2\alpha_2 \cos \omega_2 t + 2\alpha_1 \alpha_2 \cos \omega_1 t \cos \omega_2 t$$

Using Table 8.1.1 we can replace the second powers and the product of the two cosines and get, after a few steps,

$$i(t) = 1 + \frac{\alpha_1^2}{2} + \frac{\alpha_2^2}{2} + 2\alpha_2 \cos \omega_1 t + 2\alpha_2 \cos \omega_2 t + \frac{\alpha_1^2}{2} \cos 2\omega_1 t$$
$$\tag{15.4.5}$$
$$+ \frac{\alpha_2^2}{2} \cos 2\omega_2 t + \alpha_1 \alpha_2 \cos (\omega_1 - \omega_2)t + \alpha_1 \alpha_2 \cos (\omega_1 - \omega_2)t$$

The two original cosines are present, but the dc component has changed and additional signals exist as well. Their frequencies are twice the original ones, the difference of the two original frequencies and also their sum. In other words, new signals are created and the *superposition principle is not valid for nonlinear networks.* We cannot use frequency domain analysis. Network functions do not exist either. The concept of poles and zeros does not exist. We cannot apply the Laplace transform. Only two things can be done: finding the dc operating point (dc applied but no signal yet), and calculating the time domain response, once the signal has been applied.

We will not go into time domain analysis of nonlinear networks because this requires considerable knowledge which is beyond the scope of this text. However, we will present a mathematical method to obtain the operating dc point. It gives us at least the possibility of using small-signal analysis and thus applying all the theory covered in this book. To establish the principles we will use two nonlinear elements: a bipolar diode and a field-effect transistor. We will not go into physical background or details.

15.4.1. Diode

A bipolar diode is shown in Fig. 15.4.1a. Its dc characteristic is expressed by the equation

$$i = I_s(e^{v_D/V_T} - 1) \tag{15.4.6}$$

and is sketched in Fig. 15.4.1b. In Eq. (15.4.6) I_s is the *saturation* current, a constant which depends on the size of the diode, v_D is the voltage across the diode and V_T is a constant, approximately equal to 25 mV. In order to simplify the expressions we will insert the number for V_T and use

$$i = I_s(e^{40v_D} - 1) \tag{15.4.7}$$

$$v_D = v_i - v_j$$

(a)

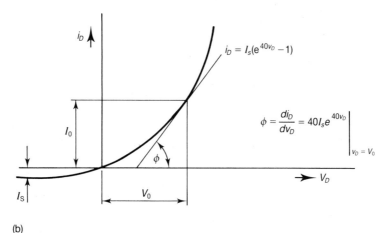

(b)

FIGURE 15.4.1   Bipolar diode: (a) Symbol. (b) Characteristic.

The current is large for $v_D > 0$ and very small, flowing in opposite direction, for $v_D < 0$. For any selected fixed voltage, $v_{D,0}$, the current will be fixed, i_0. If we increment $v_{D,0}$ by a small amount, Δv_D, the current will also increment by an amount Δi. Instead of increments we can use differentiation and obtain

$$\frac{di}{dv_D} = g(v_{D,0}) = 40 I_s s^{40 v_D} \tag{15.4.8}$$

This is an incremental conductance of the diode at the operating point $v_{D,0}$. For small variations of the voltage and current around the operating point we will use upper case letters and

$$G = g(v_{D,0}) = \frac{I}{V} \tag{15.4.9}$$

The value of G is the conductance we would use in small-signal analysis. The characteristic of the diode and the way to determine G (the tangent) are shown in Fig. 15.4.1b.

15.4.2. Field-Effect Transistor

The element symbol is Fig. 15.4.2a and a sketch of its characteristics is in b. It is described by two sets of equations: In the triode region (see the figure) the equations are

$$i_D = \beta \left[(v_{GS} - V_T) v_{DS} - \frac{1}{2} v_{DS}^2 \right] \tag{15.4.10}$$

valid for

$$v_{GS} \geq V_T$$
$$v_{DS} \leq v_{GS} - V_T \tag{15.4.11}$$

In the pinch-off, or saturation, region

$$i_D = \frac{1}{2} \beta (v_{GS} - V_T)^2 \tag{15.4.12}$$

valid for

$$v_{GS} \geq V_T$$
$$v_{DS} \geq v_{GS} - V_T \tag{15.4.13}$$

(a)

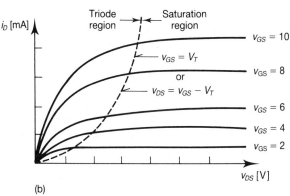

(b)

FIGURE 15.4.2   FET transistor: (a) Symbol with voltages. (b) Characteristics.

In the equations V_T is a constant which depends on technology. We will assume that $V_T = 2$. The boundary between the two regions is reached when we use the equal sign in the inequalities. Inserting $v_{DS} = v_{GS} - V_T$ into Eq. (5.4.10) results in

$$i_D = \beta \left[(v_{GS} - V_T)(v_{GS} - V_T) - \frac{1}{2}(v_{GS} - V_T)^2 \right] = \frac{\beta}{2}(v_{GS} - V_T)^2$$

which is Eq. (15.4.12). There is thus smooth transition from one set of equations to the other set.

The first question which comes into mind is how does one know which of these equations should be used? In simple networks, and with experience, one can make a reasonable guess. Without experience, or when one deals with a larger network, only one thing remains: take one set, solve the network and check whether the inequalities are satisfied. If they are not, take the other set and solve the equations again. Another way involves graphical solutions which use characteristics like those in Fig. 15.4.2b. The methods are reasonably simple for single transistor networks and are the subject of courses on electronics.

15.5. NETWORKS WITH NONLINEAR ELEMENTS

In this section we set up equations for several networks with nonlinear elements. First we consider networks with diodes, each of the diodes described by

$$i_D = I_s(e^{40v_D} - 1) \qquad (15.5.1)$$

The first network is in Fig. 15.5.1; we identify

$$v_D = v_1 - v_2 \qquad (15.5.2)$$

and write KCL for both nodes:

$$\begin{aligned} f_1 &= G_1 v_1 + i_D - J = 0 \\ f_2 &= -i_D + G_2 v_2 = 0 \end{aligned} \qquad (15.5.3)$$

The steps do not differ, in principle, from our previous ways of writing the equations. Nevertheless, there are two differences:

a. nonlinear equations *cannot* be put into a matrix form and *must* be left as separate equations.
b. The independent sources, E or J, are kept on the left side because our solution method, to be explained in Section 6, will be based on this form.

The second network, Fig. 15.5.2, is slightly more complicated. For the diode we identify

$$v_D = v_1 - v_2 \qquad (15.5.4)$$

Fig. 15.5.1

Fig. 15.5.2

FIGURE 15.5.1   Network with one diode.

FIGURE 15.5.2   Network with one diode and two current sources.

The KCL are

$$f_1 = (G_1 + G_2)v_1 - G_2v_2 + i_D - J_1 = 0$$
$$f_2 = -G_2v_1 + (G_2 + G_3)v_2 - i_D - J_2 = 0$$

$$(15.5.5)$$

The third network, Fig. 15.5.3, has two voltage sources and three diodes. One terminal of the dc sources is grounded and we can write KCL similarly as we have done in active networks by marking them with crosses and not taking them into KCL. The nodal voltages are indicated in the figure and we identify

$$v_{D1} = v_1$$
$$v_{D2} = v_1 - v_3$$
$$v_{D3} = v_3 - v_2$$

$$(15.5.6)$$

The three nodal equations are

$$f_1 = (G_1 + G_2)v_1 - G_2v_2 - E_1G_1 + i_{D1} + i_{D2} = 0$$
$$f_2 = -G_2v_1 + (G_2 + G_3)v_2 - i_{D3} = 0$$
$$f_3 = G_4v_3 - G_4E_2 - i_{D2} + i_{D3} = 0$$

$$(15.5.7)$$

A transistor can also be connected as a diode, as shown in Fig. 15.5.4. The connection makes the voltage between gate and source, v_{GS}, equal to the voltage between drain and source, v_{DS} and thus Eq. (15.4.12) simplifies to

$$i_D = \frac{\beta}{2}(v_{DS} - V_T)^2$$

$$(15.5.8)$$

Fig. 15.5.3

Fig. 15.5.4

FIGURE 15.5.3   Network with three diodes.

FIGURE 15.5.4   FET connected as a diode.

Considering the network in Fig. 15.5.5 we first identify

$$v_{DS} = v_1 - v_2 \qquad (15.5.9)$$

and then the KCL equations are

$$f_1 = 4v_1 - 5 - 2v_2 + i_D = 0$$
$$f_2 = -2v_1 + 3v_2 - i_D = 0$$

In this case we will substitute for i_D. To simplify writing, we set $V_T = -2$ and $\beta/2 = 1$. The equations are

$$f_1 = v_1^2 + v_2^2 + 2v_2 - 2v_1v_2 - 1 = 0$$
$$f_2 = -v_1^2 - v_2^2 + 2v_1 - v_2 + 2v_1v_2 - 4 = 0 \qquad (15.5.10)$$

Although the FET transistor is described by a quadratic function, even in this simple case we cannot find the unknown voltages by algebraic methods. We will derive a numerical method in the next section, but first let us consider a network where the transistor acts as an amplifier. We are asked to set up the equations for the amplifier in Fig. 15.5.6. We use Eq. (15.4.12)

$$i_D = \frac{\beta}{2} (v_{GS} - V_T)^2$$

Fig. 15.5.5

Fig. 15.5.6

FIGURE 15.5.5   Network with a FET connected as a diode.

FIGURE 15.5.6   FET amplifier.

and identify

$$v_{GS} = v_1 - v_3$$

The three KCL equations are

$$f_1 = (G_1 + G_2)v_1 - 10G_1 = 0$$
$$f_2 = G_3v_2 - 10G_3 + i_{DS} = 0$$
$$f_3 = G_4v_3 - i_{DS} = 0$$

We could substitute for i_{DS} but it would not make the result simpler and we leave the set in the above form.

15.6. NEWTON–RAPHSON ITERATIVE METHOD

We will derive a general method which can be used to solve systems of n nonlinear equations in n unknowns, but as an introduction we start with just one nonlinear equation with one variable,

$$f(x) = 0 \qquad (15.6.1)$$

We know that at a point x, we can expand such function into a Taylor series about a point which is Δx away from x. The expansion is

$$f(x + \Delta x) - f(x) + \frac{f^{(1)}(x)}{1!}\Delta x + \frac{f^{(2)}(x)}{2!}(\Delta x)^2 \ |$$

Usually we terminate the expansion after the second term and get the condition

$$f(x) + f^{(1)}(x)\,\Delta x = 0$$

or

$$f^{(1)}(x)\,\Delta x = -f(x)$$

In the case of one equation we obtain the unknown Δx by division,

$$\Delta x = -\frac{f(x)}{f^{(1)}(x)}$$

Usually we indicate the starting point and the starting increment with a superscript zero, x^0 and Δx^0; then the new x is given a superscript one,

$$x^1 = x^0 + \Delta x^0$$

The process is repeated with the new value, x^1. We can thus write an iterative scheme

$$\Delta x^k = -\frac{f(x^k)}{f^{(1)}(x^k)}$$

$$x^{k+1} = x^k + \Delta x^k$$

(15.6.2)

In many books, the first expression above is inserted into the second to obtain

$$x^{k+1} = x^k - \frac{f(x^k)}{f^{(1)}(x^k)}$$

but it is not advantageous to do so. The iterations on (15.6.2) have the advantage that we can see the sequence of Δx^k, which must be decreasing.

As an example we take the simple case

$$f(x) = x^2 - 1 = 0$$

which clearly has the solution $x = \pm 1$. The derivative is $f^{(1)} = 2x$ and $\Delta x = (1 - x^2)/2x$. To start iteration by means of Eq. (15.6.2) we need an initial point. If we have an idea what the root could be, then we should use as an initial value the best estimate. If we know nothing, then we simply select some arbitrary number. We will start with $x^0 = 2$; the sequence of iterations is shown in Table 15.6.1.

Solution of *systems* of nonlinear equations proceeds similarly; we will derive the method by using a system of two equations:

$$f_1(x_1, x_2) = 0$$

$$f_2(x_1, x_2) = 0$$

(15.6.3)

Each of these equations can be expanded into a Taylor series about points $x_1 + \Delta x_1$ and $x_2 + \Delta x_2$, with the assumption that numerical values are available

TABLE 15.6.1   Iterations on $f(x) = x^2 - 1 = 0$

k	x^k	Δx^k
0	2	-0.75
1	1.25	-0.225
2	1.025	-0.0246951
3	1.003049	-0.0003048
4	1.000000	$-4.598E-8$

for x_1 and x_2.

$$f_1(x_1 + \Delta x_1, x_2 + \Delta x_2) = f_1(x_1, x_2) + \frac{\partial f_1}{\partial x_1} \Delta x_1 + \frac{\partial f_1}{\partial x_2} \Delta x_2 + \text{higher terms} = 0$$

$$f_2(x_1 + \Delta x_1, x_2 + \Delta x_2) = f_2(x_1, x_2) + \frac{\partial f_2}{\partial x_1} \Delta x_1 + \frac{\partial f_2}{\partial x_2} \Delta x_2 + \text{higher terms} = 0$$

We neglect all higher-order terms and rewrite

$$f_1(x_1, x_2) + \frac{\partial f_1}{\partial x_1} \Delta x_1 + \frac{\partial f_1}{\partial x_2} \Delta x_2 = 0$$

$$f_2(x_1, x_2) + \frac{\partial f_2}{\partial x_1} \Delta x_1 + \frac{\partial f_2}{\partial x_2} \Delta x_2 = 0$$

Since both x_1 and x_2 are known as numbers, we can insert them into the equations and get numerical values for the functions and the partial derivatives. Transfer the function values to the right-hand side:

$$\frac{\partial f_1}{\partial x_1} \Delta x_1 + \frac{\partial f_1}{\partial x_2} \Delta x_2 = -f_1(x_1, x_2)$$

$$\frac{\partial f_2}{\partial x_1} \Delta x_1 + \frac{\partial f_2}{\partial x_2} \Delta x_2 = -f_2(x_1, x_2)$$
(15.6.4)

This is a set of two linear equations in two unknowns,

$$\begin{bmatrix} \dfrac{\partial f_1}{\partial x_1} & \dfrac{\partial f_1}{\partial x_2} \\[2mm] \dfrac{\partial f_2}{\partial x_1} & \dfrac{\partial f_2}{\partial x_2} \end{bmatrix} \begin{bmatrix} \Delta x_1 \\ \Delta x_2 \end{bmatrix} = \begin{bmatrix} -f_1(x_1, x_2) \\ -f_2(x_1, x_2) \end{bmatrix}$$
(15.6.5)

The matrix of the partial derivatives is called the *Jacobian* and is usually denoted by the letter J. Since in this book J has been used to denote independent current sources, we will use the letter M. Thus in matrix notation we have

$$\mathbf{M} = \begin{bmatrix} \dfrac{\partial f_1}{\partial x_1} & \dfrac{\partial f_1}{\partial x_2} \\[2mm] \dfrac{\partial f_2}{\partial x_1} & \dfrac{\partial f_2}{\partial x_2} \end{bmatrix}$$
(15.6.6)

and

$$\Delta \mathbf{x} = \begin{bmatrix} \Delta x_1 \\ \Delta x_2 \end{bmatrix} \tag{15.6.7}$$

and also

$$\mathbf{f} = \begin{bmatrix} f_1 \\ f_2 \end{bmatrix} \tag{(15.6.8)}$$

After solving the system we obtain new values by adding the increments to the original values of the variables. Since this must be repeated, we distinguish the steps by using a superscript. The starting values are given the superscript zero, the next value the superscript one:

$$
\begin{aligned}
x_1^1 &= x_1^0 + \Delta x_1^0 \\
x_2^1 &= x_2^0 + \Delta x_2^0
\end{aligned}
\tag{15.6.9}
$$

Equations (15.6.5) and (15.6.9) form the Newton–Raphson iterative algorithm. We usually write them in the form

$$\mathbf{M}^k \Delta \mathbf{x}^k = -\mathbf{f}^k \tag{15.6.10}$$

$$\mathbf{x}^{k+1} = \mathbf{x}^k + \Delta \mathbf{x}^k \tag{15.6.11}$$

where the superscript k starts at zero and is incremented with every iteration. We start with x_1^0 and x_2^0. If we know anything about their possible values, we use any good estimate; it will reduce the number of iterations. If we do not know anything, then we make an arbitrary guess (and hope for the best). The iterations are stopped when the increments Δx_1 and Δx_2 are sufficiently small. The method applies for any number of equations; for n equations the matrix $\mathbf{M}$ will have the size $n \times n$ and the vectors $\Delta \mathbf{x}$ and $\mathbf{f}$ will have length n.

We will use a simple example to show how the method is used. Consider two equations

$$
\begin{aligned}
f_1 &= x_1^2 + x_2^2 - 4 = 0 \\
f_2 &= x_1 - x_2 = 0
\end{aligned}
\tag{15.6.12}
$$

The first one describes a circle with radius $r = 2$ centered at the origin, the second is a straight line passing through the origin. In this simple example we can substitute the second equation into the first one and get $x_1 = x_2 = \pm\sqrt{2} = \pm 1.4142$.

The derivatives for the Jacobian are

$$\frac{\partial f_1}{\partial x_1} = 2x_1$$

$$\frac{\partial f_1}{\partial x_2} = 2x_2$$

$$\frac{\partial f_2}{\partial x_1} = 1$$

$$\frac{\partial f_2}{\partial x_2} = -1$$

Equation (15.6.10) will have the form

$$\begin{bmatrix} 2x_1 & 2x_2 \\ 1 & -1 \end{bmatrix} \begin{bmatrix} \Delta x_1 \\ \Delta x_2 \end{bmatrix} = \begin{bmatrix} -x_1^2 - x_2^2 + 4 \\ -x_1 + x_2 \end{bmatrix}$$

Since this is only a 2 × 2 system, it is easiest to invert the matrix

$$\begin{bmatrix} \Delta x_1 \\ \Delta x_2 \end{bmatrix} = \frac{1}{-2(x_1 + x_2)} \begin{bmatrix} -1 & -2x_2 \\ -1 & 2x_1 \end{bmatrix} \begin{bmatrix} -x_1^2 - x_2^2 + 4 \\ -x_1 + x_2 \end{bmatrix} \qquad (15.6.13)$$

We will use Eq. (15.6.13) repeatedly. To start we select $x_1^0 = x_2^0 = 1$:

$$\begin{bmatrix} \Delta x_1^0 \\ \Delta x_2^0 \end{bmatrix} = \frac{1}{-4} \begin{bmatrix} -1 & -2 \\ -1 & 2 \end{bmatrix} \begin{bmatrix} 2 \\ 0 \end{bmatrix} = \begin{bmatrix} 0.5 \\ 0.5 \end{bmatrix}$$

using Eq. (15.6.11) we get the next approximation to the solution

$$\begin{bmatrix} x_1^1 \\ x_2^1 \end{bmatrix} = \begin{bmatrix} x_1^0 \\ x_2^0 \end{bmatrix} + \begin{bmatrix} \Delta x_1^0 \\ \Delta x_2^0 \end{bmatrix} = \begin{bmatrix} 1.5 \\ 1.5 \end{bmatrix}$$

In the next iteration we substitute these results into Eq. (15.6.13) to obtain

$$\begin{bmatrix} \Delta x_1^1 \\ \Delta x_2^1 \end{bmatrix} = \frac{1}{-6} \begin{bmatrix} -1 & -3 \\ -1 & 3 \end{bmatrix} \begin{bmatrix} -0.5 \\ 0 \end{bmatrix} = \begin{bmatrix} -1/12 \\ -1/12 \end{bmatrix}$$

The next approximation is

$$\begin{bmatrix} x_1^2 \\ x_2^2 \end{bmatrix} = \begin{bmatrix} 1.5 - 1/12 \\ 1.5 - 1/12 \end{bmatrix} = \begin{bmatrix} 17/12 \\ 17/12 \end{bmatrix} = \begin{bmatrix} 1.4166667 \\ 1.4166667 \end{bmatrix}$$

If we make one more iteration, we will get $\Delta x_1^2 = \Delta x_2^2 = -1/408 = 0.00245098$ and $x_1^3 = x_2^3 = 577/408 = 1.4142157$, almost equal to $\sqrt{2} = 1.41421356$.

Example 1.

Apply the Newton–Raphson formula to the linear set of equations

$$f_1 = 3x_1 + 2x_2 - 7 = 0$$
$$f_2 = 2x_1 + 4x_2 - 10 = 0$$

and show that the result is obtained in one iteration starting from any initial estimate. The solution is easy to find directly: $x_1 = 1$, $x_2 = 2$.

We take an arbitrary guess $x_1^0 = x_2^0 = 10$. Equation (15.6.10) takes the form

$$\begin{bmatrix} 3 & 2 \\ 2 & 4 \end{bmatrix} \begin{bmatrix} \Delta x_1^0 \\ \Delta x_2^0 \end{bmatrix} = \begin{bmatrix} -30 - 20 + 7 \\ -20 - 40 + 10 \end{bmatrix}$$

The solution is

$$\begin{bmatrix} \Delta x_1^0 \\ \Delta x_2^0 \end{bmatrix} = \begin{bmatrix} -9 \\ -8 \end{bmatrix}$$

and inserting into Eq. (15.6.11)

$$\begin{bmatrix} x_1^1 \\ x_2^1 \end{bmatrix} = \begin{bmatrix} 10 \\ 10 \end{bmatrix} + \begin{bmatrix} -9 \\ -8 \end{bmatrix} = \begin{bmatrix} 1 \\ 2 \end{bmatrix}$$

which is the final result.

There are many more mathematical details which should be known. The Newton–Raphson method converges rapidly if the initial estimates for the variables are close to the final values. It may not converge if they are far away. It is not the purpose of this chapter to give a comprehensive treatment. What we need are the consequences of its use in the solution of nonlinear networks. They are discussed in the next section.

15.7. NONLINEAR NETWORKS

Let us now turn our attention to the solution of network equations and take for this purpose several examples. As a first one we consider Fig. 15.7.1 where the diode is described by $i_D = e^{40v} - 1$. Writing the nodal equation for the single node

$$Gv + i_D = 2$$

FIGURE 15.7.1   Network with one diode.

or, in a form suitable for Newton–Raphson solution,

$$f(v) = v + (e^{40v} - 1) - 2 = 0$$

The derivative is

$$f^{(1)}(v) = 40e^{40v} + 1$$

and thus

$$\Delta v^k = \frac{3 - v - e^{40v}}{40e^{40v} + 1}$$

The iterations are shown in Table 15.7.1.

The second case will consider the network in Fig. 15.5.2, Eq. (15.5.4) and Eq. (15.5.5), repeated here:

$$i_D = i(v_D) = I_s(e^{40v_D} - 1) \tag{15.7.1}$$

$$v_D = v_1 - v_2 \tag{15.7.2}$$

TABLE 15.7.1   Iterations for Fig. 15.7.1

k	x^k	Δx^k
0	0.2	-0.0249763
1	0.1750237	-0.0249351
2	0.1500886	-0.0248225
3	0.1252661	-0.0245168
4	0.1007493	-0.0237011
5	0.0770482	-0.0216233
6	0.0554249	-0.0169346
7	0.0384903	-0.0090729
8	0.0294174	-0.0020885
9	0.0273289	-0.0000915
10	0.0272375	-0.000002
11	0.0272373	$-8.3396E-12$

and

$$f_1 = (G_1 + G_2)v_1 - G_2v_2 + i_D - J_1 = 0$$
$$f_2 = -G_2v_1 + (G_2 + G_3)v_2 - i_D - J_2 = 0$$

(15.7.3)

The partial derivatives are

$$\frac{\partial f_1}{\partial v_1} = G_1 + G_2 + \frac{\partial i_D}{\partial v_D}\frac{\partial v_D}{\partial v_1}$$

$$\frac{\partial f_1}{\partial v_2} = -G_2 + \frac{\partial i_D}{\partial v_D}\frac{\partial v_D}{\partial v_2}$$

$$\frac{\partial f_2}{\partial v_1} = -G_2 - \frac{\partial i_D}{\partial v_D}\frac{\partial v_D}{\partial v_1}$$

$$\frac{\partial f_2}{\partial v_2} = G_2 + G_3 - \frac{\partial i_D}{\partial v_D}\frac{\partial v_D}{\partial v_2}$$

We have used the differentiation chain rule to obtain the derivatives of the current with respect to the nodal voltages. However, since we know Eq. (15.7.2), the derivatives are

$$\frac{\partial v_D}{\partial v_1} = 1$$

$$\frac{\partial v_D}{\partial v_2} = -1$$

Substituting these partial derivatives into the Jacobian

$$\mathbf{M} = \begin{bmatrix} G_1 + G_2 + \dfrac{\partial i_D}{\partial v_D} & -G_2 - \dfrac{\partial i_D}{\partial v_D} \\[2ex] -G_2 - \dfrac{\partial i_D}{\partial v_D} & G_2 + G_3 + \dfrac{\partial i_D}{\partial v_D} \end{bmatrix}$$

The last step is to substitute into this matrix the derivative of the current. From Eq. (15.7.1)

$$\frac{\partial i_D}{\partial v_D} = 40I_s e^{40v_D} = 40I_s e^{40(v_1 - v_2)}$$

From the example we conclude that the Jacobian matrix has the same form as the nodal matrix. All linear elements in the matrix are in the same positions as they were before. Every nonlinear element will also occupy the same positions as if it were linear. The number will change from iteration to iteration and will be equal to the derivative, evaluated at the voltages of that particular iteration. When the iterations finish with sufficiently small Δv_i, then this derivative represents the *small-signal* equivalent of the nonlinear element. These conclusions are valid for any other formulation method as well.

Example 1.

Write the Jacobian matrix for the network Fig. 15.5.4. Use Eq. (15.5.10), repeated here:

$$f_1 = v_1^2 + v_2^2 + 2v_2 - 2v_1 v_2 - 1 = 0$$
$$f_2 = -v_1^2 - v_2^2 + 2v_1 - v_2 + 2v_1 v_2 - 4 = 0$$

The partial derivatives are

$$\frac{\partial f_1}{\partial v_1} = 2v_1 - 2v_2$$

$$\frac{\partial f_1}{\partial v_2} = 2v_2 + 2 - 2v_1$$

$$\frac{\partial f_2}{\partial v_1} = -2v_1 + 2 + 2v_2$$

$$\frac{\partial f_2}{\partial v_2} = -2v_2 - 1 + 2v_1$$

The Jacobian matrix is

$$\mathbf{M} = \begin{bmatrix} 2v_1 - 2v_2 & 2v_2 + 2 - 2v_1 \\ -2v_1 + 2 + 2v_2 & -2v_2 - 1 + 2v_1 \end{bmatrix}$$

Example 2.

Consider the network in Fig. 15.7.2, apply Newton–Raphson iteration and find the voltages v_1 and v_2. Use the initial estimate $v_1^0 = v_2^0 = 0.1$ and let $I_s = 1$. The diodes are described by Eq. (15.7.1).

FIGURE 15.7.2   Network with two diodes.

The nodal equations are

$$f_1 = i_{D_1} + Gv_1 - Gv_2 - J = 0$$
$$f_2 = -Gv_1 + Gv_2 + i_{D_2} = 0$$

Inserting numerical values

$$f_1 = e^{40v_1} + v_1 - v_2 - 2 = 0$$
$$-v_1 + v_2 + e^{40v_2} + 1 = 0$$

The Jacobian is

$$\mathbf{M} = \begin{bmatrix} 40e^{40v_1} + 1 & -1 \\ -1 & 40e^{40v_2} + 1 \end{bmatrix}$$

Evaluation at the initial estimates provides

$$f_1 = 52.59815$$
$$f_2 = 53.59815$$

Inserting the values we must solve

$$\begin{bmatrix} 2184.926 & -1 \\ -1 & 2184.926 \end{bmatrix} \begin{bmatrix} \Delta v_1^0 \\ \Delta v_2^0 \end{bmatrix} = \begin{bmatrix} -52.59815 \\ -53.59815 \end{bmatrix}$$

The result is

$$\Delta v_1^0 = -0.0240844$$
$$\Delta v_2^0 = -0.0245419$$

and the first approximation to the nodal voltages is

$$v_1^1 = 0.0759156$$
$$v_2^1 = 0.07545810$$

The iterations were run on a computer with the sequence shown in Table 15.7.2. The final nodal voltages are $v_1 = 0.01712$ and $v_2 = 0.00041$.

15.8. THE SPICE PROGRAM

The few nonlinear networks considered in the previous sections were very primitive, but their solution still required considerable amount of work. It is, therefore, no wonder that many attempts have been made to automate the process of nonlinear network analysis and early attempts can be traced some forty years

TABLE 15.7.2   Iterations for Network Fig. 15.7.2

k	Δv_1^k	Δv_2^k	v_1^{k+1}	v_2^{k+1}
0	−0.02408	−0.02454	0.07592	0.07546
1	−0.02260	−0.02378	0.05331	0.05168
2	−0.01909	−0.02182	0.03423	0.02986
3	−0.01234	−0.01736	0.02188	0.01250
4	−0.00432	−0.00962	0.01757	0.00289
5	−0.00044	−0.00236	0.01712	0.00053
6	−0.00001	−0.00012	0.01712	0.00041
7	−0.00000	−0.00000	0.01712	0.00041

back. The work did not stop there and many programs have been written since. Large companies produced their own packages or modified available ones to their needs.

Let us briefly discuss how such programs are structured. The nodal and mesh formulations, covered in this book for hand analysis, are not suitable for programming, because of the various network transformations. The most widely used method to set up the equations is the *modified nodal* formulation. This book covers its principles in Chapter 17. It is also the method used in NETOPT and in the program SPICE, briefly mentioned below.

Because there are infinitely many nonlinear elements, it is necessary to select the most useful ones and build them into the program so that they can be automatically accessed. The user is not burdened with the models; in fact, most users do not even understand them.

Input to the program is given by a circuit description file. It does not differ much from the input style of our NETOPT program and an example will be given below. There is, however, one important difference. Nonlinear networks need for their operation the dc source and this must be given. In NETOPT, which deals with linear elements only, the dc source was not necessary.

Nonlinear equations have to be solved and the Newton–Raphson method of Section 15.6 is used, the same way as explained there. This leads to repeated linearization and solution of linear equations. Cramer's rule, used extensively in this book, is the best one for hand calculations, but is absolutely unsuitable for programming and for larger systems. The method which is used most often is called *LU* decomposition, or triangular decomposition. It is covered in courses on numerical mathematics and the student need not know them in his initial studies. However, we give in Appendix C the theoretical background and the solution routines, in FORTRAN, are available on the diskettes accompanying this book. They can handle systems of linear equations with complex coefficients, as needed for frequency domain analysis.

Commercial programs also use the *LU* decomposition, but in a special form, called *sparse matrix* solvers. These are fairly complicated programs which take advantage of the fact that matrices describing large practical networks have very

FIGURE 15.8.1   Transistor amplifier.

few nonzero entries. The trick of such solvers is not to store the zeros and not to operate on them.

As discussed above, only two things can be done with nonlinear networks: find the dc solution and the time domain solution. The first one finds the nodal voltages (and possibly some currents) for the dc source connected to the network, but with no signal applied yet. This is equivalent to what we have done in Section 15.7. Time domain responses are solved by integration, a method not covered in this book. Later in his studies, the student can find information about sparse matrix solutions and numerical integration of network equations in [1].

There are several commercially available programs for solutions of nonlinear networks or systems. We will mention here only one, because it is widely available to University students. The program is called SPICE and originated at the University of California, Berkeley. A manual must be read before its use, but many features are similar to NETOPT. We will show the SPICE input file and analysis of the nonlinear network in Fig. 15.8.1 where the nodes are given by numbers in circles. The input file for the network is

```
Common Emitter Inverter
Vcc 1 0 5
RC 1 2 1k
Rin 4 3 750
Cout 2 0 50p
Q1 2 3 0 Qnpn
Vin 4 0 pulse (0 5 10ns 10ns 10ns 50ns)
.model Qnpn npn
.trana 1.5ns 300ns
.print tran v(4) v(2)
.end
```

[1] J. Vlach and K. Singhal *Computer Methods for Circuit Analysis and Design.* Van Nostrand Reinhold 1983.

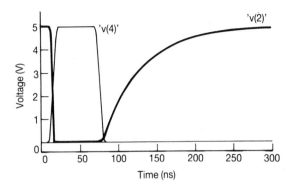

FIGURE 15.8.2   Transient response of the amplifier in Fig. 15.8.1 (bold curve) to a trapezoidal input (thin curve).

The first line is a comment, the next four lines specify the linear elements and their connections to the nodes. The line starting with Q1 gives connection of the transistor into the network while the line starting with Vin describes the input signal. It is connected between nodes 4 and 0, and it is a pulse. The numbers in the parentheses describe the voltage swing from 0 to 5 V; the pulse starts 10 ns after the beginning of simulation, has rise and fall times 10 ns each, and its width is 50 ns. The remaining lines are commands to select a transistor mode to do transient analysis, and to print the results at nodes 4 and 2. The simulation is plotted in Fig. 15.8.2; the thinner line is the input pulse (node 4), the thicker line is the output (node 2).

PROBLEMS CHAPTER 15

If you do not have access to a computer with NETOPT, redraw the networks in a form suitable for small-signal analysis and combine resistors which happen to be in parallel. If you have NETOPT, solve the linear problems by it.

P.15.1 Redraw the network for small-signal analysis by using the model on the right. The purpose of the two capacitors is to separate the dc supply voltage from the rest of the network. In this problem consider the capacitors as short circuits. Find the overall gain of the amplifier.

P.15.2 Do the same as in P.15.1.

P.15.3 Redraw the network for small-signal analysis using the model in (b) and find the transfer function.

(a)

(b)

P.15.4 The network is already redrawn for small-signal analysis (dc supply voltages are removed). Use the bipolar transistor model in (b) and find frequency domain response using NETOPT.

(a)

(b)

P.15.5 The network is already redrawn for small-signal analysis (dc supply voltages are removed). Use the model in (b) and analyze in frequency domain using NETOPT.

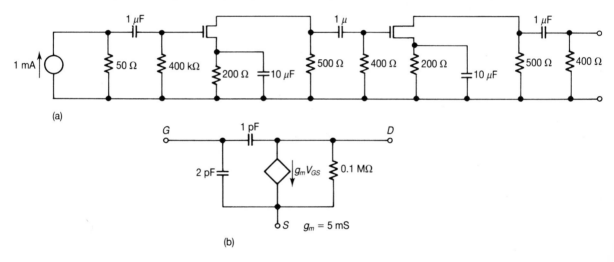

(a)

(b)

P.15.6 Redraw the network for small-signal analysis and find frequency domain response using NETOPT. Use the model in (b).

(a)

(b)

P.15.7 Do the same as in P.15.6 using the same model.

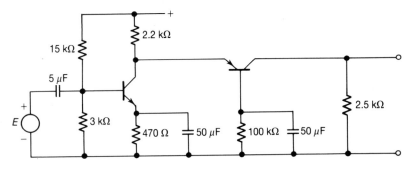

P.15.8 The nonlinear element is defined by $i = v^2$ for $v \geq 0$ and $i = 0$ for $v < 0$. (a) Find the voltage v as a solution of one quadratic equation. (b) Apply the Newton–Raphson algorithm starting with $v^0 = 1.2$.

.15.9 Use mesh and nodal formulation (after transforming the voltage source into a current source). Find e dc voltage across the diode.

P.15.10 Set up nodal equations and make two Newton–Raphson iteration steps by starting with $v_1^0 = 1$ and $v_2^0 = 0.5$.

P.15.11 Two equations are given:

$$f_1 = x_1^2 - x_2 - 2 = 0$$
$$f_2 = 2x_1 - x_2 + 3 = 0$$

(a) Eliminate one variable and find the solution as roots of one quadratic equation. (b) Start with $x_1^0 = 3.45$ and $x_2^0 = 9.9$ and make two steps in the Newton–Raphson iteration.

P.15.12 Do the same as in P.15.11 with the equations

$$f_1 = x_1 - x_2^2 = 0$$
$$f_2 = x_1 + 3x_2 - 5 = 0$$

In the iteration start with $x_1^0 = x_2^0 = 1$.

CHAPTER 16

Sensitivities

16.0. INTRODUCTION

All our previous studies assumed that the element values are accurate and do not change. This may be an ideal situation but is not true in applications. Any system which is supposed to perform a certain function is built from components which always differ somewhat from the nominal value. The differences may not be large but if all components are not what they are supposed to be, it may happen that the system eventually does not meet its specifications. It would certainly be nice to have an idea which components are critical and which are not. In electrical networks, this is achieved by the theory of sensitivities, the subject of this chapter.

Sensitivities are always difficult to obtain by hand calculations and are not available in many computer packages. They are available in NETOPT; read Chapter 18, Section 8 on how to get them.

16.1. MOTIVATION

We will use a small example to show how the change of an element influences performance of a network. The voltage divider in Fig. 16.1.1 has the transfer function

$$F = \frac{V_{out}}{E} = \frac{1}{sCR + 1} \tag{16.1.1}$$

492

FIGURE 16.1.1   Sensitivity principle on RC voltage divider.

In this chapter we will use the letter F for any function which will be subject to sensitivity studies; they need not be network functions only.

Suppose that we change the resistor by a small amount ΔR. This will influence the function and we will have

$$F + \Delta F = \frac{1}{sC(R + \Delta R) + 1} \tag{16.1.2}$$

Subtracting Eq. (16.1.1) from Eq. (16.1.2)

$$\Delta F = (F + \Delta F) - F = \frac{1}{sC(R + \Delta R) + 1} - \frac{1}{sCR + 1}$$

or

$$\Delta F = \frac{-sC\,\Delta R}{[sC(R + \Delta R) + 1][sCR + 1]}$$

Opening the brackets in the denominator we arrive at

$$\frac{\Delta F}{\Delta R} = \frac{-sC}{s^2 R^2 C^2 \left(1 + \dfrac{\Delta R}{R}\right) + sCR\left(2 + \dfrac{\Delta R}{R}\right) + 1} \tag{16.1.3}$$

For comparisons insert numerical values: $C = 1$, $R = 1$, $s = j1$ and $\Delta R/R = 0.001$. This leads to

$$\frac{\Delta F}{\Delta R} = \frac{-j}{-0.001 + j2.001}$$

and the absolute value is

$$\left|\frac{\Delta F}{\Delta R}\right| = 0.49975 \tag{16.1.4}$$

The above steps required a fair amount of algebra even though the network is so simple. If the changes are small (in fact infinitesimally small), then we get the result much faster by differentiating Eq. (16.1.1) with respect to R:

$$\frac{dF}{dR} = \frac{-sC}{(sCR + 1)^2} \tag{16.1.5}$$

Substituting the same numerical values as above, we get

$$\left| \frac{dF}{dR} \right| = 0.5 \tag{16.1.6}$$

a result which is almost the same as above, but obtained with much less work. We differentiated with respect to R but could have done so with respect to C or s, or any other variable which may appear in the function F. Such derivatives are the simplest sensitivity functions. They indicate how a small change in any variable influences the function under consideration.

16.2. DEFINITION OF SENSITIVITIES

Having established the reason why we use the derivatives to check on the influence of element changes, we now turn to the various sensitivity definitions. The simple derivative we used above is an obvious choice, but it has problems. If we change the capacitor to some other value, for instance $C = 2$, and simultaneously change the resistor to $R = 0.5$ so that their product RC does not change, we get the *same* transfer function but *another* sensitivity at the same frequency,

$$\frac{dF}{dR} = -\frac{j2}{(j + 1)^2}$$

This is quite unpleasant. In order to eliminate this problem, a normalized sensitivity has been defined:

$$S_h^F = \frac{dF}{dh} \frac{h}{F} \tag{16.2.1}$$

Here S_h^F is the symbol for this definition of sensitivity, F is the function under consideration, and h is any variable in F, for instance any network element. Let us calculate this sensitivity for the network in Fig. 16.1.1. Using Eq. (16.1.1) and Eq. (16.1.5)

$$S_R^F = \frac{-sC}{(sCR + 1)^2} \frac{R}{\dfrac{1}{sCR + 1}} = -\frac{sCR}{sCR + 1}$$

This normalized sensitivity is the same for any choice of C and R, as long as their product remains equal.

The normalized sensitivity definition Eq. (16.2.1) can be calculated by first finding the derivative and then multiplying by the ratio h/F. In some applications, it is more convenient to apply an *equivalent* definition

$$S_h^F = \frac{d \ln F}{d \ln h} \tag{16.2.2}$$

where ln denotes natural logarithm. We will see later that this definition can save us lots of work in certain situations.

The normalized sensitivity is clearly advantageous but becomes meaningless in two situations: if either F or h become zero. In such cases we apply instead two alternate normalizing definitions. If $h = 0$, then we use

$$\Sigma_1 = \frac{dF}{dh} \frac{1}{F} \tag{16.2.3}$$

simply skipping multiplication by zero. If $F = 0$, we do not divide by F and use instead

$$\Sigma_2 = \frac{dF}{dh} h \tag{16.2.4}$$

If both happen to be zero, then all we can use is the derivative itself.

It is sometimes difficult to visualize that the derivative with respect to an element can be nonzero although the element itself has zero value. To clarify this possibility we take a small example, Fig. 16.2.1, in which the capacitor C_1 actually does not exist and has a value $C_1 = 0$. Using all elements we first derive the voltage transfer function:

$$F = \frac{V_{out}}{E} = \frac{sC_1 + G}{s(C_1 + C_2) + G}$$

FIGURE 16.2.1   Sensitivity to a parasitic $C_1 = 0$.

The derivative with respect to C_1 is

$$\frac{dF}{dC_1} = \frac{s^2 C_2}{[s(C_1 + C_2) + G]^2}$$

If we now set $C_1 = 0$, we get

$$F = \frac{G}{sC_2 + G}$$

(the same as if C_1 did not exist at all), but

$$\frac{dF}{dC_1} = \frac{s^2 C_2}{[sC_2 + G]^2}$$

which is nonzero. In this situation, the use of the fully normalized sensitivity definition Eq. (16.2.1) would clearly be wrong and we must apply Σ_1.

Sensitivity with respect to a zero-valued element is a useful concept because the element does not change the network function but we can get information on how a small, actually existing, parasitic element would influence our results. This feature is available in NETOPT.

16.3. SENSITIVITIES OF TUNED CIRCUITS

A tuned circuit as in Fig. 16.3.1 is the best starting example for the study of sensitivities. Its impedance is

$$Z = \frac{1}{C} \frac{s}{s^2 + s\dfrac{G}{C} + \dfrac{1}{LC}} \tag{16.3.1}$$

FIGURE 16.3.1   Parallel tuned circuit.

In most applications the value of the conductance is chosen such that the tuned circuit has two complex conjugate poles. In such case

$$Z = \frac{1}{C} \frac{s}{s^2 + s\dfrac{\omega_0}{Q} + \omega_0^2} \tag{16.3.2}$$

Comparing the above two expressions we identify

$$\omega_0 = \frac{1}{\sqrt{LC}} \tag{16.3.3}$$

$$Q = \frac{\omega_0 C}{G} = \frac{1}{\omega_0 GL} = \frac{1}{G}\left[\frac{C}{L}\right]^{1/2} \tag{16.3.4}$$

We now have three functions, Z, ω_0 and Q, all of which describe, in some way, the properties of the tuned circuit. We can calculate sensitivities for each of them.

For instance, differentiating Eq. (16.3.1) with respect to G results in

$$\frac{dZ}{dG} = \frac{1}{C^2} \frac{-s^2}{\left(s^2 + s\dfrac{G}{C} + \dfrac{1}{LC}\right)^2}$$

and the normalized sensitivity is

$$S_G^Z - \frac{dZ}{dG}\frac{G}{Z} - \frac{-s\dfrac{G}{C}}{s^2 + s\dfrac{G}{C} + \dfrac{1}{LC}}$$

This sensitivity is a function of frequency and when we substitute $s = j\omega$, we get a complex number which is difficult to interpret. To get a meaningful result we can do two things: Either consider the absolute value of Z and find its sensitivity (we will do this later), or use Q and ω_0 because they are real variables, independent of frequency.

Suppose that we want to derive sensitivities of Q. We can differentiate with respect to, say, L, get dQ/dL and then multiply by L/Q, as required by Eq. (16.2.1). However, since Eq. (16.3.4) does not contain any additions or subtractions, we can take advantage of the normalized sensitivity definition Eq. (16.2.2) and proceed as follows: Take first the logarithm of Eq. (16.3.4)

$$\ln Q = -\ln G + \frac{1}{2}\ln C - \frac{1}{2}\ln L$$

and then its differential

$$d \ln Q = -d \ln G + \frac{1}{2} d \ln C - \frac{1}{2} d \ln L \qquad (16.3.5)$$

Since we want sensitivity with respect to L, we simply divide the equation by $d \ln L$

$$\frac{d \ln Q}{d \ln L} = -\frac{d \ln G}{d \ln L} + \frac{1}{2}\frac{d \ln C}{d \ln L} - \frac{1}{2}\frac{d \ln L}{d \ln L}$$

Comparing the left side of this equation with the definition Eq. (16.2.2), we see that it is in fact the desired S_L^Q. The elements G and C are independent of L, the ratios $d \ln G/d \ln L$ and $d \ln C/d \ln L$ are zero and $d \ln L/d \ln L = 1$. The result is

$$S_L^Q = \frac{d \ln Q}{d \ln L} = -\frac{1}{2}\frac{d \ln L}{d \ln L} = -\frac{1}{2}$$

Dividing Eq. (16.3.5) by $d \ln G$ or $d \ln C$ will result in

$$S_G^Q = -1$$

$$S_C^Q = \frac{1}{2}$$

Calculation of normalized sensitivities of ω_0 proceeds the same way. First we take the logarithm

$$\ln \omega_0 = -\frac{1}{2} \ln L - \frac{1}{2} \ln C$$

apply differentiation by writing d in front of every term and then divide by $d \ln L$ or $d \ln C$. The sensitivities are

$$S_L^{\omega_0} = \frac{d \ln \omega_0}{d \ln L} = -\frac{1}{2}$$

$$S_C^{\omega_0} = \frac{d \ln \omega_0}{d \ln C} = -\frac{1}{2}$$

Finally, since ω_0 does not depend on G, we also have

$$S_G^{\omega_0} = 0$$

The above derivations lead to one important conclusion: In the case of a tuned circuit, all $|S_h^Q|$ and $|S_h^{\omega_0}|$ are less than, or at most equal to, one. As we will discover later, these values are extremely small and no other circuit described by a second-order network function has sensitivities as low as the tuned circuit.

How can we get some practical conclusions from the sensitivity calculation? We will indicate the steps on the result $S_L^Q = -1/2$. If we take a small increment instead of the derivative, then we can write

$$S_L^Q \sim \frac{\Delta Q}{\Delta L} \frac{L}{Q} = -\frac{1}{2}$$

or

$$\frac{\Delta Q}{Q} \sim -0.5 \frac{\Delta L}{L}$$

Suppose that the value of L increases by 1%; then $\Delta L/L = 0.01$ and

$$\frac{\Delta Q}{Q} \approx -0.5 \times 0.01 = -0.005$$

If, in addition, our Q is 200, then

$$\Delta Q \approx -0.005 \times 200 = -1$$

These simple steps indicate that a 1% increase in inductor value results in an approximate 0.5% drop in Q. If the original Q was 200, the change in L will cause a reduction of Q to approximately 199. We can also check how good this approximation is. To get $Q = 200$, let us select $C = 1$, $L = 1$ and $R = 1/G = 200$. If we increase the inductance to the value $L = 1.01$, then $Q = 200/\sqrt{1.01} = 199.007\,438$. The approximation is indeed very good.

16.4. NETWORK FUNCTION SENSITIVITY

All network functions are rational functions but even if the network is not large, it is often difficult to find the function in terms of variable elements. Finding the derivatives with respect to the elements is also lengthy and any help from the theory is welcome.

Our study will start with the transfer function sensitivity of the Sallen–Key network in Fig. 16.4.1. Using the methods of Chapter 12 we set up the system equation

$$\begin{bmatrix} G_1 + G_2 + sC_1 & -G_2 - sC_1A \\ -G_2 & G_2 + sC_2 \end{bmatrix} \begin{bmatrix} V_1 \\ V_2 \end{bmatrix} = \begin{bmatrix} EG_1 \\ 0 \end{bmatrix} \qquad (16.4.1)$$

FIGURE 16.4.1   Sallen–Key second-order network.

and get the transfer function

$$F = \frac{N}{D} = \frac{AG_1G_2}{s^2C_1C_2 + s[C_1G_2(1 - A) + C_2G_1 + C_2G_2] + G_1G_2} \qquad (16.4.2)$$

If we need the normalized sensitivity with respect to, say, the gain of the amplifier, A, then we must first calculate dF/dA and then multiply this derivative by A/F. It is not possible to use the definition Eq. (16.2.2) because the logarithm cannot be applied to additions or subtractions.

We can simplify our task by first deriving a formula for the sensitivity of a rational function

$$F = \frac{N}{D}$$

Since it is only a ratio of two terms, we can use the sensitivity formula based on the natural logarithm

$$\ln F = \ln N - \ln D$$

differentiate

$$d \ln F = d \ln N - d \ln D$$

and divide by $d \ln h$

$$\frac{d \ln F}{d \ln h} = \frac{d \ln N}{d \ln h} - \frac{d \ln D}{d \ln h}$$

This is equivalent to

$$S_h^F = S_h^N - S_h^D \qquad (16.4.3)$$

The formula says that we first obtain separately normalized sensitivities of the numerator and denominator and then merely subtract them.

We will apply the steps for the above transfer function. Its numerator is

$$N = AG_1G_2$$

and its normalized sensitivity with respect to the gain is

$$S_A^N = 1$$

Differentiating the denominator

$$\frac{dD}{dA} = -sC_1G_2$$

and normalizing

$$S_A^D = \frac{A}{D}\frac{dD}{dA} = -\frac{sC_1G_2A}{D}$$

Using Eq. (16.4.3)

$$S_A^F = 1 + \frac{sC_1G_2A}{s^2C_1C_2 + s[C_1G_2(1-A) + C_2G_1 + C_2G_2] + G_1G_2} \tag{16.4.4}$$

The steps were easy but this result is rarely needed because Eq. (16.4.4) is complex for $s = j\omega$ and thus difficult to interpret. What we really need is the sensitivity of the absolute value of the function.

In order to derive the formula for $S_h^{|F|}$ we resort to a trick and start with the equivalence

$$F = |F|e^{j\phi} \tag{16.4.5}$$

Taking the logarithm and differentiating

$$d\ln F = d\ln|F| + jd\phi$$

Dividing by $d\ln h$

$$\frac{d\ln F}{d\ln h} = \frac{d\ln|F|}{d\ln h} + j\frac{d\phi}{d\ln h}$$

which is equivalent to

$$S_h^F = S_h^{|F|} + j\frac{\dfrac{d\phi}{dh}}{h}$$

On the left side is a complex number which we can write as $\text{Re } S_h^F + j \text{ Im } S_h^F$ and compare with the right-hand side. The real parts and the imaginary parts must be equal

$$S_h^{|F|} = \text{Re } S_h^F \qquad (16.4.6)$$

and

$$\text{Im } S_h^F = \frac{d\phi}{dh} h$$

If we wish to obtain the last expression as a normalized sensitivity, we divide both sides by ϕ and get

$$\frac{1}{\phi}\text{Im } S_h^F = \frac{h}{\phi}\frac{d\phi}{dh}$$

which is equivalent to

$$S_h^\phi = \frac{1}{\phi}\text{Im } S_h^F \qquad (16.4.7)$$

We will demonstrate application of Eq. (16.4.6) on the Sallen–Key network, but simplify our task by inserting numerical values: $C_1 = C_2 = 1$, $G_1 = G_2 = 1$ and $A = 2$. Using Eq. (16.4.4)

$$S_A^F = 1 + \frac{2s}{s^2 + s + 1}$$

If we select, for instance, $s = j2$, this reduces to

$$S_A^F = 1 + \frac{j4}{-3 + j2} = \frac{21}{13} - j\frac{12}{13}$$

and, as follows from Eq. (16.4.6), the sensitivity of the absolute value is

$$S_A^{|F|} = \frac{21}{13}$$

Let us summarize the steps of this section. For any frequency $s = j\omega$, the sensitivity of the transfer function, S_h^F, is obtained as an intermediate step using Eq. (16.4.3). The result is a complex number and its real part is the desired $S_h^{|F|}$.

Example 1.

For the network in Fig. 16.4.2, calculate the voltage transfer function $F = V_2/E$ by keeping the transconductance g of the VC as a variable. For the frequency $s = j2$ calculate the absolute value of F and also the sensitivity $S_g^{|F|}$.

Using the nodal formulation explained in Chapter 12 and taking numerical values for all elements except g, we arrive at the matrix equation

$$\begin{bmatrix} s+2 & -1 \\ -1+g & s+1 \end{bmatrix} \begin{bmatrix} V_1 \\ V_2 \end{bmatrix} = \begin{bmatrix} E \\ 0 \end{bmatrix}$$

The determinant is $D = s^2 + 3s + 1 + g$ and the numerator is $N_2 = (1 - g)E$. The network function is

$$F = \frac{1 - g}{s^2 + 3s + 1 + g}$$

The derivative of the numerator is $dN/dg = -1$ and the normalized sensitivity is

$$S_g^N = \frac{dN}{dg}\frac{g}{N} = \frac{-g}{1 - g}$$

The derivative of D is $dD/dg = 1$ and the normalized sensitivity is

$$S_g^D = \frac{g}{s^2 + 3s + 1 + g}$$

Since we now have all the derivatives we can substitute the numerical value for the VC, $g = 2$.

$$S_g^F = S_g^N - S_g^D = 2 - \frac{2}{s^2 + 3s + 3} = \frac{2s^2 + 6s + 4}{s^2 + 3s + 3}$$

FIGURE 16.4.2   Network with a CV.

Let us now select $s = j2$. The absolute value of F will be

$$|F| = \left| \frac{-1}{-1 + j6} \right| = \frac{1}{\sqrt{37}}$$

and the sensitivity will be

$$S_g^F = \frac{-4 + j12}{-1 + j6} = \frac{76}{37} + j\frac{12}{37}$$

According to Eq. (16.4.6) we must take only the real part to get the sensitivity of the absolute value:

$$S_g^{|F|} = \text{Re } S_g^F = \frac{76}{37}$$

Calculation of network function sensitivities by hand is an extremely laborious process because the evaluations must be done for a number of frequencies. Frequency domain sensitivities are built into NETOPT and can be calculated simultaneously for any number of elements. We will show this application in the next example.

Example 2.

Consider the network in Fig. 16.4.3 (redrawn Fig. 12.3.6). Calculate the amplitude response and the sensitivity with respect to the gain of the amplifier.

The amplitude response is given in Fig. 16.4.4a, the sensitivity in Fig. 16.4.4b. From b we see that changes in the gain of the amplifier will have their largest influence in the approximate frequency range of 0.38 to 0.75 rad/s. See Chapter 18 for instructions on how to use the program.

FIGURE 16.4.3   Network with a Twin-T.

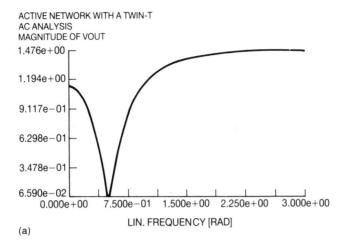

ACTIVE NETWORK WITH A TWIN-T
AC ANALYSIS
MAGNITUDE OF VOUT

(a)

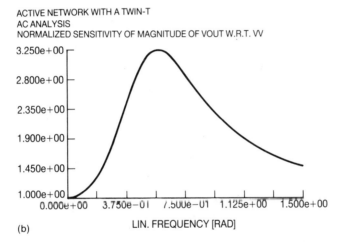

ACTIVE NETWORK WITH A TWIN-T
AC ANALYSIS
NORMALIZED SENSITIVITY OF MAGNITUDE OF VOUT W.R.T. VV

(b)

FIGURE 16.4.4   Responses of the network in Fig. 16.4.3: (a) Amplitude response. (b) Normalized sensitivity $S_A^{|F|}$.

16.5. SENSITIVITIES OF Q AND ω_0

A second-order network function is a rational function whose denominator is a polynomial of second degree and whose numerator has a degree which may be less than or equal to the degree of the denominator.

If both the numerator and denominator are polynomials of second degree and if their roots are complex conjugate, we can distinguish them by using for the denominator the notation Q_p, ω_p (p standing for the pole) and for the numerator Q_z, ω_z (z standing for the zero). We will use the notation Q and ω_0 and derive formulas for their evaluation and also for their sensitivities.

Two complex conjugate poles (or zeros) form a second-order polynomial

$$P(s) = s^2 + s\frac{\omega_0}{Q} + \omega_0^2 \qquad (16.5.1)$$

but this polynomial may still be multiplied by some element values. As an example, consider the Sallen–Key network in Fig. 16.4.1 which we have analyzed and for which we found

$$D = s^2 C_1 C_2 + s[C_1 G_2(1 - A) + C_2 G_1 + C_2 G_2] + G_1 G_2 \qquad (16.5.2)$$

In order to save us lots of unnecessary work, we will derive formulas for Q and ω_0 and for their sensitivities by assuming that the polynomial has the general form

$$P = Us^2 + Vs + W \qquad (16.5.3)$$

which we rewrite as

$$P = U\left[s^2 + \frac{V}{U}s + \frac{W}{U}\right]$$

Now we can compare the expression in the square brackets with Eq. (16.5.1) and get the following identities

$$\frac{V}{U} = \frac{\omega_0}{Q}$$

$$\frac{W}{U} = \omega_0^2$$

or

$$\omega_0 = \left[\frac{W}{U}\right]^{1/2}$$

$$\qquad (16.5.4)$$

$$Q = \frac{\sqrt{UW}}{V}$$

Applying these formulas to the polynomial Eq. (16.5.2) we identify

$$U = C_1 C_2$$
$$V = C_1 G_2(1 - A) + C_2 G_1 + C_2 G_2 \qquad (16.5.5)$$
$$W = G_1 G_2$$

and obtain

$$\omega_0 = \left[\frac{W}{U}\right]^{1/2} = \left[\frac{G_1 G_2}{C_1 C_2}\right]^{1/2}$$

$$Q = \frac{\sqrt{G_1 G_2 C_1 C_2}}{C_1 G_2 (1 - A) + C_2 G_1 + C_2 G_2}$$

(16.5.6)

More important is the fact that we can use Eq. (16.5.4) to derive general rules for sensitivities of these variables. Since Eq. (16.5.4) does not contain any additions or subtractions, we resort to the already known trick of taking the logarithm

$$\ln \omega_0 = \frac{1}{2} \ln W - \frac{1}{2} \ln U$$

differentiating and dividing by $d \ln h$:

$$\frac{d \ln \omega_0}{d \ln h} = \frac{1}{2} \frac{d \ln W}{d \ln h} - \frac{1}{2} \frac{d \ln U}{d \ln h}$$

This is equivalent to

$$S_h^{\omega_0} = \frac{1}{2} S_h^W - \frac{1}{2} S_h^U$$

(16.5.7)

Doing the same for Q:

$$\ln Q = \frac{1}{2} \ln U + \frac{1}{2} \ln W - \ln V$$

and

$$S_h^Q = \frac{1}{2} S_h^U + \frac{1}{2} S_h^W - S_h^V$$

(16.5.8)

If we want in our example the sensitivity with respect to the gain of the amplifier, we calculate first the auxiliary sensitivities

$$S_A^U = 0$$

$$S_A^V = -\frac{C_1 G_2 A}{V}$$

$$S_A^W = 0$$

and then use Eq. (16.5.7) and Eq. (16.5.8) to get

$$S_A^{\omega o} = 0$$

$$S_A^Q = \frac{C_1 G_2 A}{V} \tag{16.5.9}$$

At this point we could insert for V and leave the result. However, it is very instructive to take one more theoretical step using Eq. (16.5.4), writing $1/V = Q/\sqrt{UW}$ and inserting into S_A^Q. We get

$$S_A^Q = AQ \left[\frac{C_1 G_2}{C_2 G_1} \right]^{1/2} \tag{16.5.10}$$

This is a surprising result because the sensitivity is proportional to the product AQ! Without sensitivities we might have expected that the active network we just analyzed could serve as a tuned circuit. After all, like the tuned circuit, it is described by a second-order network function. We derived earlier that all sensitivities of the tuned circuit, S_h^Q and $S_h^{\omega o}$, are of absolute value less than or equal to one. Our active network has a small $S_A^{\omega o}$, but the sensitivity of Q is proportional to AQ. A small change in the gain of the amplifier will cause considerable change of Q and the situation will be aggravated if we need high Q.

In order to fully understand properties of the Sallen–Key (or any other) active second-order network, we must derive sensitivities with respect to all elements. We do so by first deriving the sensitivities S_h^U, S_h^V and S_h^W; they are summarized in Table 16.5.1. Next we apply Eq. (16.5.7) and Eq. (16.5.8) and obtain the sensitivities of Q and ω_0. These are summarized in Table 16.5.2. We see that *all* Q sensitivities are proportional to Q, a fairly disappointing result. Without going into any more details, we should perhaps say that all second-order active networks tend to have this disappointing property.

Sensitivity of Q and ω_0 is built into NETOPT and is available for networks of any order if the SET SENS command is in effect. See Chapter 18 Section 8 for details.

Example 1.

In the network Fig. 16.5.1 keep the two amplifiers as variables. Obtain the transfer function and select gains such that $Q = 5$. Calculate the Q sensitivities for both amplifier gains.

Using element values the system matrix becomes

$$\begin{bmatrix} s+1 & -sA_2 \\ -A_1 & s+1 \end{bmatrix} \begin{bmatrix} V_1 \\ V_2 \end{bmatrix} = \begin{bmatrix} E \\ sE \end{bmatrix}$$

TABLE 16.5.1   Auxiliary Sensitivities of the Sallen–Key Network

$S^U_{C_1} = 1$	$S^V_{C_1} = \dfrac{C_1 G_2(1-A)}{V}$	$S^W_{C_1} = 0$
$S^U_{C_2} = 1$	$S^V_{C_2} = \dfrac{C_2(G_1 + G_2)}{V}$	$S^W_{C_2} = 0$
$S^U_{G_1} = 0$	$S^V_{G_1} = \dfrac{G_1 C_2}{V}$	$S^W_{G_1} = 1$
$S^U_{G_2} = 0$	$S^V_{G_2} = \dfrac{G_2(C_2 + C_1 - C_1 A)}{V}$	$S^W_{G_2} = 1$
$S^U_A = 0$	$S^V_A = -\dfrac{C_1 G_2 A}{V}$	$S^W_A = 0$

TABLE 16.5.2   Sensitivities of the Sallen–Key Network

$S^{\omega_0}_{C_1} = -\dfrac{1}{2}$	$S^Q_{C_1} = \dfrac{1}{2} - C_1 G_2(1-A)\dfrac{Q}{\sqrt{C_1 C_2 G_1 G_2}}$
$S^{\omega_0}_{C_2} = -\dfrac{1}{2}$	$S^Q_{C_2} = \dfrac{1}{2} - C_2(G_1 + G_2)\dfrac{Q}{\sqrt{C_1 C_2 G_1 G_2}}$
$S^{\omega_0}_{G_1} = +\dfrac{1}{2}$	$S^Q_{G_1} = \dfrac{1}{2} - G_1 C_2 \dfrac{Q}{\sqrt{C_1 C_2 G_1 G_2}}$
$S^{\omega_0}_{G_2} = +\dfrac{1}{2}$	$S^Q_{G_2} = \dfrac{1}{2} - G_2(C_2 + C_1 - C_1 A)\dfrac{Q}{\sqrt{C_1 C_2 G_1 G_2}}$
$S^{\omega_0}_A = 0$	$S^Q_A = C_1 G_2 A \dfrac{Q}{\sqrt{C_1 C_2 G_1 G_2}}$

The determinant is $D = s^2 + s(2 - A_1 A_2) + 1$, $N_2 = A_2(s^2 + s + A_1)$ and the network function is

$$F = \frac{A_2(s^2 + s + A_1)}{s^2 + s(2 - A_1 A_2) + 1}$$

Comparing with Eq. (16.5.3) we identify $U = 1$, $V = 2 - A_1 A_2$ and $W = 1$. Inserting into Eq. (16.5.4) we have $\omega_0 = 1$ and $Q = 1/(2 - A_1 A_2)$. Using the steps explained above, we

FIGURE 16.5.1   Network with two amplifiers.

derive that

$$S_{A_1}^Q = S_{A_2}^Q = \frac{A_1 A_2}{2 - A_1 A_2}$$

In order to get $Q = 5$ the product $A_1 A_2$ must be equal to 9/5 and the Q sensitivity with respect to each amplifier will be 9.

Example 2.

In the network Fig. 16.5.2 keep the two capacitors as variables and find the transfer function. Calculate the Q and ω_0 sensitivities for changes in these two capacitors.

The matrix equation is

$$\begin{bmatrix} -sC_1 & 0 & -10/6 \\ -3 & sC_2 + 4 & -1 \\ 0 & 2 & -1 \end{bmatrix} \begin{bmatrix} V_1 \\ V_2 \\ V_3 \end{bmatrix} = \begin{bmatrix} 3E \\ 0 \\ 0 \end{bmatrix}$$

FIGURE 16.5.2   Network with two OPAMPs.

The determinant is $D = s^2 C_1 C_2 + 2s C_1 + 10$ and the transfer function is

$$F = \frac{-18}{s^2 C_1 C_2 + 2s C_1 + 10}$$

Following Eq. (16.5.3) we identify $U = C_1 C_2$, $V = 2C_1$ and $W = 1$. The sensitivities are $S_{C_1}^U = S_{C_2}^U = 1$, $S_{C_1}^V = 1$, $S_{C_2}^V = 0$ and W is a constant and thus does not contribute to sensitivities. Inserting into Eq. (16.5.7) and Eq. (16.5.8) we get $S_{C_1}^{\omega_0} = S_{C_2}^{\omega_0} = 1/2$ and $S_{C_1}^Q = -1/2$, $S_{C_2}^Q = 1/2$.

16.6. SENSITIVITY OF POLES AND ZEROS

The subject of this section is the sensitivity of a pole of a second-order network to any element. The derivations are also valid for the numerator of a network function with two complex conjugate zeros. For demonstration we will use the Sallen–Key network but simplify the steps by keeping only one variable, A, and by setting $C_1 = C_2 = 1$ and $G_1 = G_2 = 1$. After the necessary differentiations we will set the gain $A = 2$.

The denominator of the transfer function was derived in Eq. (16.4.2) and for the above values

$$D = s^2 + s(3 - A) + 1 \tag{16.6.1}$$

It has two complex conjugate roots. We will consider only the one in the upper half plane because results for the other one are complex conjugate. The root, denoted by r, is obtained after several steps in the form

$$r = \frac{1}{2}[A - 3 + j\sqrt{6A - A^2 - 5}] \tag{16.6.2}$$

Differentiation with respect to A results in

$$\frac{dr}{dA} = \frac{1}{2}\left[1 + j\frac{3 - A}{\sqrt{6A - A^2 - 5}}\right]$$

If we now substitute $A = 2$, we obtain

$$\frac{dr}{dA} = \frac{1}{2}\left[1 + \frac{j}{\sqrt{3}}\right]$$

Differentiation of expressions containing square roots is not all that difficult, but it is lengthy and it is easy to make a mistake. We would appreciate a method which could get us the result with less effort. This is possible if we proceed as follows. The polynomial, P, is a function of the element, h, and its root is also a function of h. The condition for finding the root can be written as

$$P[h, s(h)] = 0 \qquad\qquad (16.6.3)$$

Since we have here two variables, we take partial derivatives

$$\frac{\partial P(h, s)}{\partial h} + \frac{\partial P(h, s)}{\partial s} \frac{\partial s}{\partial h}\bigg|_{s=r} = 0$$

to get

$$\frac{\partial r}{\partial h} = \frac{\partial s}{\partial h}\bigg|_{s=r} = -\frac{\dfrac{\partial P(h, s)}{\partial h}}{\dfrac{\partial P(h, s)}{\partial s}}\bigg|_{s=r} \qquad\qquad (16.6.4)$$

Finding the sensitivity by means of Eq. (16.6.4) is easy. We first calculate the root by substituting *all* values. This leads to the polynomial

$$P = s^2 + s + 1 \qquad\qquad (16.6.5)$$

whose upper half plane root is

$$r = \frac{1}{2}[-1 + j\sqrt{3}]$$

Next we differentiate Eq. (16.6.1) with respect to A

$$\frac{\partial P(h, s)}{\partial A} = -s$$

and substitute the root for s

$$\frac{\partial P(h, s)}{\partial A} = \frac{1}{2}[1 - j\sqrt{3}]$$

Finally, we differentiate Eq. (16.6.5) with respect to s

$$\frac{\partial P(h, s)}{\partial s} = 2s + 1$$

and substitute the root for s again

$$\frac{\partial P(h, s)}{\partial s} = j\sqrt{3}$$

All entries needed for Eq. (16.6.4) are now available and

$$\frac{dr}{dA} = \frac{1}{2}\left(1 + \frac{j}{\sqrt{3}}\right)$$

We can, of course, continue to get the normalized sensitivity as well. In this case it is

$$S_A^r = \frac{dr}{dA}\frac{A}{r} = -\frac{j}{2}\left[\sqrt{3} + \frac{1}{\sqrt{3}}\right]$$

The above steps are a mixture of polynomial differentiations and substitutions of numerical values but there is no differentiation of terms with square roots. If all element values are available numerically we can use Eq. (16.6.4) for higher-order polynomials as well but for this we invariably need a computer and a polynomial root-finding routine.

NETOPT gives this sensitivity for arbitrary networks if the SET SENS command is in effect; see Chapter 18 for details.

Example 1.

Find the pole sensitivity with respect to A_1 for the network in Fig. 16.5.1. Use the transfer function derived there and the values $A_1 = 1$, $A_2 = 1.5$.

The denominator of the transfer function is

$$D = s^2 + s(2 - A_1 A_2) + 1$$

We need the derivatives

$$\frac{dD}{dA_1} = -sA_2$$

$$\frac{dD}{ds} = 2s + 2 - A_1 A_2$$

Inserting the gains, the denominator simplifies to $D = s^2 + 0.5s + 1$ and the root (in the upper half plane) is

$$s = 0.25(-1 + j\sqrt{15})$$

To evaluate Eq. (16.6.4) substitute the pole coordinate for s in the above two derivatives

and obtain

$$\frac{d(pole)}{dA_1} = 1.5 \frac{s}{2s + 0.5} = 0.75\left(1 + j\frac{1}{\sqrt{15}}\right)$$

16.7. SENSITIVITIES OF OPERATIONAL AMPLIFIERS

Ideal operational amplifiers have gains which are theoretically infinitely large. Dealing with infinity is unpleasant in theoretical work and computers cannot handle infinity at all. For these reasons we redefined the operational amplifier as a device with an inverted gain

$$B = -\frac{1}{A} \tag{16.7.1}$$

We also derived, in Chapter 12, Eq. (12.1.4), that properties of nonideal operational amplifiers are expressed by

$$V_+ - V_- + BV_o = 0 \tag{16.7.2}$$

If the amplifier is ideal, then $B = 0$ and we have an element with zero value. If we wish to find sensitivity with respect to this ideal element, we must use Eq. (16.7.2) in all our derivations, obtain the network function in terms of B, differentiate with respect to B and only afterwards insert $B = 0$.

We will demonstrate this case on the simple integrator shown in Fig. 16.7.1. Using the methods of Chapter 12 we write

$$V_1(G + sC) - sCV_{out} = EG$$
$$-V_1 + BV_{out} = 0$$

The second equation is Eq. (16.7.2), where we identified $V_1 = 0$, $V_- = V_1$ and

FIGURE 16.7.1   Network with nonideal OPAMP.

$V_o = V_{out}$. The solution is

$$F = \frac{V_{out}}{E} = \frac{G}{-sC + B(G + sC)}$$

Differentiating with respect to B

$$\frac{dF}{dB} = \frac{-G(sC + G)}{[-sC + B(sC + G)]^2}$$

At this point we can insert $B = 0$ and obtain

$$F = -\frac{G}{sC}$$

$$\frac{dF}{dB} = -\frac{G(sC + G)}{s^2 C^2}$$

Since the element has zero value, we cannot use the usual normalized sensitivity S and must apply

$$\Sigma_1 = \frac{dF}{dB}\frac{1}{F} = \frac{sC + G}{sC}$$

Hand calculation of sensitivities to changes of operational amplifiers is always time consuming because we must keep the inverted gain B in the equations and cannot assume that the input terminals are at the same potential. This increases the size of the system matrix by one row and one column per OPAMP.

Two kinds of sensitivities to OPAMP changes are built into NETOPT. One is the ideal case with $B = 0$. The other is a nonideal OPAMP whose inverted gain is defined as

$$B = -\frac{s}{GB_f}$$

GB_f being the gain–bandwidth product in Hz. See Chapter 18 for instructions.

PROBLEMS CHAPTER 16

Sensitivity analysis is fairly difficult even in small networks, but is crucial for design and for checking. We give a few problems to indicate how to proceed in sensitivity analysis; every earlier problem in which some (or all) elements were kept as variables can be used for sensitivity studies.

Sensitivity problems, even for large and higher-order networks, can be solved by NETOPT. There are no *theoretical* limitations. The only restriction is the size of the matrix of the particular NETOPT version. Do remember that if you wish to calculate sensitivity with respect to ideal operational amplifiers then you *must* select the 1-graph MNF version of the formulation method. If you ask for the amplitude response in dB, then corresponding sensitivities are semilogarithmic. Other sensitivities are normalized. Do not forget to issue the command SET SENS and follow the program instructions.

P.16.1 Find the admittance of the network in terms of variable elements. Obtain expressions for Q, ω_0, S_h^Q and $S_h^{\omega_0}$, where h stands for all elements of the network. Select unit values for all elements and get numerical values for the above expressions.

P.16.2 Find the transfer function, F, in terms of variables G and g. Calculate S_h^Q and $S_h^{\omega_0}$ where h is either g or G. Use $G = 1\,\mathrm{S}$ and such g as required to get $Q = 4$. Insert numerical values into the sensitivity expressions.

P.16.3 Find the transfer function. Select $G = 1\,\mathrm{S}$ and select g such that $Q = 1$. What is the numerical value for ω_0? Obtain S_h^Q and $S_h^{\omega_0}$ where h is either G or g and insert numerical values.

P.16.4 For the problem P.16.3 find S_g^F where F is the transfer function. Select $\omega = \omega_0$ and obtain $S_g^{|F|}$ at this frequency.

P.16.5 Using NETOPT, plot the amplitude response of P.16.3 (magnitude). Also plot the sensitivity of the amplitude response to changes of g. Use frequency range 0 to 10 rad/s. Use the NETOPT option SHOW CRIT to find at which frequency is the largest sensitivity.

P.16.6 Find the transfer function, F, in terms of variables G_1, G_2 and A. Use $G_1 = G_2$ and design for $Q = 4$ and $\omega_0 = 3$. Find the sensitivity S_A^F. Obtain $S_A^{|F|}$ at $s = j2$.

P.16.7 Derive the transfer function in terms of all G and A. Design to realize the denominator polynomial $s^2 + 2s + 2$. Find the sensitivity of S_A^F and $S_A^{|F|}$ at $s = j3$. Also derive S_A^Q. Find the sensitivity of the pole in the upper half plane to the changes of the gain A.

P.16.8 Let the operational amplifier be ideal. Derive the transfer function, F, in terms of all elements. Set afterwards $G_2 = G_3$; you should get a first-order all-pass function. Find $S_{C_1}^{|F|}, S_{G_1}^{|F|}, S_{G_2}^{|F|}$ and $S_{G_3}^{|F|}$ for $s = j3$. Find the zero and sensitivity with respect to the elements.

P.16.10 The operational amplifier is ideal. Find the transfer function in terms of variable G_1 and G_2. Calculate $S_{G_1}^{|F|}$ and $S_{G_2}^{|F|}$. Afterwards substitute $G_2 = 4G_1$ $s = j3$ to obtain $S_{G_1}^{|F|}$ and $S_{G_2}^{|F|}$.

P.16.9 Find the transfer function by keeping A as a variable. Find the sensitivity $S_A^{|F|}$ at $s = j3$. Insert $A = 2$ and obtain numerical values.

CHAPTER 17

Modified Nodal Formulation

17.0. INTRODUCTION

Until now we solved all our problems by writing either the nodal or the mesh equations. If the network had independent current sources and voltage controlled current sources, we used nodal formulation. If it had voltage sources and current-controlled voltage sources, we used mesh equations without modifications. In all other cases we had to resort to preprocessing which sometimes necessitated redrawing of the figures. We also discovered that if the network had capacitors as well as inductors, we were unable to avoid fractions of the type $1/s$, no matter which formulation we used.

The various transformations and redrawings may be acceptable if we are dealing with small problems and hand calculations, but such steps are always sources of possible errors. For larger networks, which we cannot solve by hand anyway, we need a method which avoids all transformations and can be automated. Several possibilities exist and the most practical one is the modified nodal formulation. We pay a modest price for the convenience: the matrices are larger than we could obtain with the various transformations, but are not excessively large. Solving larger systems means more work, but if we leave it to a computer, it need not worry us too much. Here we explain in detail modified nodal formulation. It is the method used in most modern programs, like SPICE; it is also the method used in NETOPT. This chapter may be considered as optional.

17.1. PASSIVE ELEMENTS

Modified nodal formulation starts as nodal formulation by first creating a $n \times n$ matrix where n is the number of ungrounded nodes. This portion handles the current sources, J, voltage-controlled current sources, VC, conductances, G and admittances of capacitances, sC, exactly the same way as we have learned in nodal formulation. We will repeat in this section how to handle the conductances and capacitances and will also introduce the concept of the *stamp*.

If a conductance, G, is connected between nodes i and j, then the value G is added to the matrix in the position (i, i) and (j, j), and is subtracted from the entries in the positions (i, j) and (j, i). This can be symbolically described by Fig. 17.1.1, the *stamp* of the conductance.

If the capacitor is connected between nodes i and j, then the product sC is added to the entries in the positions (i, i) and (j, j) and is subtracted from the entries in the positions (i, j) and (j, i). If the capacitor has an initial voltage, E_0, it can be taken into account by an impulsive current source in parallel with the capacitor; this was explained in the text related to Fig. 10.1.10. For the initial voltage, positive at node i and negative at node j, we enter $-CE_0$ in the ith row of the right-hand-side vector and $+CE_0$ in its jth row. This can be expressed schematically by the stamp of the capacitor as shown in Fig. 17.1.2. The signs of the entries of the right-hand-side vector will be interchanged if E_0 is negative at node i and positive at node j.

The first difference is encountered when we use inductors. We would like to avoid terms of the form $1/sL$ and this can be achieved by considering inductors in their impedance form. To clarify the concept we start with the example in Fig. 17.1.3 where the current through the inductor is indicated as leaving node 1 and entering node 2. Suppose that we remove the inductor and replace it by its current, I_L. This current must be added to the KCL equation of the first node and subtracted from the KCL equation of the second node:

$$V_1 sC_1 + I_L = J$$
$$V_2(sC_2 + G) - I_L = 0$$

(17.1.1)

Fig. 17.1.1

Fig. 17.1.2

FIGURE 17.1.1   Conductance and its stamp.

FIGURE 17.1.2   Capacitor and its stamp.

FIGURE 17.1.3   Network with one inductor.

In addition, we know that

$$sLI_L = V_1 - V_2$$

and since the voltages as well as the current are unknown, we transfer them to the left

$$V_1 - V_2 - sLI_L = 0 \tag{17.1.2}$$

Additional equations of this type are given the name *constitutive* equations. We will use this name throughout the text. Collecting Eq. (17.1.1) and Eq. (17.1.2) into one matrix equation

$$\begin{bmatrix} sC_1 & 0 & 1 \\ 0 & sC_2 + G & -1 \\ 1 & -1 & -sL \end{bmatrix} \begin{bmatrix} V_1 \\ V_2 \\ I_L \end{bmatrix} = \begin{bmatrix} J \\ 0 \\ 0 \end{bmatrix} \tag{17.1.3}$$

Inductors are elements which can have initial currents. As we explained in detail in Chapter 10 and as shown in Fig. 10.2.4, this initial current can be taken into account by a Dirac-impulse voltage source in series with the inductor.

Before deriving the stamp of an inductor, we take another example, Fig. 17.1.4a. The initial voltage across the capacitor is represented in Fig. 17.1.4b by the current source in parallel with C. The initial current through the inductor is taken into account by the equivalent impulsive voltage source in series with L. The set of nodal equations is written as before, now taking into account the current source due to the initial voltage across the capacitor:

$$(G_1 + sC)V_1 - sCV_2 + I_L = J - CE_0$$
$$-sCV_1 + (sC + G_2)V_2 - I_L = CE_0$$

The constitutive equation for the inductor with an initial current flowing through it is

$$V_1 - V_2 - sLI_L = -LJ_0 \tag{17.1.4}$$

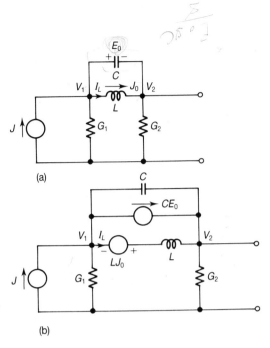

(a)

(b)

Fig. 17.1.4

Fig. 17.1.5

FIGURE 17.1.4   Network with initial conditions: (a) Original. (b) With equivalent impulse sources.

FIGURE 17.1.5   Inductor and its stamp.

and the matrix equation is

$$\begin{bmatrix} G_1 + sC & -sC & 1 \\ -sC & sC + G_2 & -1 \\ 1 & -1 & -sL \end{bmatrix} \begin{bmatrix} V_1 \\ V_2 \\ I_L \end{bmatrix} = \begin{bmatrix} J - CE_0 \\ CE_0 \\ -LJ_0 \end{bmatrix}$$

We can now draw conclusions from the results of the two examples. Every inductor *increases* the size of the system matrix by one row and one column. If the inductor is connected between nodes i and j, as shown in Fig. 17.1.5, a new column for the current I_L is added to the matrix with $+1$ in the ith row and -1 in the jth row. The constitutive equation itself is added as a new row. If the inductor has an initial current, J_0, flowing from node i to node j, $-LJ_0$ is entered in the right-hand-side column in the same row as the constitutive equation. The stamp is in Fig. 17.1.5. Here the empty matrix portion, denoted by $m \times m$, is the *previous matrix* in which the inductor makes no modifications. The additional row and column are marked symbolically by $m + 1$.

The next element we consider is a short circuit. It must be taken into account by marking an *additional* node in the network. If the short circuit is between nodes

$$
\begin{array}{c}
 I \\
 i j m+1 \\
\begin{array}{c} i \\ j \\ m+1 \end{array}
\left[
\begin{array}{c|c}
m \times m & \begin{array}{c} 1 \\ -1 \end{array} \\
\hline
1 \quad -1 &
\end{array}
\right]
\end{array}
$$

FIGURE 17.1.6   Short circuit and its stamp.

i and j and we assume, as always, that the current flows from the first node, i, to the second node, j, then I is added in row i and subtracted from row j. In addition, we know that $V_i = V_j$. Since none of these voltages is known, we write this constitutive equation as

$$
V_i - V_j = 0 \tag{17.1.5}
$$

and add it to the previous set of equations. This leads to the stamp in Fig. 17.1.6.

An example will show how we set up the modified nodal equations if we have more than one element which increases the size of the matrix. The network in Fig. 17.1.7 has three ungrounded nodes; its nodal portion will have the size 3×3. We write the nodal equations by removing the inductors and replacing them with their currents. This leads to

$$
(G_1 + sC_1)V_1 + I_{L1} = J
$$

$$
-I_{L1} + sC_2V_2 + I_{L2} = 0
$$

$$
-I_{L2} + (G_2 + sC_3)V_3 = 0
$$

Next we add the two constitutive equations

$$
V_1 - V_2 - sL_1I_{L1} = 0
$$

$$
V_2 - V_3 - sL_2I_{L2} = 0
$$

FIGURE 17.1.7   Network with two inductors.

to complete the set. In matrix form

$$\begin{bmatrix} G_1 + sC_1 & 0 & 0 & 1 & 0 \\ 0 & sC_2 & 0 & -1 & 1 \\ 0 & 0 & sC_3 + G_2 & 0 & -1 \\ 1 & -1 & 0 & -sL_1 & 0 \\ 0 & 1 & -1 & 0 & -sL_2 \end{bmatrix} \begin{bmatrix} V_1 \\ V_2 \\ V_3 \\ I_{L1} \\ I_{L2} \end{bmatrix} = \begin{bmatrix} J \\ 0 \\ 0 \\ 0 \\ 0 \end{bmatrix}$$

In the above example we first wrote the KLC equations one by one and then collected them as a matrix. This is not necessary and we can write the matrix equation directly by using the stamps. Disregard first the presence of the inductors and write the nodal 3×3 matrix. Next use the stamp for the first inductor and add one row and one column. For second inductor use the same stamp and add *again* a new row and column to the previous matrix. Now we see why we have denoted the new row and column by $m + 1$: it simply indicates an additional row and column to whatever matrix we had before.

The last element which *should not* normally be used is a linear resistor, taken in impedance form. If it is connected between nodes i and j, with the current assumed to flow from i to j, then the constitutive equation is

$$V_i - V_j - RI = 0 \tag{17.1.6}$$

In addition, current I must be added in row i and subtracted in row j. The stamp is similar to the stamp of the inductor and is in Fig. 17.1.8. It can be seen that a resistor, taken as a resistor, adds unnecessarily one row and one column to the system matrix. This is the reason why we *always* use conductances in linear networks. Even if the user gives the resistor as a resistor, we convert the element internally into G and return to the given element only in the result. This rule is implemented in NETOPT.

It is advisable to compare the stamps for the inductor, resistor and short circuit; they are quite similar and this similarity should be remembered.

FIGURE 17.1.8   Resistor and its stamp (not normally used).

FIGURE 17.2.1   Network with two independent sources.

17.2. INDEPENDENT SOURCES

The independent current source has not presented any problems in nodal formulation and we use it exactly as before. The voltage source was causing problems in nodal formulation and we always transformed it into a current source. In modified nodal formulation we take its current into account and do not perform any network transformations.

To introduce the concept we start with a simple network having one voltage source and one current source, see Fig. 17.2.1. Assign voltages to all three nodes. At the beginning disregard the presence of the voltage source but take into account its current I_E (in the direction from $+$ to $-$). The three nodal equations are

$$I_E + sC_1V_1 - sC_1V_2 = 0$$
$$-sC_1V_1 + (sC_1 + G_1 + G_2)V_2 - G_2V_3 = 0$$
$$-G_2V_2 + (G_2 + sC_2)V_3 = J$$

At this point we have three equations and four únknowns: V_1, V_2, V_3 and I_E. However, we have not yet used the fact that the voltage V_1 is known and equal to E,

$$V_1 = E$$

Now we have the necessary number of equations; in the matrix we first write the nodal portion and account for the voltage source in the fourth row and column:

$$\begin{bmatrix} sC_1 & -sC_1 & 0 & 1 \\ -sC_1 & sC_1 + G_1 + G_2 & -G_2 & 0 \\ 0 & -G_2 & G_2 + sC_2 & 0 \\ 1 & 0 & 0 & 0 \end{bmatrix} \begin{bmatrix} V_1 \\ V_2 \\ V_3 \\ I_E \end{bmatrix} = \begin{bmatrix} 0 \\ 0 \\ J \\ E \end{bmatrix}$$

The steps do not change if the voltage source is floating (not connected to ground), as in Fig. 17.2.2. First we disregard the voltage source, keep its current

Fig. 17.2.2

Fig. 17.2.3

FIGURE 17.2.2   Network with a floating voltage source.

FIGURE 17.2.3   Independent current source and its stamp.

and write the nodal equations:

$$(G_1 + G_2)V_1 - G_2 V_2 + 0V_3 + I_E = 0$$

$$-G_2 V_1 + (G_2 + G_3 + G_4)V_2 - G_4 V_3 = 0$$

$$0V_1 - G_4 V_2 + (G_4 + G_5)V_3 - I_E = 0$$

In addition, we know the constitutive equation

$$V_1 - V_3 = E$$

which is the last equation needed. In matrix form

$$\begin{bmatrix} G_1 + G_2 & -G_2 & 0 & 1 \\ -G_2 & G_2 + G_3 + G_4 & -G_4 & 0 \\ 0 & -G_4 & G_4 + G_5 & -1 \\ 1 & 0 & -1 & 0 \end{bmatrix} \begin{bmatrix} V_1 \\ V_2 \\ V_3 \\ I_E \end{bmatrix} = \begin{bmatrix} 0 \\ 0 \\ 0 \\ E \end{bmatrix}$$

These equations give us enough information to write the stamps for both independent sources. Let the current source be as in Fig. 17.2.3, connected to nodes i and j with current flowing away from node i. The stamp is in the same figure and follows the rules established for nodal formulation.

If the voltage source is connected as in Fig. 17.2.4, with the + sign at node i and the − sign at j, then the current I is added in row i and subtracted in row j. The constitutive equation is

$$V_i - V_j = E \qquad (17.2.1)$$

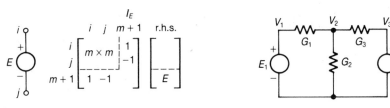

Fig. 17.2.4   Fig. 17.2.5

FIGURE 17.2.4   Independent voltage source and its stamp.

FIGURE 17.2.5   Network with two voltage sources.

and the stamp is as shown; it requires addition of one row and one column to the previously defined matrix.

If the network has several independent voltage sources, then we must add one row and one column for each. For instance, for the network in Fig. 17.2.5 we get the matrix equation

$$
\begin{bmatrix}
G_1 & -G_1 & 0 & 1 & 0 \\
-G_1 & (G_1 + G_2 + G_3) & -G_3 & 0 & 0 \\
0 & -G_3 & G_3 & 0 & 1 \\
1 & 0 & 0 & 0 & 0 \\
0 & 0 & 1 & 0 & 0
\end{bmatrix}
\begin{bmatrix}
V_1 \\
V_2 \\
V_3 \\
I_{E1} \\
I_{E2}
\end{bmatrix}
=
\begin{bmatrix}
0 \\
0 \\
0 \\
E_1 \\
E_2
\end{bmatrix}
$$

17.3. CONTROLLED SOURCES

The four dependent sources in Chapter 4 were already presented in a suitable form. There is no change in the use of the voltage-controlled current source, VC. If the controlling voltage is taken between nodes i (non-inverting) and node j (inverting) and the output is connected to nodes k and l with the current flowing from node k to l, then the currents

$$I_k = gV_i - gV_j$$
$$I_l = -gV_i + gV_j$$
$$(17.3.1)$$

are added to the kth and lth nodal equation, respectively. This means that the stamp of the VC is shown in Fig. 17.3.1. As examples use Fig. 4.1.2 through Fig. 4.1.6 in Chapter 4 without any modification.

The voltage-controlled voltage source, VV is shown in Fig. 17.3.2. The output voltage is

$$V_k - V_l = \mu(V_i - V_j)$$

FIGURE 17.3.1 VC and its stamp.

but since none of these voltages is known, we transfer all terms to the left and get the constitutive equation

$$-\mu V_i + \mu V_j + V_k - V_l = 0 \tag{17.3.2}$$

A current also flows through the voltage source, in the positive direction at node k and in the negative direction at node l. The stamp is in the same figure. As another example consider the network in Fig. 17.3.3. Similarly as in the case of the independent source, we first "remove" the voltage source but consider its current, I_{VV}. The nodal equations are

$$(G_1 + G_2)V_1 - G_2V_2 + I_{VV} = J$$

$$-G_2V_1 + (G_2 + G_3 + G_4)V_2 - G_4V_3 = 0$$

$$-G_4V_2 + (G_4 + G_5)V_3 - I_{VV} = 0$$

We also use the constitutive equation, Eq. (17.3.2), in which we identify $V_i = V_2$, $V_j = V_3$, $V_k = V_1$ and $V_l = V_3$:

$$\mu V_2 + \mu V_3 + V_1 - V_3 = 0$$

Fig. 17.3.2

Fig. 17.3.3

FIGURE 17.3.2 VV and its stamp.

FIGURE 17.3.3   Network with a floating VV.

This results in

$$
\begin{bmatrix}
(G_1 + G_2) & -G_2 & 0 & 1 \\
-G_2 & (G_2 + G_3 + G_4) & -G_4 & 0 \\
0 & -G_4 & (G_4 + G_5) & -1 \\
1 & -\mu & \mu - 1 & 0
\end{bmatrix}
\begin{bmatrix}
V_1 \\
V_2 \\
V_3 \\
I_{VV}
\end{bmatrix}
=
\begin{bmatrix}
J \\
0 \\
0 \\
0
\end{bmatrix}
$$

The next controlled source is the current-controlled current source, CC, Fig. 17.3.4. The controlling port is a short circuit and we must increase the number of nodes by one and actually mark such a node in the circuit. The constitutive equation which we add is the same one as for a short circuit,

$$V_i - V_j = 0 \tag{17.3.3}$$

We must also add the current I to the ith nodal equation and subtract it from the jth equation. In addition, the output of the CC adds current αI at node k and subtracts the same current at node l. These explanations should now be sufficient to derive the stamp; it is shown in Fig. 17.3.4. For better understanding we will explain application on the network in Fig. 17.3.5 for which we write the nodal

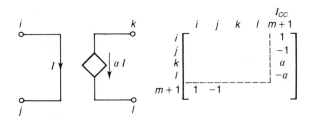

FIGURE 17.3.4 CC and its stamp.

FIGURE 17.3.5   Network with a floating CC.

equations and attach to them Eq. (17.3.3):

$$G_1V_1 + I + \alpha I = J$$
$$-I + V_2G_2 - V_3G_2 = 0$$
$$-V_2G_2 + (G_2 + G_3 + G_4)V_3 - G_4V_4 = 0$$
$$-G_4V_3 + (G_4 + G_5)V_4 - \alpha I = 0$$
$$V_1 - V_2 = 0$$

In matrix form

$$\begin{bmatrix} G_1 & 0 & 0 & 0 & (1 + \alpha) \\ 0 & G_2 & -G_2 & 0 & -1 \\ 0 & -G_2 & (G_2 + G_3 + G_4) & -G_4 & 0 \\ 0 & 0 & -G_4 & (G_4 + G_5) & -\alpha \\ 1 & -1 & 0 & 0 & 0 \end{bmatrix} \begin{bmatrix} V_1 \\ V_2 \\ V_3 \\ V_4 \\ I \end{bmatrix} = \begin{bmatrix} J \\ 0 \\ 0 \\ 0 \\ 0 \end{bmatrix}$$

The last element of this section is a current-controlled voltage source, CV, shown in Fig. 17.3.6. The short circuit at its input port requires one additional row and column. The voltage source at its output port also requires one more row and column. Thus the set of equations must be increased by two and we must add two variables and two constitutive equations. The constitutive equation of the short circuit is

$$V_i \quad V_j - 0 \tag{17.3.4}$$

For the voltage source we have $V_k - V_l = rI_1$. Since neither the voltages nor the current are known, we transfer all terms to one side and the constitutive equation becomes

$$V_k - V_l - rI_1 = 0 \tag{17.3.5}$$

FIGURE 17.3.6   *CV* and its stamp.

FIGURE 17.3.7   Network with a floating CV.

The current I_1 modifies the ith and jth nodal equations, the current of the voltage source, I_2, modifies also the kth and lth nodal equations. This leads to the stamp in Fig. 17.3.6.

To clarify the stamp, we also take the example in Fig. 17.3.7. The nodal equations are

$$G_1 V_1 + I_1 + I_2 = J$$

$$-I_1 + G_2 V_2 - G_2 V_3 = 0$$

$$-G_2 V_2 + (G_2 + G_3)V_3 - I_2 = 0$$

To these equations we add the constitutive equations

$$V_1 - V_2 = 0$$

$$V_1 - V_3 - rI_1 = 0$$

and in matrix form

$$
\begin{bmatrix}
G_1 & 0 & 0 & 1 & 1 \\
0 & G_2 & -G_2 & -1 & 0 \\
0 & -G_2 & (G_2 + G_3) & 0 & -1 \\
1 & -1 & 0 & 0 & 0 \\
1 & 0 & -1 & -r & 0
\end{bmatrix}
\begin{bmatrix}
V_1 \\
V_2 \\
V_3 \\
I_1 \\
I_2
\end{bmatrix}
=
\begin{bmatrix}
J \\
0 \\
0 \\
0 \\
0
\end{bmatrix}
$$

17.4. SPECIAL ELEMENTS

Several new elements were introduced in Chapter 13 and we give here their stamps. Before doing so, it is useful to show how nodal formulation relates to the two-port definitions. The boxes in Fig. 17.4.1a and b denote the *same* network. In a is the two-port definition, in b is the nodal definition with voltages measured with respect

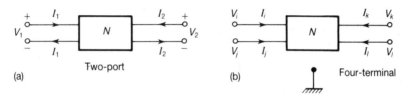

FIGURE 17.4.1   A network and its: (a) Two-port description. (b) Four-terminal description.

to ground. Comparison immediately reveals the following equivalences:

$$V_i - V_j = V_1$$
$$V_k - V_l = V_2$$
$$I_i = I_1$$
$$I_j = -I_1$$
$$I_k = I_2$$
$$I_l = -I_2$$

(17.4.1)

We will use them to derive the stamps.

17.4.1. Ideal Operational Amplifier

The ideal operational amplifier was already used in Chapter 12 but here we will make it more general and allow both output terminals to be floating, see Fig. 17.4.2. The constitutive equation is

$$(V_i - V_j)A = V_k - V_l$$

(17.4.2)

Since we wish to have one stamp for both ideal as well as nonideal OPAMPs, we cannot keep A; instead we again introduce the inverted gain

$$B = -\frac{1}{A}$$

(17.4.3)

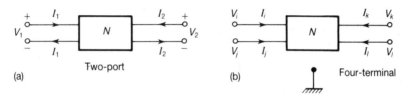

FIGURE 17.4.2   Operational amplifier and its stamp.

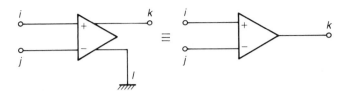

FIGURE 17.4.3   Equivalent symbols for an OPAMP.

and modify Eq. (17.4.2)

$$V_i - V_j + BV_k - BV_l = 0 \qquad (17.4.4)$$

This is the constitutive equation to be used in the stamp. The ideal operational amplifier does not draw any current at its input terminals; only the kth and lth nodes will be influenced by the output current. The stamp thus has the form given in Fig. 17.4.2. If terminal l is grounded (the usual case), then the floating operational amplifier symbol reduces to Fig. 17.4.3 and V_l becomes zero.

We will use the network in Fig. 17.4.4 to write modified nodal equations and also the matrix equation. For the three ungrounded nodes we have

$$GV_1 - GV_2 + I_E = 0$$
$$-GV_1 + (sC + G)V_2 - sCV_3 = 0$$
$$-sCV_2 + sCV_3 + I_{OP} = 0$$

and the constitutive equations for the voltage source and operational amplifier are, respectively,

$$V_1 = E$$
$$-V_2 + BV_3 = 0$$

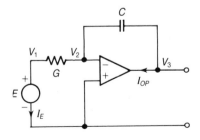

FIGURE 17.4.4   Integrator.

Collected together

$$
\begin{bmatrix}
G & -G & 0 & 1 & 0 \\
-G & G + sC & -sC & 0 & 0 \\
0 & -sC & sC & 0 & 1 \\
1 & 0 & 0 & 0 & 0 \\
0 & -1 & B & 0 & 0
\end{bmatrix}
\begin{bmatrix}
V_1 \\ V_2 \\ V_3 \\ I_E \\ I_{OP}
\end{bmatrix}
=
\begin{bmatrix}
0 \\ 0 \\ 0 \\ E \\ 0
\end{bmatrix}
$$

This matrix equation can also be written directly by inspection. Set up first the nodal portion; it has the dimension 3×3 because there are three ungrounded nodes. Next attach the constitutive equation for the voltage source and also modify the first row by adding one column with unit entry in the first row. This takes care of the independent voltage source. Next attach the constitutive equation of the operational amplifier and add one column with a unit in row three, because the current of the OPAMP output influences the current balance of the third node.

As a reminder, we repeat that the gain of a practical OPAMP changes with frequency and we often assume that

$$
A = \frac{A_0 \omega_0}{s + \omega_0}
$$

Refer to Chapter 16 and 18 for additional details.

17.4.2. Gyrator

The symbol for a gyrator is in Fig. 17.4.5 and the equations were given in Chapter 13,

$$
I_1 = -g_2 V_2
$$
$$
I_2 = g_1 V_1
$$

To make them suitable for modified nodal formulation, all we have to do is convert

FIGURE 17.4.5   Gyrator and its stamp.

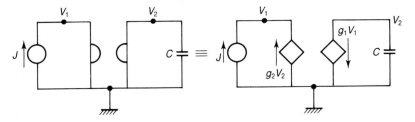

FIGURE 17.4.6   Gyrator loaded by a capacitor.

the two-port expressions into terminal expressions using Eq. (17.4.1):

$$I_i = -g_2(V_k - V_l)$$
$$I_j = +g_2(V_k - V_l)$$
$$I_k = +g_1(V_i - V_j)$$
$$I_l = -g_1(V_i - V_j)$$

The stamp in Fig. 17.4.5 follows immediately. As an example we consider Fig. 17.4.6 with only two ungrounded nodes. Only two nodal equations are needed:

$$-g_2 V_2 = J$$
$$g_1 V_1 + sC V_2 = 0$$

and if we wish to use matrix form

$$\begin{bmatrix} 0 & -g_2 \\ g_1 & sC \end{bmatrix} \begin{bmatrix} V_1 \\ V_2 \end{bmatrix} = \begin{bmatrix} J \\ 0 \end{bmatrix}$$

17.4.3. Negative Impedance Convertor

The negative impedance convertor, *NIC*, is in Fig. 17.4.7 and its two-port equations, already introduced in Chapter 13, are

$$V_1 = k_1 V_2$$
$$I_1 = -k_2 I_2$$

For modified nodal formulation this is adjusted to

$$V_i - V_j - k_1 V_k + k_1 V_l = 0$$
$$I_1 + k_2 I_2 = 0$$

(17.4.7)

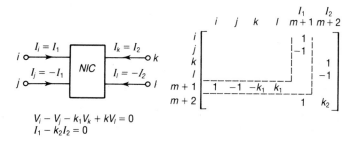

$$V_i - V_j - k_1 V_k + k V_l = 0$$
$$I_1 - k_2 I_2 = 0$$

FIGURE 17.4.7   Negative impedance convertor and its stamp.

Since I_1 influences the sum of currents at nodes i and j and I_2 modifies the currents at nodes k and l, the stamp must increase the size of the previous matrix by two rows and two columns; it is shown in Fig. 17.4.7.

When we consider the example in Fig. 17.4.8, we first realize that there are only two ungrounded nodes and the KCL at these nodes are

$$I_1 = J$$

$$GV_2 + I_2 = 0$$

In addition, we must add the constitutive equations for the negative impedance convertor:

$$V_1 - k_1 V_2 = 0$$

$$I_1 + k_2 I_2 = 0$$

This results in the matrix equation

$$\begin{bmatrix} 0 & 0 & 1 & 0 \\ 0 & G & 0 & 1 \\ 1 & -k_1 & 0 & 0 \\ 0 & 0 & 1 & +k_2 \end{bmatrix} \begin{bmatrix} V_1 \\ V_2 \\ I_1 \\ I_2 \end{bmatrix} = \begin{bmatrix} J \\ 0 \\ 0 \\ 0 \end{bmatrix}$$

FIGURE 17.4.8   Negative impedance convertor with a conductance.

An ideal transformer, being a special form of a convertor, has the same type of stamp.

17.4.4. Transformer with Two Coils

We derived, in Section 14.1, that a transformer with initial currents through the two coils is described by the equations

$$V_1 = sL_1I_1 + sMI_2 - L_1J_{10} - MJ_{20}$$
$$V_2 = sMI_1 + sL_2I_2 - MJ_{10} - L_2J_{20}$$

If we substitute from Eq. (17.4.1) for V_1 and V_2, we get two constitutive equations

$$V_i - V_j - sL_1I_1 - sMI_2 = -L_1J_{10} - MJ_{20}$$
$$V_k - V_l - sMI_1 - sL_2I_2 = -MJ_{10} - L_2J_{20}$$

(17.4.8)

Nodes i and j have their currents modified by I_1, nodes k and l by I_2; the stamp is in Fig. 17.4.9. The right-hand-side vector is also influenced if the initial currents are nonzero.

As an example, consider the network in Fig. 17.4.10. Although we will set up the equations, the reader should attempt to write the matrix equation directly. The equations are

$$sC_1V_1 - sC_1V_2 + I_1 = J$$
$$-sC_1V_1 + sC_1V_2 + I_2 = 0$$
$$sC_2V_3 - I_1 - I_2 = 0$$
$$V_1 - V_3 - sL_1I_1 - sMI_2 = 0$$
$$V_2 - V_3 - sMI_1 - sL_2I_2 = 0$$

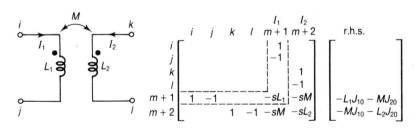

FIGURE 17.4.9   Transformer and its stamp.

FIGURE 17.4.10   Network with a transformer.

and the matrix equation is

$$
\begin{bmatrix}
sC_1 & -sC_1 & 0 & 1 & 0 \\
-sC_1 & sC_1 & 0 & 0 & 1 \\
0 & 0 & sC_2 & -1 & -1 \\
1 & 0 & -1 & -sL_1 & -sM \\
0 & 1 & -1 & -sM & -sL_2
\end{bmatrix}
\begin{bmatrix}
V_1 \\
V_2 \\
V_3 \\
I_1 \\
I_2
\end{bmatrix}
=
\begin{bmatrix}
J \\
0 \\
0 \\
0 \\
0
\end{bmatrix}
$$

Note how simple the handling of the transformer in this formulation really is.

17.5. COMPUTER APPLICATIONS

We are adding this section for readers interested in understanding how the theories presented in this chapter can be applied to computer applications. For additional information see J. Vlach and K. Singhal *Computer Methods for Circuit Analysis and Design* Van Nostrand Reinhold 1983, Chapter 4.

In all examples presented in this chapter we always wrote one system matrix with all its components in it. For instance, for the network in Fig. 17.1.7 we obtained the system equation

$$
\begin{bmatrix}
G_1 + sC_1 & 0 & 0 & 1 & 0 \\
0 & sC_2 & 0 & -1 & 1 \\
0 & 0 & sC_3 + G_2 & 0 & -1 \\
1 & -1 & 0 & -sL_1 & 0 \\
0 & 1 & -1 & 0 & -sL_2
\end{bmatrix}
\begin{bmatrix}
V_1 \\
V_2 \\
V_3 \\
I_{L1} \\
I_{L2}
\end{bmatrix}
=
\begin{bmatrix}
J \\
0 \\
0 \\
0 \\
0
\end{bmatrix}
$$

Denote the matrix by $\mathbf{T}$, the vector of unknowns by $\mathbf{X}$ and the right-hand side by $\mathbf{W}$,

$$\mathbf{TX} = \mathbf{W}$$

For computer application, it is convenient to decompose the matrix $\mathbf{T}$ into two real matrices which we will call $\mathbf{GM}$ and $\mathbf{CM}$:

$$\mathbf{T} = \mathbf{GM} + s\mathbf{CM}$$

The two matrices are:

$$\mathbf{GM} = \begin{bmatrix} G_1 & 0 & 0 & 1 & 0 \\ 0 & 0 & 0 & -1 & 1 \\ 0 & 0 & G_2 & 0 & -1 \\ 1 & -1 & 0 & 0 & 0 \\ 0 & 1 & -1 & 0 & 0 \end{bmatrix}$$

$$\mathbf{CM} = \begin{bmatrix} C_1 & 0 & 0 & 0 & 0 \\ 0 & C_2 & 0 & 0 & 0 \\ 0 & 0 & C_3 & 0 & 0 \\ 0 & 0 & 0 & -L_1 & 0 \\ 0 & 0 & 0 & 0 & -L_2 \end{bmatrix}$$

The right-hand-side vector is unchanged,

$$\mathbf{W} = [J \quad 0 \quad 0 \quad 0 \quad 0]^T$$

To save space we have written it here in transposed form.

Let us take another network, Fig. 17.5.1. When writing the $\mathbf{GM}$ matrix we simply disregard all capacitors and inductors, as if they were erased from the figure. The network has five ungrounded nodes and the voltage source with the operational amplifier will increase the size of the overall matrix to 7×7. We can fill the nodal portion of the matrix by the methods we learned earlier, or element by element: for instance, G_1 will be positive in positions $(1, 1)$ and $(2, 2)$ and negative in positions $(1, 2)$ and $(2, 1)$. G_5 will be positive in positions $(1, 1)$ and $(4, 4)$, and negative in positions $(1, 4)$ and $(4, 1)$. Any grounded conductance will appear only once: G_2 occurs in only one place, the entry $(2, 2)$.

When the nodal portion has been prepared, we add entries for the other two elements. The voltage source is connected to node 1 and its current will influence row 1; 1 comes into the $(1, 6)$ position. The constitutive equation for the source is $V_1 = E$ and 1 is entered into the position $(6, 1)$. The operational amplifier

FIGURE 17.5.1   Second-order active network.

influences the balance of currents at node 5; this is taken care of by an additional, seventh, column with 1 in the (5, 7) position. Finally we attach the constitutive equation for the operational amplifier, in this case

$$V_3 - V_4 + BV_5 = 0$$

and the entries are -1 in position (7, 3), $+1$ in the position (7, 4) and B in the position (7, 5). This completes writing of the **GM** matrix:

$$\mathbf{GM} = \begin{bmatrix} G_1 + G_5 & -G_1 & 0 & -G_5 & 0 & 1 & 0 \\ -G_1 & G_1 + G_2 & 0 & 0 & 0 & 0 & 0 \\ 0 & 0 & G_3 + G_4 & 0 & -G_3 & 0 & 0 \\ -G_5 & 0 & 0 & G_5 + G_6 + G_7 & -G_7 & 0 & 0 \\ 0 & 0 & -G_3 & -G_7 & G_3 + G_7 & 0 & 1 \\ 1 & 0 & 0 & 0 & 0 & 0 & 0 \\ 0 & 0 & 1 & -1 & B & 0 & 0 \end{bmatrix}$$

The **CM** matrix must have the same size as **GM**, 7×7. C_1 will be positive in positions (2, 2) and (3, 3), negative in (2, 3) and (3, 2). C_2 will be taken positive in

(1, 1) and (5, 5), negative in (1, 5) and (5, 1). All other entries are zero,

$$\mathbf{CM} = \begin{bmatrix} 0 & 0 & 0 & 0 & 0 & 0 & 0 \\ 0 & C_1 + C_2 & -C_1 & 0 & -C_2 & 0 & 0 \\ 0 & -C_1 & C_1 & 0 & 0 & 0 & 0 \\ 0 & 0 & 0 & 0 & 0 & 0 & 0 \\ 0 & -C_2 & 0 & 0 & C_2 & 0 & 0 \\ 0 & 0 & 0 & 0 & 0 & 0 & 0 \\ 0 & 0 & 0 & 0 & 0 & 0 & 0 \end{bmatrix}$$

The right-hand-side vector length must be the same as the size of the matrices, 7, and only the voltage source will have an entry in this vector. Since the constitutive equation for the source was taken care of by the sixth row of the **GM** matrix, the only nonzero entry will be the sixth one. The vector will be

$$\mathbf{W} = [0 \quad 0 \quad 0 \quad 0 \quad 0 \quad E \quad 0]^T$$

written horizontally in transposed form to save space.

In programming, the matrices **GM** and **CM** and the vector **W** can be prepared by first setting zeros into all entries and then writing the nonzero entries as they are given above. This can be done by a small program segment:

```
C...SET ZEROS INTO MATRICES GM, CM AND VECTOR W
     DO 1 I = 1, 7
     W(I) = 0.
     DO 1 J = 1, 7
     GM(I,J) = 0.
1    CM(I,J) = 0.
C...START NONZERO ENTRIES
     GM(1,1) = G1 + G5
     GM(1,2) = -G1
     GM(1,6) = 1.
     . . . .
```

where we have filled the first row of the matrix **GM**.

The above steps are done only once and the matrices are reused. For instance, if we wish to solve the network in frequency domain, we must substitute $s = j\omega$. In FORTRAN this is done by the statement S = CMPLX(0., OMEGA). Next we create a complex matrix **T**

$$\mathbf{T} = \mathbf{GM} + s\mathbf{CM} \tag{17.5.1}$$

which in programming is done simply by

```
     DO 2 I = 1, 7
     DO 2 J = 1, 7
2    T(I,J) = GM(I,J) + S*CM(I,J)
```

Whenever we change the frequency, all we have to do is change OMEGA, define a new S and repeat this three-line segment of the program. For each value of S we must solve the equation

$$\mathbf{TX} = \mathbf{W}$$

which means that we must have routines which work with complex numbers. They are on the diskettes accompanying this book and the appropriate theory is in Appendix C. A FORTRAN compiler which can operate with complex numbers and a linker must be available to the user.

Example 1.

Write the **GM**, **CM** matrices and the **W** vector for the network in Fig. 17.5.2.

The network has three ungrounded nodes, one inductor and one voltage source and thus the overall size of the matrices will be 5×5. The matrices are

$$\mathbf{GM} = \begin{bmatrix} G_1 & -G_1 & 0 & 0 & 1 \\ -G_1 & G_1 & 0 & 1 & 0 \\ 0 & 0 & G_2 & -1 & 0 \\ 0 & 1 & -1 & 0 & 0 \\ 1 & 0 & 0 & 0 & 0 \end{bmatrix}$$

$$\mathbf{CM} = \begin{bmatrix} 0 & 0 & 0 & 0 & 0 \\ 0 & C_1 & 0 & 0 & 0 \\ 0 & 0 & C_2 & 0 & 0 \\ 0 & 0 & 0 & -L & 0 \\ 0 & 0 & 0 & 0 & 0 \end{bmatrix}$$

FIGURE 17.5.2   LCG network.

and the right-hand-side vector is

$$\mathbf{W} = [0 \quad 0 \quad 0 \quad 0 \quad E]^T$$

PROBLEMS CHAPTER 17

Every problem given in this book can be set up in modified nodal formulation. Here we give only a few simple networks to test the skills.

NETOPT can use two formulations; the default formulation is different from the one explained in this book, but you can test your results by NETOPT, with some caution. You *must* request the 1-graph formulation to get results corresponding to the material of this chapter. In the element list give first all elements which enter the nodal portion of

the matrix. Afterwards give the elements in the same sequence as you would do on paper. You can issue the command SHOW MATR, which will display the information on the screen. If it runs over more than one screen, you can send the information into the listing file by the command LIST MATR, but then you must exit NETOPT and display the file NETOPT.LST by your editor. Let us take the problem P.17.1; the file will display the following information (all elements have value 1).

```
INTERNAL GROUND NODE IS NUMBERED      6

USER NODE = INTERNAL NODE
     1 = 1           2 = 2          3 = 3

STRUCTURE OF NETWORK MATRICES

ELEMENT      SC.VAL      MATRIX       +        -        -        +
  1    G1     1.000        G       1, 1    1, 2    2, 1    2, 2
  2    C1     1.000        C       2, 2
  3    G2     1.000        G       2, 2    2, 3    3, 2    3, 3
  4    E1     1.000        W         4
  5    A1     0.

TOPOLOGICAL UNIT ENTRIES IN MATRIX G

                                  4, 1

                                  1, 4

                                  5, 3

                                  3, 5
```

The *topological units* are unit values outside of the nodal portion of the matrix. The file has additional information which we did not reproduce here. Note especially the line under the text: USER NODE = INTERNAL NODE. If the numbers on both sides of the equal signs are the same, like in this example, then the matrix will be as you would get it on paper. If the numbers are interchanged, then the program renamed the nodes for efficient sparse-matrix solution. If this happens, it is perhaps not worth trying to trace back the interchanges. Also note that ground is renumbered, in this

example to 6. You can disregard this information, your matrix will be 5 × 5.

In each of the following problems determine first the size of the resulting matrix. The nodes are marked by numbers in circles and voltages at these nodes will have the same subscripts.

P.17.1 through P.17.11 Write the modified nodal matrix equations.

P.17.1

P.17.2

P.17.3

P.17.4

P.17.5

P.17.6

P.17.7

P.17.8

P.17.9

P.17.10

P.17.11

P.17.15 through P.17.18 Write the modified nodal matrix equations.

P.17.15

P.17.12 through P.17.14 Normally we convert all resistors into conductances, to keep the size of the matrix small. In these examples take the resistors as resistors and write the modified nodal matrix equation. The matrix grows by one row and one column for each such resistor. Write also the matrix equations by converting the resistors into conductances.

P.17.16

P.17.12

P.17.17

P.17.13

P.17.14

P.17.18

CHAPTER 18

NETOPT Program

18.0. INTRODUCTION

NETOPT is a program for simulation and design of linear networks. It is available for various computers, including IBM-PC or equivalents, and is fully interactive, with plotting. In the frequency domain it can calculate and plot the following responses:

1. Amplitude (magnitude or dB)
2. Phase in degrees
3. Real and imaginary part
4. Group delay
5. Sensitivity of the above to changes of any element.

In addition, it can calculate

6. Poles and zeros
7. Q and ω_0 of poles and zeros
8. Sensitivities of poles and zeros
9. Sensitivities of Q and ω_0 of poles and zeros
10. Symbolic functions with up to five variable elements

Higher-level options are also available:

545

11. Tolerance analysis
12. Monte Carlo analysis (with plotting)
13. Interactive optimization of amplitude and group delay
14. Yield optimization

All results can be tabulated, directed to the terminal or to a listing file. Plot directives can be saved in a file for a hardcopy.

The educational IBM-PC version of NETOPT has all features of the full program with certain limitations: The sizes of networks are limited and optimization can be performed with fewer elements. The version also works on computers which do not have a mathematical coprocessor (but will utilize its presence).

Instructions given in this chapter refer only to the theories covered in this book or to features useful to the student. Detailed instructions on how to use the full capabilities of the program are in a separate manual.

Before you run the program for the first time, prepare a file with the name GRAPHICS.DEF to specify which graphics equipment you have. The only information in the file will be a single line with the following:

NO if you do not have any graphics equipment
CGA if you have IBM CGA interface and graphics monitor
EGAM if you have EGAM monochrome monitor
EGAC for EGA or VGA color monitor
HERC if you have Hercules graphics board and appropriate monitor.

The program looks for this file in current directory and then in the root directory. If you give wrong information, the results are unpredictable. You can print the graphs by installing a hardcopy driver and using the Shift PrtScr command.

18.1. NETWORK DESCRIPTION

A network to be solved by the program must be prepared similarly as in Fig. 18.1.1 (taken from Chapter 1, Fig. 1.2.10). Each node is given a number and ground is given the number zero. Any short circuit in the network must be marked by an additional node, see nodes 3 and 4 in Fig. 18.1.1.

The command NETOPT loads the program. It comes with questions and suggests actions. As in the example below, you may see something like [Y/N D:N].

FIGURE 18.1.1   Network with various elements.

This means that you have the choice YES or NO and if you hit return, the Default D is NO. If you agree with what the program suggests, just hit return. Data can be entered either in lower or upper case. The following is what will appear on the screen of your computer as you enter the data for the network in Fig. 18.1.1.

```
NETOPT V2.0                 15-May-90 09:01:27
Copyright (C) 1987 by J.Vlach and E.Christen
University of Waterloo, Waterloo, Ontario, Canada
   Network description from file   [Y/N D:N] : (hit return)
   Want to enter element tolerance [Y/N D:N] : (hit return)
   Want to enter element bounds    [Y/N D:N] : (hit return)
   Enter title ( <= 72 char.) :
   FIGURE 18.1.1     (this is the title you type in)
   Enter network description or HELP
   Enter name of element no. 1 : el 1 0 5   (return)
   Enter name of element no. 2 : rl 1 2 10 (return, etc.)
   Enter name of element no. 3 : cl 2 0 0.5
   Enter name of element no. 4 : r2 2 3 20
   Enter name of element no. 5 : cc 3 4 0 4 3
   Enter name of element no. 6 : ll 4 0 1
   Enter name of element no. 7 : c2 4 0 2
   Enter name of element no. 8 : v4 4 0      (voltmeter)
   Enter name of element no. 9 : done       (ends input)
   Listing output: Terminal (T) or File (F) [D:T] : hit return

      FIGURE 18.1.1

      CIRCUIT DESCRIPTION
```

	ELEMENT	INTERCONNECTION				VALUE
1	E1	1	0			5.0000000d+00
2	R1	1	2			1.0000000d+01
3	C1	2	0			5.0000000d-01
4	R2	2	3			2.0000000d+01
5	CC	3	4	0	4	3.0000000d+00
6	L1	4	0			1.0000000d+00
7	C2	4	0			2.0000000d+00
8	V2	4	0			0.0000000d+00

Next the program asks whether you wish to use manual or automatic scaling. For small networks *always* select manual and accept the suggested values by hitting returns.

```
Element scaling Manual (M) or Automatic (A) [D:A] : m
Enter frequency scaling factor FSCALE [D:1/2PI] : hit return
Enter impedance scaling factor ZSCALE [D: 1.DO] : hit return
FREQUENCY SCALING FACTOR FSCALE = 1.59154943d-01
IMPEDANCE SCALING FACTOR ZSCALE = 1.00000000d+00
```

The program asks which formulation you wish to use and suggests 1-graph or 2-graph. Normally we use 2-graph, the default. It makes the matrices somewhat smaller and thus runs faster. The 1-graph method is needed in sensitivity calculations with respect to ideal operational amplifiers and is equivalent to the theory explained in Chapter 17. Accept the indicated defaults by hitting returns until you get the main menu:

```
Select command:
NET ,SPEC,AC  ,POLE,TIME,SYMB,TOL ,MC  ,OPT ,
PLOT,SAVE,SET ,SHOW,LIST,SYS ,HELP,QUIT     :
```

At this point *always* issue the command SAVE. To simplify your job, the program will suggest a file name [NETWORK.DAT] which you can either accept by hitting return, or give another name. If you first SAVE and then quit NETOPT using the QUIT command, you will have your work available next time you load NETOPT. The command

```
netopt network.dat
```

will first load NETOPT which, in turn, will automatically read the file network.dat.

If you make an error when entering data, the program complains. Look carefully for what seems to be the problem. You must give a correction or at least something acceptable. The program refuses to continue until you give acceptable input. Continue input of your data until the end and SAVE what you have done. Also issue the command SHOW NET, to check what exactly is in the program. If all is fine, you can continue. Should there be an error, then QUIT the program and using any editor available on your computer correct the network file. During corrections by your editor you *MUST* give the correct values in the same format (with the same spacing) as the program writes them. At least one source and at least one output (voltmeter or ammeter) must be given. If you forget, the program will remaind you.

When you look at the file saved by NETOPT, you will see a somewhat different picture than above:

```
FIGURE 18.1.1
El  1  0      5.00000000d+00   0.0000d+00   0.0000d+00   0.0000d+00
R1  1  2      1.00000000d+00   0.0000d+00   0.0000d+00   0.0000d+00
C1  2  0      5.00000000d-01   0.0000d+00   0.0000d+00   0.0000d+00
R2  2  3      2.00000000d+00   0.0000d+00   0.0000d+00   0.0000d+00
CC  3  4  0  4  3.00000000d+00   0.0000d+00   0.0000d+00   0.0000d+00
L1  4  0      1.00000000d+00   0.0000d+00   0.0000d+00   0.0000d+00
C2  4  0      2.00000000d+00   0.0000d+00   0.0000d+00   0.0000d+00
V4  4  0      0.00000000d+00   0.0000d+00   0.0000d+00   0.0000d+00
```

The last three columns are needed for optimization and higher-level functions which we will not use; the program will insert zeros.

Each element name can occupy only four positions. The easiest is to give the name of the element, as summarized below, followed by a number which distinguishes the element from other similar ones. This was done in the above file. Elements available in NETOPT are denoted the same way as in the text. The dots below indicate spaces where additional information (numbers or letters) can be given:

```
Two-terminal elements:

G...    conductor       value in siemens
R...    resistor        value in ohms
C...    capacitor       value in farads
L...    inductor        value in henrys

Independent sources:

E...    voltage source   value in volts
J...    current source   value in amperes

Output selectors:

V...    voltmeter        no value
A...    amperemeter      no value

Operational amplifiers:

O...    ideal opamp      no value
N...    non-ideal opamp  value is gain-bandwidth product in Hz

VV..    voltage-controlled voltage source  value dimensionless
VC..    voltage-controlled current source  value in mhos
CV..    current-controlled voltage source  value in ohms
CC..    current-controlled current source  value dimensionless
```

All these elements are summarized in Fig. 18.1.2 and the sequence of data for each of them is under the element. The dots following the element name indicate how many additional letters you can give. If you give a nonideal operational amplifier (N) and enter zero for the GB product, then the element automatically converts into an ideal operational amplifier.

The writing of large or small element values is simplified by attaching, after the number, a letter (letters) denoting the multiplicative factor. For instance, a capacitor value of 1.52 nanofarad can be given as 1.52n. The symbols differ from standard notations because we cannot type Greek letters with a computer keyboard. The symbols are in Table 18.1.1.

The program works with two sets of menus. The main menu is

```
NET , SPEC, AC  , POLE, TIME, SYMB, TOL , MC  , OPT ,
PLOT, SAVE, SET , SHOW, LIST, SYS , HELP, QUIT        :
```

If you issue the command NET, you can start solving a new network without quitting NETOPT. Some of the options are not described in this chapter and can

FIGURE 18.1.2   NETOPT elements.

be found only in the NETOPT manual (TOL, MC, OPT). The command SYS gets you to the computer system (DOS). The commands SET, SHOW or LIST switch you to a secondary menu which looks like this:

```
NET , MATR, FILE, MNF , SIZE, DEG , SCAL, SENS, VARS, CRIT,
SPEC, GOAL, PIVO, GD  , PREF, MARG, LIM , HELP, QUIT      :
```

We will use some of its options (GD, SENS), some are described only in the manual. It is important for your work to be able to get back to the main menu which controls all operations. This is achieved by typing QUIT in this second menu.

TABLE 18.1.1	Prefixes for Electrical Quantities in NETOPT	
Prefix	Name	Value
P	pico	10^{-12}
N	nano	10^{-9}
U or MU	micro	10^{-6}
M	milli	10^{-3}
K	kilo	10^{+3}
ME	mega	10^{+6}
G	giga	10^{+9}
T	tera	10^{+12}

18.2. RESISTIVE NETWORKS

As an example consider the network in Fig. 18.2.1, (used in Chapter 2, Fig. 2.1.6). We wish to find all three nodal voltages. The network data are

```
 1  J1     0    1      1.0000000d+00
 2  G1     1    0      1.0000000d+00
 3  G2     1    2      2.0000000d+00
 4  G3     2    0      1.0000000d+00
 5  G4     2    3      2.0000000d+00
 6  G5     1    3      3.0000000d+00
 7  J2     0    3      2.0000000d+00
 8  J3     1    3      3.0000000d+00
 9  V1     1    0      0.0000000d+00
10  V2     2    0      0.0000000d+00
11  V3     3    0      0.0000000d+00
```

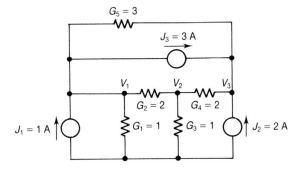

FIGURE 18.2.1   Resistive network with three current sources.

The actions we take are:

```
NET ,SPEC,AC   ,POLE,TIME,SYMB,TOL ,MC   ,OPT ,
PLOT,SAVE,SET ,SHOW,LIST,SYS ,HELP,QUIT      : ac

Enter number of frequency points : 1
Enter frequency : 0     (corresponds to dc)
Frequencies in Hertz (H) or Radians (R) [D:H] :
  (return)
Listing output: Terminal (T), File (F) or No output
  (N) [D:N] : t
Enter response types. Finish with DONE:
  DB (DB), MAgnitude (MA), PHase (PH),
  REal part (RE), IMaginary part (IM) [D:DB] : re

NETWORK FIGURE 2.1.6

AC ANALYSIS

FREQUENCY        V1           V2           V3
  [HZ]       REAL PART    REAL PART    REAL PART

0.00000d+00  1.43243d+00  1.56757d+00  2.48649d+00
```

18.3. SYMBOLIC FUNCTIONS FOR RESISTIVE NETWORKS

NETOPT can find network functions in terms of variable elements. This is called symbolic analysis and up to five elements can be kept as variables (letters) in the result. Numerical values are given for all elements and it is imperative that you *select manual scaling with the values suggested by the program.* A symbolic function automatically assumes that the source (only one) has unit value. *Give all resistors as conductances.* Nothing happens if you give them as resistors, but the program automatically switches to conductances to keep the system matrices small and the symbolic function theory cannot switch back. The result uses conductances.

As an example we take the network in Fig. 18.3.1 (used in Chapter 5, Fig. 5.2.1). For computer solution we must add one node to take into account the controlling

FIGURE 18.3.1   Network with a *CC*.

current of the *CC*. The input data are

```
ELEMENT      INTERCONNECTION           VALUE

1 E1       1   0                2.0000000d+00
2 G1       1   2                2.0000000d+00
3 G2       3   0                3.0000000d+00
4 G3       3   4                1.0000000d+00
5 G4       4   0                2.0000000d+00
6 CC       2   3    0    4      2.0000000d+00
7 V4       4   0                0.0000000d+00
```

The steps you take depend on how many elements you wish to take as variables. The largest number is five, which we will use below, but anything less, including none, works also.

```
Select command:
NET ,SPEC,AC  ,POLE,TIME,SYMB,TOL ,MX  ,OPT ,
PLOT,SAVE,SET ,SHOW,LIST,SYS ,HELP,QUIT      : symb
Coeff. derivatives (D) or Symbolic function (S)
  [D:S] : s
Enter explicit elements (not more than 5)
Enter element name or DONE : g1
Enter element name or DONE : g2
Enter element name or DONE : g3
Enter element name or DONE : g4
Enter element name or DONE : cc
Enter element name or DONE : done
Enter network degree (<= 150) : 0
```

At this point it is useful to send the result to a file because it might be longer than one screen. To read the file you must quit NETOPT by the command QUIT and use the editor of your computer. The result is in the file NETOPT.LIS:

```
FIGURE 18.3.1

SYMBOLIC ANALYSIS

SYMBOLIC FUNCTION IN TERMS OF ELEMENTS

INPUT AT E1  ,   OUTPUT AT V4

EXPLICIT        POWER  NUMERATOR          DENOMINATOR
ELEMENTS        OF S

G1    G3

                       MULTIPLIER          1.00000000D+00
                0      -1.00000000d+00    -1.00000000d+00
```

(program continues on next page)

EXPLICIT ELEMENTS			POWER OF S	NUMERATOR	DENOMINATOR
G2 G3					
				MULTIPLIER	1.00000000D+00
			0	0.00000000d+00	-1.00000000d+00
G1		G4			
				MULTIPLIER	1.00000000D+00
			0	0.00000000d+00	-1.00000000d+00
	G2	G4			
				MULTIPLIER	1.00000000D+00
			0	0.00000000d+00	-1.00000000d+00
	G3 G4				
				MULTIPLIER	1.00000000D+00
			0	0.00000000d+00	-1.00000000d+00
G1 G2		CC			
				MULTIPLIER	1.00000000D+00
			0	-1.00000000d+00	0.00000000d+00
G1	G3	CC			
				MULTIPLIER	1.00000000D+00
			0	-1.00000000d+00	-1.00000000d+00

The multiplier is always equal to 1 if you have chosen manual scaling as suggested above; you simply ignore it. The power of s is always zero for a network without capacitors and inductors. The table says, on the top, that the product G_1G_3 will be in the numerator as well as the denominator, multiplied by minus one in both cases. Lower in the table, the product $G_1G_2\alpha$ (CC standing for α to indicate the current controlled current source) will be in the numerator multiplied by minus one, but will be absent in the denominator (multiplied by zero). The product $G_1G_3\alpha$ will be in the numerator and denominator, multiplied by minus one. Thus after a few simplifying steps the symbolic function is

$$F = \frac{G_1G_3(1 + \alpha) + G_1G_2\alpha}{G_1G_3(1 + \alpha) + G_2G_3 + (G_1 + G_2 + G_3)G_4}$$

18.4. FREQUENCY DOMAIN ANALYSIS

To demonstrate this type of analysis we use the network Fig. 18.4.1; it is a Cauer-parameter filter and its element values are in the input file:

ELEMENT		INTERCONNECTION		VALUE
1 E1	1	0		1.0000000d+00
2 G1	1	2		1.0000000d+00

(program continues on next page)

FIGURE 18.4.1   Cauer-parameter LC filter.

ELEMENT	INTERCONNECTION		VALUE
3 C1	2	0	1.2526200d+00
4 C2	2	3	5.5110000d-02
5 L1	2	3	1.2898500d+00
6 C3	3	0	1.9828300d+00
7 C4	3	4	1.4819000d-01
8 L2	3	4	1.1845000d+00
9 C5	4	0	1.1718300d+00
10 G2	4	0	1.0000000d+00
11 V4	4	0	0.0000000d+00

From the main menu the choices are

```
NET ,SPEC,AC   ,POLE,TIME,SYMB,TOL ,MC   ,OPT ,
PLOT,SAVE,SET ,SHOW,LIST,SYS ,HELP,QUIT      : ac
```

```
Enter number of frequency points : 100
Enter start and end frequencies : 0 5
Frequencies in Hertz (H) or Radians (R) [D.H] : r
Frequency scaling LInear (LI) or LOgarithmic (LO)
   [D:LO] : li
Listing output: Terminal (T), File (F) or No output
   (N) [D:N] :
```

The above selection of radians is suitable for normalized networks, like ours. Real networks are usually studied with frequencies expressed in Hertz. After the calculations the main menu appears again and we can issue the command PLOT. The program responds with the following questions and your answers.

```
Enter file name [no default] : ac.tmp
Enter response type:
   DB (DB), MAgnitude (MA), PHase (PH),
      REal (RE) or IMaginary part (IM) [D:DB] : return
      Y axis scaling Manual (M) or Automatic (A) [D:A] :
      (return)
      Plot output to file for hardcopy [Y/N D:N] :
      (return)
```

FIGURE 18.4.2   Amplitude (dB) of network Fig. 18.4.1.

Plot of the response is in Fig. 18.4.2. Another hit of the return key gets us back to the main menu for further actions.

18.5. SYMBOLIC FUNCTIONS OF GENERAL NETWORKS

Networks with capacitors and/or inductors lead to rational functions in *s* with coefficients determined by the elements and their values. NETOPT can find such functions in terms of *s* with up to five elements kept as variables. We will demonstrate this application on the network Fig. 18.5.1 (used in Chapter 7, Fig. 7.4.4). The input data are

```
1 E1      1   0              1.0000000d+00
2 L       1   2              3.0000000d+00
3 G1      3   0              1.0000000d+00
4 C       3   4              2.0000000d+00
5 G2      4   0              2.0000000d+00
6 CC      2   3    0    4    2.0000000d+00
7 V4      4   0              0.0000000d+00
```

FIGURE 18.5.1   Network with a *CC*.

We will demonstrate two types of symbolic analysis. In the first one we ask for the function in terms of *s* only. The steps are

```
NET ,SPEC,AC  ,POLE,TIME,SYMB,TOL ,MC  ,OPT ,
PLOT,SAVE,SET ,SHOW,LIST,SYS ,HELP,QUIT      : symb
Coeff. derivatives (D) or Symbolic function (S) [D:S] :
Enter explicit elements (not more than 5)
Enter element name or DONE : done
Enter network degree (<= 150) : 2
Listing output: Terminal (T) or File (F) [D:T] :
```

The program asks for the network's degree. For a beginner this is difficult to answer but, fortunately, giving a number larger than needed is harmless, while giving a number less than needed leads to a wrong result. A rule of thumb is to give a degree equal to the sum of capacitors and inductors, in this case 2. The result is

EXPLICIT ELEMENTS	POWER OF S	NUMERATOR	DENOMINATOR
		MULTIPLIER	1.00000000D+00
	0	-2.00000000d+00	-2.00000000d+00
	1	-6.00000000d+00	-1.20000000d+01
	2	0.00000000d+00	-1.80000000d+01

which gives the function

$$F = \frac{6s + 2}{18s^2 + 12s + 2}$$

If we select up to five (as in the following case) or fewer elements, then the steps are

```
NET ,SPEC,AC  ,POLE,TIME,SYMB,TOL ,MC  ,OPT ,
PLOT,SAVE,SET ,SHOW,LIST,SYS ,HELP,QUIT      : symb
Coeff. derivatives (D) or Symbolic function (S) [D:S] :
  symb
Enter explicit elements (not more than 5)
Enter element name or DONE : 1
Enter element name or DONE : g1
Enter element name or DONE : c
Enter element name or DONE : g2
Enter element name or DONE : cc
Enter element name or DONE : done
Enter network degree (<= 150) : 2
Listing output: Terminal (T) or File (F) [D:T] : f
```

Note that in this case it is advantageous to send the result to a file because it may be longer than one screen page. After quitting NETOPT, we read the file NETOPT.LIS by means of any editor. The result is

```
EXPLICIT          POWER   NUMERATOR              DENOMINATOR
ELEMENTS          OF S
           C
                          MULTIPLIER             1.00000000D+00
                  1       -1.00000000d+00        -1.00000000d+00

  L G1 C
                          MULTIPLIER             1.00000000D+00
                  2       0.00000000d+00         -1.00000000d+00

       G2
                          MULTIPLIER             1.00000000D+00
                  0       0.00000000d+00         -1.00000000d+00

  L G1    G2
                          MULTIPLIER             1.00000000D+00
                  1       0.00000000d+00         -1.00000000d+00

  L    C G2
                          MULTIPLIER             1.00000000D+00
                  2       0.00000000d+00         -1.00000000d+00

    G1        CC
                          MULTIPLIER             1.00000000D+00
                  0       -1.00000000d+00        0.00000000d+00

       C    CC
                          MULTIPLIER             1.00000000D+00
                  1       -1.00000000d+00        -1.00000000d+00
```

If we have used manual normalization with the suggested values, the multiplier will always be 1 and is disregarded. Reading the output from top down, we get the following terms of the symbolic function:

$$F = \frac{-sC - G_1\alpha - sC\alpha}{-sC - s^2LCG_1 - G_2 - sLG_1G_2 - s^2LCG_2 - sC\alpha}$$

Trace the above table with this equation and you will understand how to read the result. Rearranging terms and multiplying both numerator and denominator by -1 gets us the symbolic function:

$$F = \frac{sC(1 + \alpha) + G_1\alpha}{s^2LC(G_1 + G_2) + s[C(1 + \alpha) + LG_1G_2] + G_2}$$

Run this example again by requiring only one or only two elements.

In simple networks it may sometimes happen that a given choice of network degree does not work. The remedy is very simple: select a higher degree. Should this not work for several choices, then it is almost certain that you have an error in your data.

18.6. POLES AND ZEROS

NETOPT has two possibilities for calculating poles and zeros:

1. Using the QZ algorithm.
2. Obtaining first the network function and then finding the roots of the numerator and denominator polynomials.

QZ is a very accurate algorithm and is easier to use because it automatically finds the right number of poles and zeros. The interpolation option needs the order of the filter.

As an example we will take the filter in Fig. 18.4.1 and use both options. For the QZ algorithm we get:

POLE ZERO ANALYSIS OF V4

NETWORK ZEROS

REAL PART [RAD]	IMAG.PART [RAD]	ZERO FREQ. [RAD]	ZERO Q	
0.00000000d+00	-3.75072472d+00	3.75072472d+00	0.00000000d+00	Z
0.00000000d+00	3.75072472d+00	3.75072472d+00	0.00000000d+00	Z
9.25090915d-17	-2.38683841d+00	2.38683841d+00	-3.87580035d-17	Z
-5.50204702d-16	2.38683841d+00	2.38683841d+00	2.30516108d-16	Z

NETWORK POLES

REAL PART [RAD]	IMAG.PART [RAD]	POLE FREQ. [RAD]	POLE Q
-1.29752204d-01	-1.05027231d+00	1.05825684d+00	4.07799180d+00
-1.29752204d-01	1.05027231d+00	1.05825684d+00	4.07799180d+00
-5.07120006d-01	0.00000000d+00	5.07120006d-01	5.00000000d-01
-3.81129314d-01	-6.86198581d-01	7.84938243d-01	1.02975318d+00
-3.81129314d-01	6.86198581d-01	7.84938243d-01	1.02975318d+00

Some interpretation is needed. If the number is of a similar magnitude as the real part of the third zero (9.250d − 17) then clearly the real part can be considered to be zero. It is not exactly zero because the elements are given with only six decimal digits. All poles are in the negative half plane (negative real parts) and the filter is stable (as it must be, because there are no active elements). The terms ZERO FREQ. or POLE FREQ. refer to ω_0 like in Eq. (8.4.5) and are actually the distances of the zeros or poles from the origin. Notice also the Q of the zeros. Since the zeros are either exactly or almost on the imaginary axis, their Q would be infinitely large and the computer cannot handle infinity. The symbol Z indicates that the number actually printed is $\zeta = 1/2Q$, and expression frequently used in control theory.

If we take the interpolation option, the network function is first printed, followed by the poles and zeros (they may be in different sequence). Certain amount of work is again needed to correctly interpret very small numbers: anything smaller than, say, 10^{-15} can be replaced by zero.

POWER OF S	NUMERATOR	DENOMINATOR
0	-1.00000000d+00	-2.00000000d+00
1	1.18423789d-15	-6.88163000d+00
2	-2.46614689d-01	-1.13983826d+01
3	-8.88178420d-16	-1.40147980d+01
4	-1.24773852d-02	-8.73856991d+00
5	5.92118946d-16	-5.71565625d+00

NETWORK ZEROS

REAL PART [RAD]	IMAG. PART [RAD]	ZERO FREQ. [RAD]	ZERO Q
-2.18952885d-47	2.38683841d+00	2.38683841d+00	9.17334345d-48 Z
0.00000000d+00	-2.38683841d+00	2.38683841d+00	0.00000000d+00 Z
-1.09476443d-46	3.75072472d+00	3.75072472d+00	2.91880772d-47 Z
1.97057597d-46	-3.75072472d+00	3.75072472d+00	-5.25385389d-47 Z

NETWORK POLES

REAL PART [RAD]	IMAG. PART [RAD]	POLE FREQ. [RAD]	POLE Q
-5.07120006d-01	-1.36845553d-48	5.07120006d-01	5.00000000d-01
-3.81129314d-01	6.86198581d-01	7.84938243d-01	1.02975318d+00
-3.81129314d-01	-6.86198581d-01	7.84938243d-01	1.02975318d+00
-1.29752204d-01	1.05027231d+00	1.05825684d+00	4.07799180d+00
-1.29752204d-01	-1.05027231d+00	1.05825684d+00	4.07799180d+00

18.7. GROUP DELAY

NETOPT has a very powerful theoretical method to calculate group delay for any network. Since this requires certain amount of additional computations which are not otherwise needed, the option must be requested by the SET GD command. We will demonstrate the computation and plotting of the group delay for the filter

NETWORK FIGURE 18.4.1
AC ANALYSIS
DELAY [MS] OF VOUT

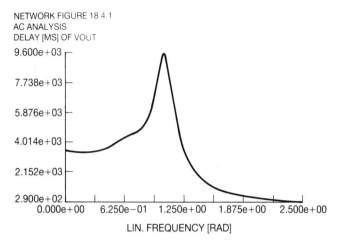

LIN. FREQUENCY [RAD]

FIGURE 18.7.1   Group delay of network Fig. 18.4.1.

Fig. 18.4.1. In the menu we request the computation as follows:

```
NET ,SPEC,AC  ,POLE,TIME,SYMB,TOL ,MC  ,OPT ,
PLOT,SAVE,SET ,SHOW,LIST,SYS ,HELP,QUIT    : set gd
```

The computer returns afterwards to the same menu and we select the choice AC, like for amplitude responses. When the program has finished and has displayed the same menu, we can request plotting and select the option GD:

```
Enter file name [no default] : ac.tmp
Enter response typo:
   DB (DB), MAgnitude (MA), PHase (PH), Group Delay
      (GD), REal (RE) or IMaginary part (IM) [D:DB] : gd
         Y axis scaling Manual (M) or Automatic (A) [D:A] :
```

The response is in Fig. 18.7.1.

18.8. SENSITIVITIES

Calculation of sensitivities of network functions by hand is extremely time consuming. Even worse, pole and zero sensitivities can be found by hand for network functions of at most second order. NETOPT uses special methods which work for high-order networks. The sensitivities are available for all responses, for the poles, zeros, their Q and ω_0. You can calculate the sensitivity to changes of as many elements as needed, even all of them. One important item: if you wish to calculate sensitivity with respect to ideal operational amplifiers, you *must* select the 1-graph option.

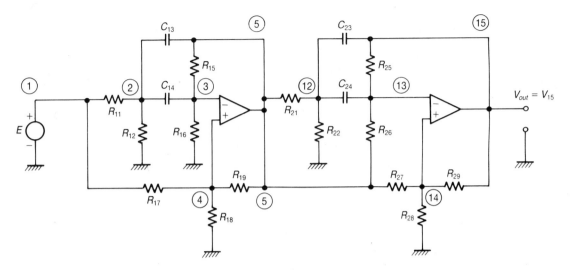

FIGURE 18.8.1   Active band-pass filter.

To demonstrate sensitivity calculations we take the active network in Fig. 18.8.1. It is a realistic example of a bandpass filter with the passband from 1 kHz to 1.2 kHz, maximum passband ripple of 0.3 dB and stopbands below 800 Hz and above 1.5 kHz. Minimum attenuation in the stopband is 22 dB, relative to the maximum gain in the passband. The input data are

```
4-TH ORDER BANDPASS FILTER
EIN      1      0                  1.00000000D+00
R11      1      2                  2.69670000D+04
R12      2      0                  7.25990000D+04
C13      2      5                  2.20000000D-09
C14      2      3                  2.20000000D-09
R15      3      5                  1.58478000D+05
R16      3      0                  2.26403000D+05
R17      1      4                  2.83330000D+04
R18      4      0                  4.25000000D+04
R19      4      5                  1.00000000D+05
N1       4      3      5      0    0.00000000D+00
C1A      3      0                  0.00000000D+00
C1B      4      0                  0.00000000D+00
R21      5     12                  2.69440000D+04
R22     12      0                  9.23800000D+04
C23     12     15                  2.20000000D-09
C24     12     13                  2.20000000D-09
R25     13     15                  2.13682000D+05
R26     13      5                  1.49834000D+05
R27      5     14                  1.50000000D+04
```

R28	14	0		6.00000000D+04
R29	14	15		1.00000000D+05
N2	14	13	15 0	0.00000000D+00
C2A	13	0		0.00000000D+00
C2B	14	0		0.00000000D+00
VOUT	15	0		0.00000000D+00

Note that in the file are also four capacitors (C1A, C1B, C2A, C2B) with zero values. We have not drawn them in Fig. 18.8.1. They do not influence the response but we can evaluate and plot sensitivities with respect to these elements and thus get an estimate of how they would influence the response if they were nonzero but very small. We suggest that you do this yourself. Operational amplifiers can be ideal, O, in which case no value is given, or nonideal, N, in which case the value is the gain–bandwidth product in Hz. If this value is set equal to zero, then the nonideal operational amplifier converts to an ideal one; this was used in the example. Figure 18.8.2 shows the overall frequency domain response of the filter, Fig. 18.8.3 is the plot of the passband only, both in dB. The group delay is plotted in Fig. 18.8.4. In order to calculate sensitivity due to changes in the operational amplifiers, we must request the 1-graph option. Figure 18.8.5 shows the sensitivity of the amplitude in dB with respect to N1. Note the trailing P in the description of the response. It indicates that the value of an element is zero (P, generally standing for parasitic). Similarly, a trailing P will appear for the above four-valued capacitors. Finally Fig. 18.8.6 shows sensitivity of the group delay (in milliseconds) to changes of N1.

Sensitivities of poles, zeros, their Q and their ω_0 are also available, but the printout is easy to interpret and we do not give it here.

FIGURE 18.8.2   Amplitude response (dB), network Fig. 18.7.1.

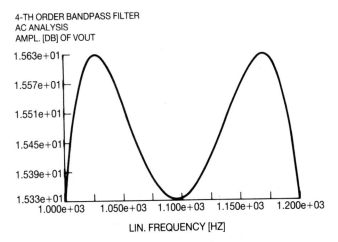

FIGURE 18.8.3   Pass-band response (dB), network Fig. 18.7.1.

FIGURE 18.8.4   Group delay, network Fig. 18.7.1.

4-TH ORDER BANDPASS FILTER
AC ANALYSIS
SEMI-NORMALIZED SENSITIVITY OF AMPL. [DB] OF VOUT W.R.T. N1 P

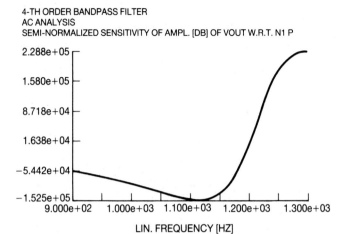

FIGURE 18.8.5   Sensitivity to the OPAMP, network Fig. 18.7.1.

4-TH ORDER BANDPASS FILTER
AC ANALYSIS
SEMI-NORMALIZED SENSITIVITY OF DELAY [MS] OF VOUT W.R.T. N1 P

FIGURE 18.8.6   Group delay sensitivity, network Fig. 18.7.1.

There is one very powerful sensitivity option built into NETOPT: the program can give maximum sensitivities in the whole frequency band for all selected elements (or all elements) and order them in magnitude. Thus one can immediately see which of the elements are critical and will perhaps require tighter element tolerances to build. Suppose that we want sensitivity with respect to all resistors. This is achieved as follows:

```
Select command:
NET ,SPEC,AC  ,POLE, TIME,SYMB,TOL ,MC  ,OPT ,
PLOT,SAVE,SET ,SHOW,LIST,SYS ,HELP,QUIT      : set sens

Want to enter element tolerance [Y/N D:Y] : n
Enter element name, HELP or DONE : r* (requests ALL
  resistors)
Enter element name, HELP or DONE : done
```

Run ac analysis, for instance between 800 Hz and 1.2 kHz. When the main menu appears again, give the command SHOW CRIT. The following will appear on the screen:

```
4-TH ORDER BANDPASS FILTER

AC ANALYSIS

SORTED MAXIMUM SENSITIVITIES OF MAGNITUDE OF VOUT
BETWEEN 8.00000d+02 HZ AND 1.20000d+03 HZ
```

ELEMENT	FREQUENCY [HZ]	MAGNITUDE	SENSITIVITY
R25	8.00000d+02	4.79957d-01	1.2303d+01
R15	1.16364d+03	6.04506d+00	6.1916d+00
R26	8.00000d+02	4.79957d-01	-5.5619d+00
R27	8.00000d+02	4.79957d-01	5.4862d+00
R21	8.00000d+02	4.79957d-01	4.6539d+00
R29	9.69697d+02	4.89597d+00	-4.3703d+00
R19	1.20000d+03	5.84377d+00	-4.1346d+00
R11	1.20000d+03	5.84377d+00	-3.0393d+00
R28	8.00000d+02	4.79957d-01	-2.6996d+00
R17	1.20000d+03	5.84377d+00	2.0808d+00
R18	1.20000d+03	5.84377d+00	2.0539d+00
R16	1.17172d+03	6.04851d+00	-1.6610d+00
R22	8.00000d+02	4.79957d-01	1.3582d+00
R12	1.20000d+03	5.84377d+00	-1.1290d+00

The largest sensitivity is due to resistor R25 at frequency 800 Hz, equal to 12.303, the next largest sensitivity occurs at 1.1636 kHz due to the changes in R15 and so on.

18.9. TIME RESPONSES

NETOPT can calculate time domain responses of networks to the Dirac impulse, to the step and to a pulse with specified width and infinitely steep sides. There are no approximations. In other available programs, like in SPICE, nonzero rise-time or fall-time must be given. The method does not need poles and residues of the network and is explained in the book by J. Vlach and K. Singhal *Computer Methods for Circuit Analysis and Design* Van Nostrand Reinhold 1982, Chapter 10.

Two choices are implemented, denoted by 0/2 or 8/10. We do not want to go into theoretical details and thus our suggestion is to test several runs with various step sizes. If two runs with different step sizes have identical responses, then the result is correct. A good starting point is the method 0/2 with 100 points for the time length you need. The 8/10 is accurate for considerably longer step sizes than the method 0/2 but takes longer to evaluate and there is no clear rule which is faster or better. The method cannot evaluate the starting point at $t = 0$ and the first available time point is at the first step.

For time domain analysis select the command TIME from the main menu and follow the suggested options. As an example we give in Fig. 18.9.1 the step response of the LC filter in Fig. 18.4.1.

It is also possible to give initial conditions. A positive value for a capacitor means that the initial voltage is $+$ at the first node given in the input data. If you give a negative value, then the polarity is reversed. A similar rule applies for the initial current of the inductor: a positive value means that the initial current flows from the first given node to the second. A negative value will represent reversed direction.

There is one more important point to mention in time domain analysis. For any analysis, the program always insists that there must be at least one external

NETWORK FIGURE 18.4.1
TIME DOMAIN ANALYSIS (0/2 APPROXIMATION)
STEP RESPONSE OF VOUT

FIGURE 18.9.1   Step response, filter Fig. 18.4.1.

source. This is reasonable in most cases because it prevents meaningless runs, but is not suitable when the network has only initial conditions and no external source. In such a case, to fool the program and still get a correct result, just add a current source across any two terminals and set its value equal to zero.

CHAPTER 19

Fourier Series and Transform

19.0. INTRODUCTION

Books on basic network theory usually include the Fourier series and Fourier integral theories, but the subject is not strongly related to network analysis. Both Fourier methods are always taught in courses on mathematics and we assume that the student either has been, or will be, exposed to them elsewhere. We cover only those points which have some importance for the material of this book.

In the previous chapters we encountered two types of signals. The first were purely sinusoidal and we discussed them in Chapter 8. We have seen that linear networks influence the amplitude and phase of sinusoidal signals passing through them. The other type were signals suddenly starting at $t = 0$ and we considered them in time domain, Chapter 10. In every example the output signal was different from the input signal and the networks had different responses to the same signals. We would like to know why this is so, and perhaps what can be done to either avoid such distortions, or to create distortions according to our wishes. To be able to do that we must have more information about the signals; Fourier series and Fourier integral are the tools which help in answering the above questions. Fourier series deals with periodic but non-sinusoidal signals. The Fourier integral gives answers for signals of arbitrary shapes, occurring only once. Both methods are the subject of this brief chapter.

19.1. FOURIER SERIES

The Fourier-series method tells us what types of components are present in a periodic but non-sinusoidal signal. As we know from Chapter 8, a periodic function satisfies the condition

$$f(t) = f(t + nT) \tag{19.1.1}$$

where T is the time of the period and n is an arbitrary integer. We also defined the frequency,

$$f = \frac{1}{T} \tag{19.1.2}$$

and the angular frequency

$$\omega = 2\pi f = \frac{2\pi}{T} \tag{19.1.3}$$

The theory of Fourier series states that any periodic function is composed of components having various amplitudes and phases, but *all* of them having frequencies which are integer multiples of the fundamental frequency, given by Eq. (19.1.2). The most frequently used form of the Fourier series is defined by the equations

$$f(t) = \frac{a_0}{2} + \sum_{n=1}^{\infty} (a_n \cos n\omega t + b_n \sin n\omega t) \tag{19.1.4}$$

where the coefficients are obtained by solving the formulas

$$a_0 = \frac{2}{T} \int_{t_0}^{t_0 + T} f(t)\, dt \tag{19.1.5}$$

$$a_n = \frac{2}{T} \int_{t_0}^{t_0 + T} f(t) \cos n\omega t\, dt \qquad n = 1, 2, \ldots \tag{19.1.6}$$

$$b_n = \frac{2}{T} \int_{t_0}^{t_0 + T} f(t) \sin n\omega t\, dt \qquad n = 1, 2, \ldots \tag{19.1.7}$$

Of special importance in the above equations is the integration interval from t_0 to $t_0 + T$. We can use an arbitrary point on the periodic function and let it be t_0. If we select $t_0 = 0$, then we will integrate from 0 to T. We can equally well set $t_0 = -T/2$ and integrate from $-T/2$ to $+T/2$. The name *fundamental harmonic* was given to the component with $n = 1$. The other components are called second, third ... harmonic, or just *higher harmonics*.

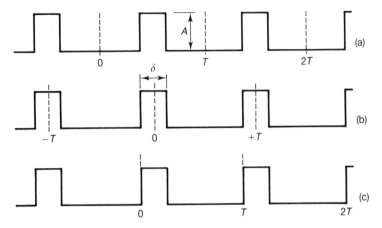

FIGURE 19.1.1   Three choices of origin, (a)–(c), for a periodic pulse.

Fourier series have very few practical restrictions, and we quote them for completeness:

a. $f(t)$ is a single-valued function.
b. The integral $\displaystyle\int_{t_0}^{t_0+T} |f(t)|\, dt$ is finite.
c. $f(t)$ has a finite number of discontinuities.

None of these conditions will be violated for any practical function.

Figure 19.1.1a–c shows an impulse of duration δ, repeated with period T. In a we selected the interval from the middle of the lower portion. In b we placed the origin in the middle of the impulse, while in c we placed it at the start of the impulse. This freedom of choice leads to considerable simplifications if the functions are even or odd. As is known from mathematics, an even function satisfies the equation

$$f(t) = f(-t)$$

and choice a or b makes the function in Fig. 19.1.1 an even function. An odd function satisfies

$$f(t) = -f(-t)$$

Consider next the two cases in Fig. 19.1.2. The choice of the origin in a makes the function an odd function. The choice in b makes it an even function.

We give here so much importance to these choices for one very practical reason. An even function will have all b_n coefficients equal to zero and we need not calculate them. Similarly, an odd function will have all a_n terms equal to zero and

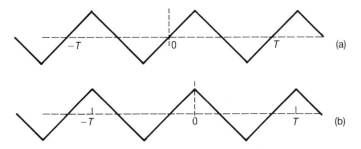

FIGURE 19.1.2   Periodic triangle: (a) As an odd function. (b) As an even function.

we can skip the calculations again. We also note that the coefficient $a_0/2$ is the average of the functions over the period T. Thus the functions in Fig. 19.1.2 will have $a_0 = 0$, because the area above the axis is the same as the area below. The integration to obtain the coefficients of even or odd functions simplifies as well: it is sufficient to calculate the integral over one half of the period and take the result twice.

As an example of a Fourier expansion consider Fig. 19.1.1b. The function has height A and is nonzero from $-\delta/2$ to $+\delta/2$. The coefficients will be

$$a_0 = \frac{2A}{T} \int_{-\delta/2}^{+\delta/2} dt = \frac{2A\delta}{T} \tag{19.1.8}$$

$$a_n = \frac{2A}{T} \int_{-\delta/2}^{+\delta/2} \cos n\omega t \, dt = \frac{2A}{T} \left[\frac{\sin n\omega t}{n\omega} \right]_{-\delta/t}^{\delta/2}$$

$$= \frac{4A}{T} \frac{\sin n\omega \dfrac{\delta}{2}}{n\omega} = \frac{2A\delta}{T} \frac{\sin n\omega \dfrac{\delta}{2}}{n\omega \dfrac{\delta}{2}} \tag{19.1.9}$$

The expansion is

$$f(t) = \frac{A\delta}{T} + \sum_{1}^{\infty} \frac{2A\delta}{T} \frac{\sin n\omega \dfrac{\delta}{2}}{n\omega \dfrac{\delta}{2}} \cos n\omega t \tag{19.1.10}$$

We will consider four cases with $\delta = T/2$, $T/4$, $T/6$ and $T/8$. For these cases the dc term, $a_0/2$, is equal to 1/2, 1/4, 1/6 and 1/8, respectively. The coefficients are given by

$$a_n = \frac{2A}{\pi} \frac{\sin \dfrac{\pi}{T} \delta}{n}$$

and all four cases are plotted in Fig. 19.1.3a–d for $A = 1$. The coefficients of the harmonic components are indicated as lines with circles at their ends. An envelope can be drawn by connecting the points. It will have the form $\sin x/x$, compare with

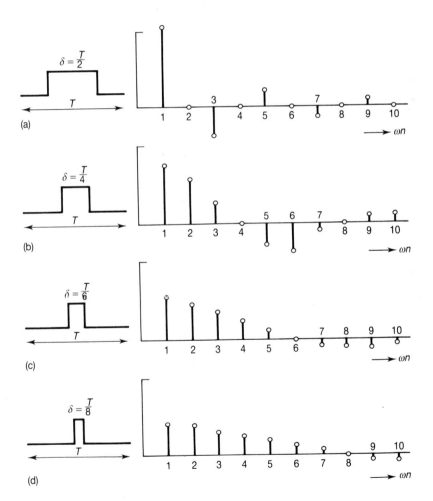

FIGURE 19.1.3   Spectra for the function in Fig. 19.1.1:

(a) $\delta = \dfrac{T}{2}$.   (b) $\delta = \dfrac{T}{4}$.   (c) $\delta = \dfrac{T}{6}$.   (d) $\delta = \dfrac{T}{8}$.

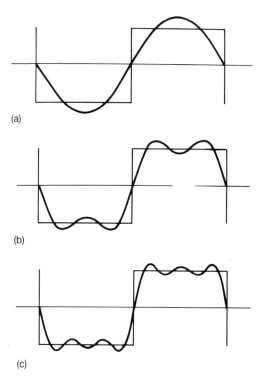

(a)

(b)

(c)

FIGURE 19.1.4   One period of a square wave and its: (a) First harmonic. (b) First two harmonics. (c) First three harmonics.

Eq. (19.1.10). Figure 19.1.4 shows approximation of the square wave with $\delta = T/2$ by a, one, b, two, and c, three harmonics.

Example 1.

Find the Fourier coefficients and the expansion for the waveform in Fig. 19.1.2a.

The function is an odd function, the areas above the horizontal axis are the same as those below and thus the coefficient $a_0 = 0$. In addition, the choice of the origin makes the function an odd one and only b_n coefficients need be considered. We will integrate only from 0 to $T/4$ and take the result twice.

For the interval from 0 to $T/4$ we have the function

$$y = kt \tag{19.1.11}$$

with the slope

$$k = \frac{4A}{T} \tag{19.1.12}$$

For the interval from $T/4$ to $T/2$ we have the function

$$y = -kt + 2A \qquad (19.1.13)$$

with k given by Eq. (19.1.12). Altogether we have to solve

$$b_n = 2 \times \frac{2}{T}\left[\int_0^{T/4} t \sin n\omega t + \int_{T/4}^{T/2} (-kt + 2A) \sin n\omega t\right] \qquad (19.1.14)$$

where the first coefficient, 2, comes from the fact that we treat the function as an odd function and integrate only over half of the period. In any book on mathematical formulas we find the solution for the indefinite integrals

$$\int t \sin n\omega t = \frac{\sin n\omega t}{n^2\omega^2} - \frac{t \cos n\omega t}{n\omega}$$

and

$$\int \sin n\omega t = -\frac{\cos n\omega t}{n\omega}$$

Somewhat lengthy but simple evaluations for the bounds in Eq. (19.1.14) lead to the result

$$b_n = \frac{8A}{n^2\pi^2} \sin n\frac{\pi}{2}$$

All even coefficients are zero. The odd coefficients are

$$b_n = \frac{8A}{n^2\pi^2}(-1)^{(n-1)/2} \qquad n = 1, 3, 5, 7, \ldots \qquad (19.1.15)$$

and the Fourier expansion is

$$f(t) = \frac{8A}{\pi^2}\left[\frac{1}{1}\sin \omega t - \frac{1}{9}\sin 3\omega t + \frac{1}{25}\sin 5\omega t - \frac{1}{49}\sin 7t + \cdots\right] \qquad (19.1.16)$$

Fourier expansions of periodic functions lead to evaluations of integrals with trigonometric functions and such evaluations are always time consuming. It is not the purpose of this book to practice such integrations. All we wish to achieve is to give some understanding without placing too much emphasis on the mathematical details.

Decomposition of any non-sinusoidal periodic signal into its components is not only a mathematical abstraction; the harmonic components do exist and can be measured, or heard. If different musical instruments play the same steady tone, we hear somewhat different sounds. The difference is caused by different amplitudes of the higher harmonic components generated by the various instruments.

Another important point is the speed with which the coefficients decrease. Our first case, Fig. 19.1.1, has the coefficients given in Eq. (19.1.10). The signal has abrupt changes (is not continuous) and amplitudes of the coefficients decrease slowly. Our second example, Fig. 19.1.2, is a continuous signal and its coefficients decrease much more rapidly, see Eq. (19.1.16). As a rule of thumb we will remember that abrupt changes of the signal result in slowly decreasing coefficients. We will have to take more terms of the expansion if we wish to approach the shape of the original function.

Another point worth noticing is the following. Suppose that we have a signal of constant shape, for instance like the impulses in Fig. 19.1.1. The width of the impulse will be constant but we will start increasing the time of the period, T. The amplitude will be influenced, but more importantly the spectral lines will start getting closer if the period is increased. If the period becomes very long, the spectral lines will get very close. We can expect that if the period T stretches to infinity, the spectral lines will merge into one continuous spectrum. This is, indeed, the case and the appropriate mathematical theory is the subject of the next section.

19.2. FOURIER TRANSFORM

The Fourier transform studies cases of individual signals which do not repeat. It can be considered as a coupling link between the Fourier series and the Laplace transform theory. Let us first discuss informally what is involved. Similarly to the Laplace transform, the Fourier transform is defined by a pair of integrals. The time function is denoted by $f(t)$ and the Fourier transform, also called the *spectrum*, by $S(\omega)$. The two functions are related by the expressions

$$S(\omega) = \int_{-\infty}^{+\infty} f(t)e^{-j\omega t}\, dt \tag{19.2.1}$$

$$f(t) = \frac{1}{2\pi} \int_{-\infty}^{+\infty} S(\omega)e^{+j\omega t}\, d\omega \tag{19.2.2}$$

In all applications relating to this book the signal starts at $t = 0$ and thus the lower limit of the integral in Eq. (19.2.1) can be changed to 0. If we accept this restriction, then the definition of the Laplace transform, $F(s)$, Eq. (9.1.3), and the Fourier transform, Eq. (19.2.1), differ only in the exponent of the exponential: the first one uses s while the second uses $j\omega$. There are some subtle differences but we will not go into such details. Other Laplace transform formulas also remain valid. For instance, time shift of a function by an amount T is expressed by

$$f(t - T) = e^{-j\omega T} S(\omega)$$

like in Eq. (9.5.16).

Considerable theoretical differences exist between the inverse transformations Eq. (9.1.4) and Eq. (19.2.2), but we never used the inverse Laplace transformation and we will not need the inverse of the Fourier transformation either. In this book we give them only for the sake of completeness. Because of the similarities of the

FIGURE 19.2.1   Spectrum of the Dirac impulse.

first formulas, we already have spectra for most practical signals: All we have to do is use our Laplace transform derivations and substitute $j\omega$ for s.

Take as the first example the Dirac impulse. Its Laplace transform was $F_\delta(s) = 1$, a constant. The spectrum, shown in Fig. 19.2.1, will also be a constant and thus *all* frequencies, up to infinity, are present with the same amplitude.

For the unit step, $u(t)$, we derived the Laplace transform $F_u(s) = 1/s$; substituting $s = j\omega$ we get the spectrum

$$S_u(\omega) = \frac{1}{j\omega} = -j\frac{1}{\omega}$$

All components are shifted by $-90°$ (due to the $-j$). The absolute value

$$|S_u(\omega)| = \left|\frac{1}{\omega}\right| \tag{19.2.3}$$

gives amplitudes of the components at the various frequencies and is plotted in Fig. 19.2.2. It is a continuous curve indicating that all frequencies exist, theoretically up to infinity. Low frequencies are prominently present and the amplitudes get smaller as we move to higher frequencies.

Consider next the pulse, $p(t)$, for which we have derived in Eq. (9.2.3)

$$F_p(s) = \frac{1 - e^{-sT}}{sT}$$

Substituting $s = j\omega$

$$S_p(\omega) = \frac{1 - e^{-j\omega T}}{j\omega T} = \frac{\sin \omega T}{\omega T} + j\frac{\cos \omega t - 1}{\omega T}$$

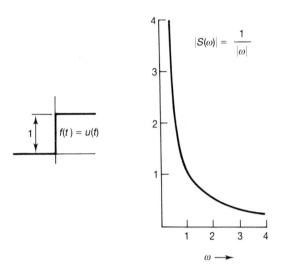

FIGURE 19.2.2   Spectrum of the unit step.

The absolute value is

$$|S_p(\omega)| = \left| \frac{1 - \cos \omega T + j \sin \omega T}{j\omega T} \right| = \left| \frac{[2(1 - \cos \omega T)]^{1/2}}{\omega T} \right| = \left| \frac{\sin \omega \dfrac{T}{2}}{\omega \dfrac{T}{2}} \right|$$

The plot of the absolute value is in Fig. 19.2.3. Note that frequencies for which $\sin (\omega T/2) = 0$ will not appear in the spectrum. For $\omega = 0$ we find the value of 1 by applying the L'Hospital rule.

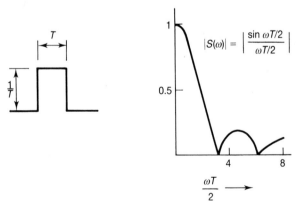

FIGURE 19.2.3   Spectrum of an impulse having width T and height $1/T$.

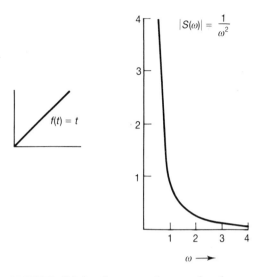

FIGURE 19.2.4   Spectrum of a ramp function.

The Laplace transform of the ramp function was found to be $F_r(s) = 1/s^2$ and thus the absolute value of the spectrum is

$$|S_r(\omega)| = \frac{1}{\omega^2} \qquad (19.2.5)$$

The spectrum is shown in Fig. 19.2.4; the amplitudes of its components decrease faster than those for the unit step, but otherwise the curve is similar.

Suppose that we take the difference of two ramp functions, one starting at $t - 0$ and one delayed by the amount T. For $r(t) = t/T$, this combined function will have a linear edge, followed by a constant value equal to 1, see Fig. 19.2.5. Its Laplace transform is

$$F(\omega) = \frac{1 - e^{-sT}}{Ts^2}$$

Substituting $s = j\omega$ we get the spectrum

$$|S(\omega)| = \left| \frac{1}{\omega} \frac{\sin \omega \dfrac{T}{2}}{\omega \dfrac{T}{2}} \right| \qquad (19.2.6)$$

The curve decreases faster than in the case of a unit step, because the time function

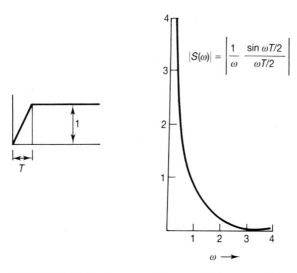

FIGURE 19.2.5   Spectrum of a step with linear transition.

is now continuous. This observation can be extended to a general rule, similarly as in the case of Fourier series: the smoother the curve of the signal, the faster the decrease of its spectral function at higher frequencies.

We will consider two more simple signals, first the function $\sin dt$, suddenly applied at $t = 0$. The Laplace transform was derived in Eq. (9.2.10). Substituting $s = j\omega$ and taking the absolute value,

$$|S(\omega)| = \left| \frac{d}{d^2 - \omega^2} \right| \tag{19.2.7}$$

The spectrum is sketched in Fig. 19.2.6 for $d = 4$.

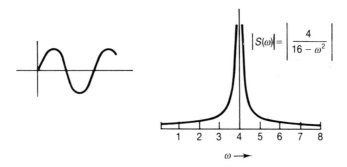

FIGURE 19.2.6   Spectrum of a switched sine function.

For cos dt we derived Eq. (9.2.11) and the spectrum is

$$|S(\omega)| = \left| \frac{\omega}{d^2 - \omega^2} \right| \qquad (19.2.8)$$

It is similar to Fig. 19.2.5, but is zero at $\omega = 0$.

Let us now discuss what conclusions we can draw from the above results. First of all we see that all suddenly applied non-periodic signals have a continuous spectrum of frequencies. This is particularly striking in the case of a sine or cosine signal. If these functions are not switched, that is if they started at $t \to -\infty$ and continued for all t unperturbed, their spectrum is just one single line. Once they are suddenly switched, there is a complete spectrum of frequencies, as seen from Eq. (19.2.7) or Eq. (19.2.8). Compare similarly the spectra of the unit step and of the signal composed of the two ramp functions. The smoother change of the signal (see Fig. 19.2.5) resulted in smaller spectral components at higher frequencies.

As was the case in periodic functions, the frequencies in the spectrum actually exist. An imperfect but useful explanation can be found in nature by considering lightning. It is a short pulse of electromagnetic energy. If we have tuned our radio to an AM station, we will hear those spectral components, no matter what station we are tuned to. If we tune to the FM channel, we may not hear them. The FM frequency band is in the high-frequency range where the spectral components may already be so small that they are lost below the noise level.

One additional practical question is still to be answered: what properties should a network have in order *not* to distort the incoming signal? Rather obviously we would ask for the same transfer of all amplitudes, but this is not enough. We must also delay all frequencies by the same amount of time and this corresponds to a phase response which is a straight line passing through the origin of the axes at $\omega = 0$. Since the group delay is the derivative of the phase response and the derivative of a straight line is a constant, it follows that a non-distorting network must also have a constant group delay for all frequencies. No network composed of the components discussed in this book can simultaneously satisfy the conditions of constant amplitude and constant group delay. We can only approach such a situation by a design which satisfies the requirements for a significant part of the spectrum.

PROBLEMS CHAPTER 19

P.19.1 Find the Fourier series. The function has a nonzero a_0 but all other a_n coefficients will be zero. Explain.

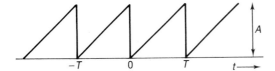

P.19.2 Find the Fourier series for the one-phase rectifier.

P.19.3 Find the Fourier series for the two-phase rectifier.

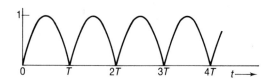

P.19.4 Calculate the spectrum of $f(t) = e^{-at}u(t)$ and plot its absolute value for $a = 1$.

P.19.5 Calculate the spectrum of $f(t) = e^{-at}\cos\omega t$ and plot its absolute value for $a = 1$ and $\omega = 2$.

APPENDIX A

Solution of Linear Systems

The program LINSYS.EXE solves systems of linear equations, the largest size being 20. Suppose that you wish to solve the system

$$\begin{bmatrix} 3 & 3 & 0 \\ 0 & 2 & 3 \\ 5 & 0 & 2 \end{bmatrix} \begin{bmatrix} x_1 \\ x_2 \\ x_3 \end{bmatrix} = \begin{bmatrix} 2 \\ 1 \\ 2 \end{bmatrix}$$

Input data can be given interactively or from a file. In the interactive option just follow the suggestions of the program. If the input is from a file, then you must first prepare it using any editor available on your computer. The file can have a name with up to 20 characters. On the first line is the size of the matrix, on the following lines are entries of the matrix by rows. The last line represents the right-hand-side vector. For the above example the file will have the form

```
3
3.  3.  0.
0.  2.  3.
5.  0.  2.
2.  1.  2.
```

The command LINSYS loads the program. The result will appear on the screen

of your terminal and will also be written into the file RESULT. For the above
example it will be

```
***THE MATRIX IS:
        3.00000        3.00000        0.00000
        0.00000        2.00000        3.00000
        5.00000        0.00000        2.00000
***THE RIGHT HAND SIDE VECTOR IS:
        2.00000        1.00000        2.00000
SOLUTION
********
*- THE DETERMINANT IS:        57.00000

*- THE NUMERATORS ARE:
        20.00000        18.00000        7.00000

*- THE SOLUTION VECTOR IS:
        0.35088        0.31579        0.12281
```

The program uses an efficient numerical method with full pivoting but prints
the solution as if it has been obtained by Cramer's rule: first comes the determinant
of the system (the denominator), then all numerators and finally all solution
variables. For the above example

$$x_1 = \frac{20}{57} = 0.35088$$

$$x_2 = \frac{18}{57} = 0.31579$$

$$x_3 = \frac{7}{57} = 0.12281$$

You can use the program repeatedly for new systems without exiting but if you
need to inspect or print the file RESULT, you must exit LINSYS and do so
before running another problem. It is advantageous to use the file version for
larger problems because a typing error cannot be corrected interactively and you
must start the problem again.

APPENDIX B

Responses of Network Functions

The program MAIN.EXE calculates and plots time domain and frequency domain responses of networks given either by their zeros and poles or by their network functions. You can rename the program, if you wish. The input can be either interactive or from a file. Suppose that we have a network function given in two forms:

$$F = \frac{(s + j3)(s - j3)}{(s + 1)(s + 1 + j2)(s + 1 - j2)} = \frac{s^2 + 9}{s^3 + 3s^2 + 7s + 5}$$

The numerator has degree 2, the denominator degree 3. We can prepare files with names having up to eight characters. If we call the file for the poles and zeros "pz", then this file will have the following form for the above example:

```
2           (degree of the numerator)
0   3       (first zero)
0  -3       (its complex conjugate)
3           (degree of the denominator)
-1   0      (real pole)
-1   2      (complex pole)
-1  -2      (its complex conjugate)
```

The text in brackets is for explanation only and is not part of the file. For the network function we select another file name, for instance "tr". The file will have the form

```
2            (degree of the numerator)
9 0 1        (coefficients)
3            (degree of the denominator)
5 7 3 1      (coefficients)
```

The coefficients must be given in the sequence of increasing powers of *s*.

The command **MAIN** loads the program which displays the first menu:

```
       TIME AND FREQUENCY RESPONSE
          FROM POLES AND ZEROS
       OR FROM THE NETWORK FUNCTION
     * * * * * * * * * * * * * * * * * * * * * * * * * * * * * * * * * *
YOU HAVE THE FOLLOWING INPUT OPTIONS
(1) Poles and zeros given interactively
(2) Poles and zeros given in a file
(3) Network function given interactively
(4) Network function given in a file
(5) Exit
     Select the Option (1-5):
```

The interactive options (1) and (3) give questions which we simply answer. Make sure that complex conjugate poles or zeros are *exactly* complex conjugate, to the last digit. Option (2) can use the above file "pz", option (4) the file "tr". After this the program displays input data (for checking) and the second menu appears:

```
     PLOT MENU
     * * * * * * * *

     TIME RESPONSE
     = = = = = = = = = = = = =
(1) Impulse Response
(2) Unit Step Response
(3) Sinusoidal Response
(4) Cosinusoidal Response
     FREQUENCY RESPONSE
     = = = = = = = = = = = = = = = = = = -
(5) Decibel (dB)
(6) Magnitude
(7) Group Delay
(8) Phase
(9) Return to Main Menu
     Select the Option (1-9):
```

Select option and follow instructions displayed on the screen.

The program does not accept multiple poles but this can be remedied quite simply by giving slightly different poles. For instance, should there be a double pole at $s = -1$ then we can give one of the poles as $s = -1$ and the other one as $s = -1.00001$.

Plotting requires information about the hardware actually installed in the computer. It is necessary to prepare a file GRAPHICS.DEF, placed into the same directory as the program MAIN. This file contains one of the following letters:

```
NO        (for PC without graphics equipment)
CGA       (for IBM CGA interface)
EGAM      (for EGA interface monochrome monitor)
EGAC      (for EGA or VGA interface color monitor)
HERC      (for Hercules graphics card and monitor)
```

It is the same file as you must prepare for running NETOPT.

Each of the nine options of the second menu also write the numerical results to the disk. The names of the files are: (1) impulse, (2) step, (3) sine, (4) cosine, (5) dB, (6) magnitude, (7) delay, (8) phase. The files are overwritten if you run another problem. To get them you must exit (option 5 of the main menu) and inspect or print them before loading MAIN for another problem.

The program has many checks which prevent you from entering unacceptable data and gives interactive instructions on how to use it.

APPENDIX C

LU Decomposition

Numerical solution of linear systems is usually done by the **LU** or *triangular* decomposition. It may take several forms but we will consider only one. The method is especially useful when we have a problem with one matrix but with several right-hand-side vectors. The expensive operations on the matrix are done only once while the cheaper forward and back substitutions are repeated for different right-hand sides.

Consider a general linear system

$$\mathbf{Ax} = \mathbf{b} \tag{C.1}$$

The **LU** decomposition will crease a product of two matrices, $\mathbf{A} = \mathbf{LU}$, and thus modify Eq. (C.1) into

$$\mathbf{LUx} = \mathbf{b} \tag{C.2}$$

For a 3 × 3 matrix the decomposition will be

$$\mathbf{L} = \begin{bmatrix} l_{11} & 0 & 0 \\ l_{21} & l_{22} & 0 \\ l_{31} & l_{32} & l_{33} \end{bmatrix} \tag{C.3}$$

and

$$\mathbf{U} = \begin{bmatrix} 1 & u_{12} & u_{13} \\ 0 & 1 & u_{23} \\ 0 & 0 & 1 \end{bmatrix} \tag{C.4}$$

We break the problem in Eq. (C.2) into the solution of two triangular systems. First we define

$$\mathbf{Ux} = \mathbf{z} \tag{C.5}$$

At this moment we cannot solve this system because neither $\mathbf{x}$ nor $\mathbf{z}$ are known. However, we can replace $\mathbf{Ux}$ in Eq. (C.2) by $\mathbf{z}$

$$\mathbf{Lz} = \mathbf{b} \tag{C.6}$$

This equation can be solved to obtain the auxiliary vector $\mathbf{z}$. Once it is known, we go to Eq. (C.5) and solve this second triangular system to get $\mathbf{x}$. The steps in Eq. (C.6) are called the *forward* substitution while those in Eq. (C.5) are the *back* substitution.

The decomposition operates on the original matrix and overwrites entries of this matrix. We will use a small example to derive the algorithm. First we multiply the two matrices to get the product

$$\mathbf{A} = \mathbf{LU} \tag{C.7}$$

or

$$\begin{bmatrix} a_{11} & a_{12} & a_{13} \\ a_{21} & a_{22} & a_{23} \\ a_{31} & a_{32} & a_{33} \end{bmatrix} = \begin{bmatrix} l_{11} & l_{11}u_{12} & l_{11}u_{13} \\ l_{21} & l_{21}u_{12} + l_{22} & l_{21}u_{13} + l_{22}u_{23} \\ l_{31} & l_{31}u_{12} + l_{32} & l_{31}u_{13} + l_{32}u_{23} + l_{33} \end{bmatrix}$$

Comparison of both sides reveals that $a_{i1} = l_{i1}$ and nothing has to be done with the first column. Consider next the first row. Clearly, if we divide the entries (1, 2) and (1, 3) by l_{11}, then the first row will already have the terms of the decomposed matrix. Since we are operating on the original matrix, this is written as the instruction

$$a_{1j} = \frac{a_{1j}}{a_{11}} \qquad j = 2, 3, \ldots$$

As a result of this operation, our intermediate matrix will have the form

$$\begin{bmatrix} l_{11} & u_{12} & u_{13} \\ l_{21} & l_{21}u_{12} + l_{22} & l_{22}u_{13} + l_{22}u_{23} \\ l_{31} & l_{31}u_{12} + l_{32} & l_{31}u_{13} + l_{32}u_{23} + l_{33} \end{bmatrix}$$

At this point look at the entries which were not yet changed. In each of them is a term which is the product of the top element in the same column and the left-most element in the same row. If we form the product of these entries and subtract it from the entry in the matrix, we simplify the term. For instance, taking u_{12} and l_{21} and subtracting their product from the (2, 2) entry will result in l_{22}. Doing so for the whole remaining matrix we will modify the matrix to

$$\begin{bmatrix} l_{11} & u_{12} & u_{13} \\ l_{21} & l_{22} & l_{22}u_{23} \\ l_{31} & l_{32} & l_{32}u_{23} + l_{33} \end{bmatrix}$$

All these operations are equivalent to

$$a_{ij} = a_{ij} - a_{1j}a_{i1} \qquad i = 2, 3, \ldots \qquad j = 2, 3, \ldots$$

In order to continue it is now advantageous to prepare a piece of paper with such a cut-out that we can cover the first row and first column of the matrix but see the rest. The covered first row and column will not change any more. The process is now repeated with the remaining matrix exactly as above, only now it is a smaller matrix and the subscripts are shifted. First we divide the entry (2, 3) by l_{22} to obtain the intermediate result

$$\begin{bmatrix} l_{11} & u_{12} & u_{13} \\ l_{21} & l_{22} & u_{23} \\ l_{31} & l_{32} & l_{32}u_{23} + l_{33} \end{bmatrix}$$

Now with the first row and column covered with the cut-out we go to the top and to the left, form the product $l_{32}u_{23}$ and subtract it from the (3, 3) entry, with the result

$$\begin{bmatrix} l_{11} & u_{12} & u_{13} \\ l_{21} & l_{22} & u_{23} \\ l_{31} & l_{32} & l_{33} \end{bmatrix}$$

For a 3 × 3 system the decomposition is finished. It should be stressed again that this is a special way of writing the two matrices in place of the original one, which is destroyed. The units of the **U** matrix are not stored; they will be taken care of later. The student should at this moment write a few larger **L** and **U** matrices, multiply them and apply the above steps as a practice.

We now summarize the algorithm:

1. Set $k = 1$
2. For $j = k + 1, k + 2, \ldots, n$ calculate $a_{kj} = a_{kj}/a_{kk}$
3. For $i = k + 1, k + 2, \ldots, n$ and $j = k + 1, k + 2, \ldots, n$ calculate $a_{ij} = a_{ij} - a_{ik}a_{kj}$
4. If $j = n$ stop; else set $k = k + 1$ and go to 2.

In FORTRAN this decomposition has the code

```
C...LU DECOMPOSITION
      NM1=N-1
      DO 1 K=1,NM1
      KP1=K+1
      DO 1 J=KP1,N
      A(K,J)=A(K,J)/A(K,K)
      DO 1 I=KP1,N
1     A(I,J)=A(I,J)-A(I,K)*A(K,J)
```

This simple algorithm does not work always. We will have a problem if $a_{kk} = 0$ when we are in point 2 of the algorithm: division by zero is prohibited. The remedy is to select *any lower row* in which the same column does not have a zero, and *interchange* these two *complete* rows. The right-hand side must also be interchanged. We often select for the interchange a row which has the largest number (in absolute value) in the same column. This choice increases accuracy and is called *partial pivoting*. Should we select the largest number (in absolute value) in the whole remaining matrix, then we would perform *full pivoting*. Full pivoting increases accuracy and makes the algorithm robust. The subroutines on the diskette use full pivoting.

After the decomposition the determinant is also available: it is the product of entries on the main diagonal:

$$|\mathbf{A}| = \prod_{i=1}^{n} a_{ii} \tag{C.8}$$

Programming for the determinant is done in one DO loop:

```
C...CALCULATE THE DETERMINANT
      DET=1.
      DO 10 I=1,N
10    DET=DET*A(I,I)
```

Once the **LU** decomposition is finished, we must first solve the system Eq. (C.6).

In our example this will be

$$
\begin{bmatrix} l_{11} & 0 & 0 \\ l_{21} & l_{22} & 0 \\ l_{31} & l_{32} & l_{33} \end{bmatrix} \begin{bmatrix} z_1 \\ z_2 \\ z_3 \end{bmatrix} = \begin{bmatrix} b_1 \\ b_2 \\ b_3 \end{bmatrix}
$$

Starting from the top we get $z_1 = b_1/l_{11}$. When we move to the second equation, we already know z_1 and thus the term $l_{21}z_1$ can be transferred to the right side, with the result

$$
z_2 = (b_2 - l_{21}z_1)/l_{22}
$$

Since now both z_1 and z_2 are known, the third equation can be solved as

$$
z_3 = (b_3 - l_{31}z_1 - l_{32}z_2)/l_{33}
$$

The whole algorithm for an arbitrary size of the matrix will have the form:

$$
z_1 = \frac{b_1}{l_{11}}
$$

$$
z_i = \left(b_i - \sum_{j=1}^{i-1} l_{ij}z_j \right) \bigg/ l_{ii} \qquad i = 2, 3, \ldots, n
$$

This process is called the forward substitution. Since we have done all operations on the original matrix, we should modify the algorithm to

$$
z_1 = \frac{b_1}{a_{11}}
$$

$$
z_i = \left(b_i - \sum_{j=1}^{i-1} a_{ij}z_j \right) \bigg/ a_{ii} \qquad i = 2, 3, \ldots, n
$$

(C.9)

In FORTRAN, the algorithm is

```
C...FORWARD SUBSTITUTION
      Z(1)=B(1)/A(1,1)
      DO 2 I=2,N
      IM1=I-1
      SUM=0.
      DO 3 J=1,IM1
3     SUM=SUM+A(I,J)*Z(J)
2     Z(I)=(B(I)-SUM)/A(I,I)
```

At this moment we know the vector **z** and we turn our attention to Eq. (C.5). For our example

$$\begin{bmatrix} 1 & u_{12} & u_{13} \\ 0 & 1 & u_{23} \\ 0 & 0 & 1 \end{bmatrix} \begin{bmatrix} x_1 \\ x_2 \\ x_3 \end{bmatrix} = \begin{bmatrix} z_1 \\ z_2 \\ z_3 \end{bmatrix}$$

Note that we are now considering the units on the main diagonal, although they are not stored in the decomposed matrix **A**. Starting with the *last* equation we have $x_3 = z_3$. Moving up one row we have $x_2 + u_{23}x_3 = z_2$ and since we already know x_3, this can be rewritten as

$$x_2 = z_2 - u_{23}x_3$$

Going up to the first equation we have

$$x_1 + u_{12}x_2 + u_{13}x_3 = z_1$$

and since we now know x_2 and x_3, we can rewrite as

$$x_1 = z_1 - u_{12}x_2 - u_{13}x_3$$

It is easy to establish the algorithm

$$x_n = z_n$$

$$x_i = z_i - \sum_{j=i+1}^{n} u_{ij}x_j \qquad i = n-1, n-2, \ldots, 1$$

Because the decomposition overwrote the original entries, we modify the formula to

$$x_n = z_n$$

$$x_i = z_i - \sum_{j=i+1}^{n} a_{ij}x_j \qquad i = n-1, n-2, \ldots, 1 \tag{C.10}$$

In FORTRAN the algorithm is

```
C...BACK SUBSTITUTION
      DO 4 I=NM1,1,-1
      X(N)=Z(N)
      SUM=0.
      IP1=I+1
      DO 5 J=IP1,N
5     SUM=SUM+A(I,J)*X(J)
4     X(I)=Z(I)-SUM
```

The routines on the diskette, LUCFUL and SOCFUL, work with complete pivoting and are written for matrices with complex entries. First call LUCFUL (LU decomposition, Complex, Full pivoting). A is the matrix, N its size, DET is the returned value of the determinant, PROW and PCOL are vectors needed in the following solution phase, IER contains information about the matrix and IA is the size of the largest problem the user intends to use. If IER = 0, the matrix was decomposed successfully, if IER = 1 then the matrix is singular (the determinant is zero). Afterwards call SOCFUL (Solution, Complex, Full pivoting). In this routine A is the decomposed matrix coming from LUCFUL, Y is the right-hand-side vector, X is the solution vector. N, PROW, PCOL and IA have the same meaning as above.

Solutions to Problems

SOLUTIONS CHAPTER 1

It is a good idea to work with fractions and not with a calculator, since this avoids round-off errors. It is preferable to leave the results as fractions. Use the calculator only in the last stage, if necessary.

P.1.1 (a) $R = \dfrac{8}{15}\,\Omega$. (b) $R = 2\,\Omega$. (c) $G = \dfrac{14}{3}\,S$ or $R = \dfrac{3}{14}\,\Omega$. (d) $G = \dfrac{9}{7}\,S$ or $R = \dfrac{7}{9}\,\Omega$. (e) $R = \dfrac{8}{31}\,\Omega$.

(f) $R = \dfrac{13}{8}\,\Omega$. (g) $R = 6\,k\Omega$. (h) $R = \dfrac{23}{6}\,\Omega$.

(i) $R = \dfrac{41}{9}\,\Omega$.

P.1.2 (a) $R = \dfrac{3}{2}\,\Omega$. (b) $R = \dfrac{21}{4}\,\Omega$. (c) $R = \dfrac{13}{3}\,\Omega$.

(d) $R = \dfrac{9}{17}\,\Omega$. (e) $R = \dfrac{20}{9}\,\Omega$.

P.1.3 (a) $R = \dfrac{31}{5}\,\Omega$. (b) $R = 800\,\Omega$. (c) same as (b).

(d) $R = 625\,\Omega$. (e) same as (d).

P.1.4 (a) 7 V, the connection is possible. (b) 4 A, the connection is possible. (c) 3 A. (d) 7 V.

P.1.5 $V = 4$ V.

P.1.6 (a) $I_{R1} = 2.5$ A, $I_{R2} = 4$ A, $V_{R1} = 5$ V, $V_{R2} = 12$ V. (b) $I_{R1} = 3$ A, $I_{R2} = 1$ A, $I_{R3} = 4$ A, $V_{R1} = 6$ V, $V_{R2} = 2$ V, $V_{R3} = 12$ V.

P.1.7 $I = 0.4$ A.

P.1.8 $V = 40$ V.

P.1.9 $I = 0.5$ A, the left source delivers, the right consumes power.

P.1.10 $I_R = 2$ A, $P_R = 8$ W, $P_E = 12$ W (consumed), $P_J = 20$ W (delivered).

P.1.11 (a) $V_R = 6$ V, $I_R = 2$ A, $P_R = 12$ W, $P_E = 8$ W (delivered), $P_J = 4$ W (delivered). (b) $V_R = 6$ V, $P_R = 12$ W, $P_E = 8$ W (delivered), $P_J = 4$ W (delivered).

P.1.12 $I_R = \dfrac{1}{2}$ A, $P_R = \dfrac{5}{2}$ W, $I_E = 3.5$ A (pointing

down), $P_E = 3.5 \times 5 = 17.5$ W   (consumed),
$P_J = 4 \times 5 = 20$ W (delivered).

P.1.13 $I_R = 4$ A, $P_R = 160$ W, $P_{J1} = 40$ W,
$P_{J2} = 120$ W.

P.1.14 (a) $P_J = 300$ W (delivered), $P_{R1} = 20$ W,
$P_{R2} = 250$ W, $P_E = 30$ W (consumed). (b) $V = 12$ V,
$V_{R1} = 15$ V, $V_{R2} = 12$ V, $V_{R3} = 12$ V, $P_{R1} = 45$ W,
$P_{R2} = 72$ W, $P_{R3} = 36$ W, $P_{J1} = 81$ W, $P_{J2} = 72$ W
(both deliver power).

P.1.15 $V = \dfrac{4}{5}$ V, $I_{G1} = \dfrac{8}{5}$ A, $I_{G2} = \dfrac{12}{5}$ A, $P_{G1} =$

$\dfrac{32}{25}$ W, $P_{G2} = \dfrac{48}{25}$ W, $P_{J1} = \dfrac{12}{5}$ W, $P_{J2} = \dfrac{8}{5}$ W

(both deliver power), $P_{J3} = \dfrac{4}{5}$ W (consumed).

P.1.16 $I = \dfrac{6}{5}$ A, $V_{R1} = \dfrac{12}{5}$ V, $V_{R2} = \dfrac{18}{5}$ V,

$P_{R1} = \dfrac{72}{25}$ W, $P_{R2} = \dfrac{108}{25}$ W, $P_{E1} = \dfrac{18}{5}$ W, $P_{E2} =$

$\dfrac{24}{5}$ W (both deliver power), $P_{E3} = \dfrac{6}{5}$ W (consumes

power).

P.1.17 $I_E = 6$ A, $P_E = 156$ W, $V_{R1} = 18$ V,
$V_{R2} = V_{R3} = 8$ V, $I_{R1} = 6$ A, $I_{R2} = 4$ A, $I_{R3} = 2$ A,
$P_{R1} = 108$ W, $P_{R2} = 32$ W, $P_{R3} = 16$ W.

P.1.18 $I_{R1} = 2.5$ mA, $I_{R2} = 5$ mA, $I_{R3} = 2$ mA.

P.1.19 Without the device $V = 8$ V. With the
device $V = 7.90514$ V.

P.1.20 (a) Without the device $V = 41.6667$ V. With
the device $V = 38.4615$ V. (b) Without the device
$V = 5$ V. With the device $V = 4.54545$ V.

P.1.21 (a) Since the source is a current source,
it does not matter what we connect, the current
will still be the same. (b) Without the device
$I = 0.5$ mA. With the device $I = 0.4762$ mA.

P.1.22 Without the device $I = 4$ mA. With the
device $I = 3.99$ mA.

P.1.23 Let the resistance of the bottom resistor
be R.   Then   the   total   resistance   is
$(1 + 3 + 1 + 2 + 3)R = 10R$. $I = 10/10R = 1/R$.
The largest resistor is $3R$, the voltage across it will
be $V_{max} = 3R/R = 3$ V. Maximum power will be
$P_{max} = V_{max}I = 3/R$ and this should be equal to
1 W or less. We select exactly $R = 3 \, \Omega$ and the
resistors will be, from bottom up: $3 \, \Omega$, $9 \, \Omega$, $3 \, \Omega$,
$6 \, \Omega$, $9 \, \Omega$.

P.1.24 Let the top resistor be the internal resistor.
Using the same notation as in P.1.23, $R_5 = 10 \, \Omega =$
$3R$ where $R = 10/3 \, \Omega$. Total resistance is
$R_T = 10R$, the current $I = 10/10R = 1/R$. Power
consumed in the largest resistor is $P_{R5} = I^2 R_5 =$
$(1/R^2)3R = 3/R$. Since $R = 10/3 \, \Omega$, this power will
be $P_{R5} = 9/10 < 1$ W and we satisfy the condition
on largest consumed power.

SOLUTIONS CHAPTER 2

P.2.1 $I = \dfrac{5}{10}$ A, $V = \dfrac{15}{10}$ V.

P.2.2 $R_T = 200 \, \Omega$, $I = \dfrac{7}{200} = 35$ mA.

P.2.3 $V = 1$ V.

P.2.4 $V = 2$ V.

P.2.5 $I_2 = \dfrac{18}{1500}$ A, $V = \dfrac{18}{5}$ V.

P.2.6 $V = 0.588235$ V, $I = 0.029412$ A, $V_s =$
1.588235 V.

P.2.7 $V_A = 1$ V, $I = \dfrac{1}{6}$ A, $V = \dfrac{1}{12}$ V.

P.2.8 $V_1 = \dfrac{18}{11}$ V, $V_2 = \dfrac{18}{33}$ V.

P.2.9 $I_1 = I_2 = 0.070125$ A.

P.2.10 $I_1 = \dfrac{15}{47}, I_2 = -\dfrac{7}{47}.$

P.2.11 $V = \dfrac{10}{3}$ V.

P.2.12 $V_1 = \dfrac{5}{11}$ V, $V_2 = \dfrac{13}{11}$ V.

P.2.13 $I_1 = 1$ A, $I_2 = -1$ A, $V_1 = 0$ V, $V_2 = -2$ V.

P.2.14 $V_1 = \dfrac{2}{7}$ V, $V_2 = -\dfrac{4}{7}$ V.

P.2.15 $V_1 = \dfrac{12}{19}$ V, $V_2 = \dfrac{15}{19}$ V.

P.2.16 $V_1 = \dfrac{119}{591}$ V, $V_2 = \dfrac{54}{591}$ V, $V_3 = \dfrac{62}{591}$ V.

P.2.17 $V_{R1} = \dfrac{8}{3}$ V, $V_{R2} = \dfrac{40}{9}$ V, $V_{R3} = \dfrac{8}{9}$ V.

P.2.18 $I = -\dfrac{1}{5}$ A, the current actually flows in the opposite direction. $V_{R1} = \dfrac{2}{5}$ V, $V_{R2} = \dfrac{3}{5}$ V.

P.2.19 $I_1 = -1$ A, $I_2 = -2$ A. The voltage is 2 V (and can be read directly from the picture).

P.2.20 $I_1 = -2$ A, $I_2 = -1$ A, $V = 5$ V.

P.2.21 $I_1 = \dfrac{5}{275}$ A, $I_2 = -\dfrac{40}{275}$ A, $V = -\dfrac{40}{275} \times 15 =$ -2.1818 V.

P.2.22 $V_{R1} = \dfrac{51}{47}$ V, $V_{R2} = -\dfrac{192}{47}$ V, $V_{R3} = \dfrac{325}{47}$ V,

$P_{R1} = \dfrac{867}{2209}$ W, $P_{R2} = \dfrac{9216}{2209}$ W, $P_{R3} = \dfrac{21125}{2209}$ W,

$P_{E1} = \dfrac{34}{47}$ W (delivered), $P_{E2} = -\dfrac{51}{47}$ W (consumed),

$P_{E3} = \dfrac{96}{47}$ W (delivered), $P_{E4} = \dfrac{260}{47}$ W (delivered),

$P_{E4} = \dfrac{325}{47}$ W (delivered).

P.2.23 $V_1 = 1$ V, $V_2 = 0$ V.

P.2.24 $I_1 = -\dfrac{5}{11}$ A, $I_2 = \dfrac{15}{11}$ A, $V = \dfrac{15}{11}$ V.

P.2.25 $V_1 = \dfrac{9}{19}$ V, $V_2 = \dfrac{5}{19}$ V, $V_J = V_1 + \dfrac{2}{3} V.$

P.2.26 $I_1 = \dfrac{57}{117}$ A, $I_E = I_1 + \dfrac{3}{2}$ A.

SOLUTIONS CHAPTER 3

P.3.1 (a) $|A| = 6, x_1 = 13/3, x_2 = -1, x_3 = -4/3.$
(b) $|A| = -120, x_1 = 2, x_2 = 0.2, x_3 = 0, x_4 = 0.8.$
(c) $|A| = 12$, $x_1 = 45.5$, $x_2 = -3.166667$, $x_3 = -30, x_4 = 11.$ (d) $|A| = 12, x_1 = 1, x_2 = x_3 = 0,$ $x_4 = 1$, $x_5 = 0.$ (e) $|A| = 1$, $x_1 = -3$, $x_2 = 2,$ $x_3 = -2, x_4 = -1, x_5 = 4.$ (f) $|A| = 46050.00391,$ $x_1 = 3.15961, x_2 = 2.47557, x_3 = -1.14007, x_4 = -1.71824.$ (g) $|A| = 24$, $x_1 = 1.66667$, $x_2 = -2.5,$ $x_3 = 2.5$, $x_4 = -1.66667.$ (h) $|A| = 6$, $x_1 = 9,$ $x_2 = -6, x_3 = -9, x_4 = 3.$

P.3.2 $I_1 = \dfrac{810}{135}$ A, $I_2 = \dfrac{405}{135}$ A, $I_3 = \dfrac{135}{135} = 1$ A, $V = 4I_3 = 4$ V.

P.3.3 $V_2 = \dfrac{124}{14.48} = 8.563536$ V, $I = 0.68508$ A.

P.3.4 $V_1 = -\dfrac{51}{139}$ V, $V_2 = \dfrac{54}{139}$ V, $V_3 = -\dfrac{56}{139}$ V.

P.3.5 $I_1 = \dfrac{80}{92}$ A, $I_2 = -\dfrac{18}{92}$ A, $I_3 = -\dfrac{76}{92}$ A. $V_1 = 3 - I_1 \times 1 = 2.13043$ V, $V_2 = (I_2 - I_3)4 = 2.52174$ V.

P.3.6 $I_1 = -\dfrac{40}{44}$ A, $I_2 = -\dfrac{60}{44}$ A, $I_3 = -\dfrac{44}{44} = -1$ A.

P.3.7 $V_1 = \dfrac{27}{39}$ V, $V_2 = \dfrac{12}{39}$ V, $V_3 = \dfrac{15}{39}$ V.

P.3.8 $V_1 = \dfrac{8}{7}$ V, $V_2 = -\dfrac{2}{7}$ V, $V_3 = \dfrac{11}{7}$ V.

P.3.9 $V_1 = \dfrac{28}{50}$ V, $V_2 = -\dfrac{18}{50}$ V, $V_3 = \dfrac{26}{50}$ V.

P.3.10 $I_1 = \dfrac{555}{618}$ A, $I_2 = \dfrac{120}{618}$ A, $I_3 = \dfrac{30}{618}$ A.

P.3.11 $I_1 = \dfrac{64}{56}$ A, $I_2 = \dfrac{40}{56}$ A, $I_3 = \dfrac{24}{56}$ A.

P.3.12 $V_1 = \dfrac{15}{75}$ V, $V_2 = \dfrac{40}{75}$ V,

$V_3 = \dfrac{105}{75}$ V, $V_4 = \dfrac{215}{75}$ V.

P.3.13 $V_1 = \dfrac{59}{55}$ V, $V_2 = \dfrac{35}{55}$ V, $V_3 = \dfrac{28}{55}$ V, $V_4 = \dfrac{18}{55}$ V,

$V_5 = \dfrac{9}{55}$ V.

P.3.14 $I_1 = 2.86840$ A, $I_2 = 2.34141$ A, $I_3 = 1.92095$ A, $I_4 = 0.16431$ A.

P.3.15 $V_1 = 0.30238$ V, $V_2 = 0.25$ V, $V_3 = 0.39524$ V.

P.3.16 $V_1 = \dfrac{15}{12.5} = 1.2$ V, $V_2 = \dfrac{55}{12.5} = 4.4$ V,

$I = 4V_1 = \dfrac{60}{12.5} = 4.8$ A, $V = V_1 + \dfrac{4}{10} = 1.6$ V.

P.3.17 $D = 11R$, $N_2 = 33$, $I_2 = \dfrac{33}{11R}$,

$V_{R4} = \dfrac{33R}{11R} = 3$ V.

P.3.18 $D = 36G + 31$, $N_3 = -56$.

P.3.19–P.3.22 Refer to results of previous problems after inserting the same values.

P.3.23 $D = 31G$, $N_2 = 217$. The required value is $G = 7/2$.

SOLUTIONS CHAPTER 4

P.4.1 $I_1 = \dfrac{96}{620}$ A, $I_2 = \dfrac{51}{620}$ A.

P.4.2 $I_1 = \dfrac{200}{160}$ A, $I_2 = \dfrac{30}{160}$ A.

P.4.3 $I_1 = \dfrac{12}{31}$ A, $I_2 = \dfrac{2}{31}$ A.

P.4.4 $I_1 = \dfrac{3200}{660}$ A, $I_2 = \dfrac{1600}{660}$ A, $I_3 = \dfrac{1200}{660}$ A.

P.4.5 $V_1 = \dfrac{15}{20}$ V, $V_2 = \dfrac{17}{20}$ V.

P.4.6 $V_1 = \dfrac{12}{14}$ V, $V_2 = \dfrac{8}{14}$ V.

P.4.7 $I_1 = \dfrac{6}{11}$ A, $I_2 = -\dfrac{4}{11}$ A.

P.4.8 $I_1 = \dfrac{1260}{495}$ A, $I_2 = \dfrac{1500}{495}$ A, $I_3 = \dfrac{600}{495}$ A.

P.4.9 $V_1 = -\dfrac{70}{16}$ V, $V_2 = -\dfrac{120}{16}$ V.

P.4.10 $V_1 = \dfrac{60}{95}$ V, $V_2 = \dfrac{50}{95}$ V, $V_3 = \dfrac{40}{95}$ V.

P.4.11 $V_1 = \dfrac{315}{141}$ V, $V_2 = \dfrac{549}{141}$ V,

$V_3 = \dfrac{354}{141}$ V, $V_{out} = V_3$.

P.4.12 $I_1 = \dfrac{10}{14}$ A, $I_2 = -\dfrac{30}{14}$ A,

$V_{out} = (I_1 - I_2)3 = \dfrac{120}{14} = 8.57143$ V.

P.4.13 $I_1 = \dfrac{110}{64}$ A, $I_2 = -\dfrac{30}{64}$ A,

$V_{out} = \dfrac{540}{64} = 8.4375$ V.

P.4.14 $V_1 = \dfrac{46}{126}$ V, $V_2 = \dfrac{36}{126}$ V, $V_{out} = V_2$.

P.4.15 $V_1 = \dfrac{256}{120}$ V, $V_2 = \dfrac{352}{120}$ V, $V_3 = \dfrac{272}{120}$ V,

$V_{out} = V_3$.

P.4.16 $I_1 = \dfrac{400}{128}$ A, $I_2 = \dfrac{320}{128}$ A, $I_3 = 0$,

$V_{out} = 4I_2 + 0 = 10$ V.

P.4.17 $V_1 = \dfrac{0.4}{0.42} = 0.95238$ V,

$V_2 = \dfrac{4}{0.42} = 9.5238$ V.

P.4.18 $I_1 = \dfrac{156}{62}$ A, $I_2 = \dfrac{12}{62}$ A,

$V_1 = 0.5(I_1 - I_2) = \dfrac{36}{31}$ V, $V_2 = V_1 - 2I_2 = \dfrac{24}{31}$ V.

P.4.19 $V_1 = \dfrac{720}{840} = 0.857$ V, $V_2 = \dfrac{1440}{840} = 1.7142$ V.

P.4.20 $V_1 = V_2 = \dfrac{22}{22} = 1$ V.

P.4.21 $V_1 = \dfrac{22}{7}$ V, $V_2 = \dfrac{25}{7}$ V.

P.4.22 $I_1 = \dfrac{50}{4}$ A, $V = -\dfrac{100}{4}$ V,

$V_{out} = 2 V = -50$ V.

P.4.23 $I_1 - \dfrac{50}{22}$ A, $V_{out} = 2I_1 = \dfrac{50}{11}$ V.

P.4.24 We need only $I_3 = \dfrac{170}{187}$ A to calculate the output voltage $V_{out} = 10 - 3I_3 = 7.27273$ V.

P.4.25 We need only $I_2 = \dfrac{90}{77}$ to get

$V_{out} = 4\dfrac{90}{77} = 4.67532$ V.

P.4.26 We need only $I_2 = \dfrac{50}{41}$ A to get

$V_{out} = 4I_2 = 4.87805$ V.

P.4.27 $V_1 = \dfrac{10}{3}$ V, $I = \dfrac{2}{3}$ A,

$V_2 = 2I \times 4 = \dfrac{16}{3} = 5.3333$ V.

P.4.28 $V_1 = 10$ V, $V_2 = V_{out} = 0$.

P.4.29 We need only $I_2 = -\dfrac{20}{9}$ A to get

$V_{out} = -\dfrac{80}{9} = -8.88889$ V.

P.4.30 We need only $I_2 = \dfrac{70}{96}$ A to get

$V_{out} = 4I_2 = 2.9167$ V.

P.4.31 $V_{out} = V_2 = \dfrac{30}{51}$ V.

P.4.32 $V_{out} = V_2 = \dfrac{16}{37}$ V.

SOLUTIONS CHAPTER 5

P.5.1 $V_1 = \dfrac{14}{23} = 0.6087$ V.

P.5.2 $V_1 = \dfrac{900}{620} = 1.4516$ V.

P.5.3 $I = \dfrac{30}{160}$ A, $V_1 = \dfrac{600}{160} = 3.75$ V.

P.5.4 $V_1 = 1$ V.

P.5.5 $V_A = \dfrac{20}{31} = 0.6452$ V.

P.5.6 $V_1 = \dfrac{5}{11} = 0.454545$ V, $V_2 = \dfrac{2}{11} = 0.18182$ V.

P.5.7 $I_A = \dfrac{6}{7} = 0.8571$ A.

P.5.8 $I = 1$ A.

P.5.9 $V_1 = \dfrac{34}{9} = 3.77778$ V, $V_2 = \dfrac{55}{9} = 6.1111$ V.

P.5.10 $V_1 = \dfrac{2}{7}$, $V_2 = -\dfrac{4}{7}$,

$V_{G2} = V_1 - V_2 = \dfrac{6}{7} = 0.85714$ V.

P.5.11 $V_{R3} = \dfrac{18}{11} = 1.63636$ V, with the right end of the resistor at a higher potential.

P.5.12 $V_{R3} = -\dfrac{60}{44} = -1.363636$ V.

P.5.13 $V_{R2} = \dfrac{20}{3} = 6.66667$ V.

P.5.14 $V_1 = \dfrac{135}{179}$ V, $V_2 = \dfrac{134}{179}$ V and $V_3 = \dfrac{234}{179}$ V.

P.5.15 $E_{EQ} = 2$ V, $R_{EQ} = 1\,\Omega$, $J_{EQ} = 2$ A.

P.5.16 $E_{EQ} = \dfrac{27}{423}$ V, $R_{EQ} = \dfrac{51}{423}\,\Omega$, $J_{EQ} = \dfrac{27}{51}$ A.

P.5.17 $E_{EQ} = 1$ V, $R_{EQ} = \dfrac{1}{3}\,\Omega$, $J_{EQ} = 3$ A.

P.5.18 $J_{EQ} = \dfrac{3}{7}$ A, $G_{EQ} = \dfrac{2}{7}$ S, $E_{EQ} = \dfrac{3}{2}$ V.

P.5.19 $J_{EQ} = \dfrac{4}{5}$ V, $G_{EQ} = \dfrac{2}{5}$ S, $E_{EQ} = 2$ V.

P.5.20 $E_{EQ} = \dfrac{6}{7}$ V, $R_{EQ} = \dfrac{2}{7}\,\Omega$, $J_{EQ} = 3$ A.

P.5.21 $V_1 = \dfrac{16 - 10\alpha}{39 - 10\alpha}$, $V_2 = \dfrac{6 + 2\alpha}{39 - 10\alpha}$.

P.5.22 $E_{EQ} = \dfrac{20}{29}$ V, $R_{EQ} = \dfrac{35}{29}\,\Omega$.

P.5.23 $E_{EQ} = \dfrac{14}{3}$ V, $R_{EQ} = \dfrac{4}{3}\,\Omega$, $J_{EQ} = \dfrac{7}{2}$ A.

P.5.24 $V_{out} = 40$ V.

P.5.25 $V_{out} = -\dfrac{600}{37} = -16.216$ V.

P.5.26 $V_{out} = \dfrac{170}{7} = 24.2857$ V.

P.5.27 $V_{out} = \dfrac{130}{11} = 11.8181$ V.

P.5.28 $E_{EQ} = \dfrac{3}{7}$ V, $R_{EQ} = \dfrac{4}{21}\,\Omega$, $J_{EQ} = \dfrac{9}{4}$ A.

P.5.29 $E_{EQ} = \dfrac{15}{11}$ V, $J_{EQ} = \dfrac{15}{6}$ A, $G_{EQ} = \dfrac{11}{6}$ S.

P.5.30 $V_1 = \dfrac{12G_3 + 25}{3G_3 + 2}$, $V_2 = \dfrac{14G_3 + 15}{3G_3 + 2}$.

P.5.31–P.5.33 To check your result, insert the same values as earlier. If you use the symbolic option in NETOPT, give variable resistors as conductances.

P.5.34 The currents in terms of r are $I_1 = \dfrac{2300 - 400r}{450 - 50r}$, $I_2 = \dfrac{1300 - 400r}{450 - 50r}$, the current $I = \dfrac{20}{9 - r}$. The condition is $3I = 10$ and the resulting $r = 3$.

P.5.35 $V_1 = \dfrac{7}{5}$ V, $V_2 = \dfrac{3}{5}$ V.

P.5.36 $V_{out} = \dfrac{14}{57}$ V.

P.5.37 $V_1 = V_2 = \dfrac{5}{3}$ V.

P.5.38 $V_{out} = \dfrac{38}{90} = 0.42222$ V.

P.5.39 $V_{out} = -\dfrac{5}{37} = -0.135135$ V.

P.5.40 $V_{out} = -\dfrac{20}{19} = -1.05263$ V.

P.5.41 $V_1 = \dfrac{102}{69} = 1.47826$ V, $V_2 = -\dfrac{60}{69} = -0.86957$ V.

P.5.42 $V_{out} = \dfrac{2700}{178} = 15.1685$ V.

P.5.43 $V_3 = \dfrac{12}{41} = 0.292683$ V.

P.5.44 $V_{out} = \dfrac{204}{141} = 1.44681$ V.

P.5.45 $V_1 = -\dfrac{13}{5}$, $V_2 = -\dfrac{17}{5}$, $V_{out} = -\dfrac{32}{5} = -6.4$ V.

P.5.46 $V_{out} = \dfrac{22}{23} = 0.95652$ V.

P.5.47 $V_{out} = \dfrac{G_3 + G_4}{G_1 G_4 - G_2 G_3} J.$

P.5.48 $V_{out} = \dfrac{J}{G}.$

P.5.49 $V_{out} = (G_1 + G_2)/(G_2 - 2G_1).$ For the output to be twice the input we must select $G_2 = 5G_1$, for instance $G_1 = 1$ S and $G_2 = 5$ S.

SOLUTIONS CHAPTER 6

P.6.1 $i_C = 45 \cos 5t.$

P.6.2 $v_C = -\dfrac{1}{5} \cos 5t$

P.6.5a $C_{EQ} = \dfrac{C_1 C_2 + C_1 C_3 + C_2 C_3}{C_2 + C_3}$

P.6.5b $C_{EQ} = \dfrac{107}{40} = 2.675$ F.

P.6.5c $C_{EQ} = \dfrac{59}{11} = 5.3636$ F.

P.6.5d $C_{EQ} = \dfrac{162}{101} = 1.60396$ F.

P.6.7 $v_L = 24 \cos 4t.$

P.6.8 $i_L = -\dfrac{3}{8} \cos 4t.$

P.6.11a $L_{EQ} = \dfrac{L_1 L_2 + L_1 L_3 + L_1 L_4 + L_2 L_3 + L_2 L_4}{L_2 + L_3 + L_4}$

P.6.11b $L_{EQ} = \dfrac{60}{17} = 3.5294$ H.

P.6.11c $L_{EQ} = \dfrac{20}{29} = 0.68965$ H.

P.6.12 $L_{EQ} = \dfrac{348}{75} = 4.64$ H.

SOLUTIONS CHAPTER 7

P.7.1a $V_2 = V_{out} = \dfrac{2s}{4s + 1}.$

P.7.1b $V_2 = V_{out} = \dfrac{2}{4s + 1}.$

P.7.1c $V_{out} = \dfrac{2}{2s^2 + 5s + 2}.$

P.7.1d $V_1 = \dfrac{2s + 4}{10s + 11}, V_2 = \dfrac{2s + 1}{10s + 11}$

P.7.2 $V_1 = \dfrac{5s - 1}{6s^2 + 19s - 3}, V_2 = \dfrac{2s}{6s^2 + 19s - 3}.$

P.7.3 $V_{out} = \dfrac{-2}{6s^2 + s + 2}.$

P.7.4 $V_{out} = \dfrac{2s - 1}{s - 11}.$

P.7.5 $V_{out} = \dfrac{3}{17s + 30}.$

P.7.6 $V_{out} = \dfrac{12s}{24s + 2}.$

P.7.7 $V_{out} = E_{EQ} = \dfrac{7}{2s^2 + 9s - 12},$
$Z_{EQ} = \dfrac{s + 3}{2s^2 + 9s - 12}$

P.7.8 $E_{EQ} = 1, Y_{EQ} = 2s + 2, J_{EQ} = 2s + 2.$

P.7.9 $E_{EQ} = 2, Y_{EQ} = \dfrac{2s + 3}{2}, J_{EQ} = 2s + 3.$

P.7.10 $V_{out} = \dfrac{6s}{5s - 1}.$

P.7.11 $V_1 = \dfrac{2s + 3}{2s + 4}, V_2 = \dfrac{3}{2s + 4}.$

P.7.12 $V_{out} = \dfrac{12}{8s + 7}$.

P.7.13 $V_{out} = \dfrac{3}{5s + 8}$.

P.7.14 $V = \dfrac{5}{3s + 8}$.

P.7.15 $V_{out} = \dfrac{-12s^2}{2s + 1}$.

P.7.16 $V_{out} = \dfrac{s - 1}{8s - 1}$.

P.7.17 $V_1 = \dfrac{8s - 2}{4s + 5}$, $V_2 = \dfrac{-4s - 2}{4s + 5}$.

P.7.18 $V_1 = \dfrac{8s + 1}{8s + 3}$, $V_2 = \dfrac{4s + 1}{8s + 3}$.

P.7.19 $E_{EQ} = \dfrac{7s}{s + 1}$, $Y_{EQ} = \dfrac{s + 1}{2s}$, $J_{EQ} = \dfrac{7}{2}$.

P.7.20 $V_{out} = \dfrac{8}{20s - 3}$.

P.7.21 $V_{out} = \dfrac{8s^2}{8s^2 + 14s + 5}$.

P.7.22 $V_{out} = \dfrac{4s + 60}{500s^2 + 1504s - 20}$.

P.7.23 $V_{out} = \dfrac{15s + 3}{10s^2 + 17s + 5}$.

P.7.24 $V = \dfrac{-1}{s^2 - s - 1}$.

P.7.25 $V_{out} = \dfrac{-4s^2 + 2s}{12s^2 + 2s + 2}$.

P.7.26 $V_{out} = \dfrac{2s - 2}{6s^2 + 2s + 2}$.

P.7.27 $E_{EQ} = \dfrac{14s}{12s^2 - 22s + 3}$, $Z_{EQ} = \dfrac{4s^2 + 6s}{12s^2 - 22s + 3}$.

P.7.28 $V_{out} = \dfrac{-10}{20s^2 + 92s + 9}$.

SOLUTIONS CHAPTER 8

P.8.1 (a) $2 + j6$.
(b) $2 - j5$.
(c) $-2 - j3$.
(d) $-11 + j10$.
(e) $-3.0588 + j0.7647$.
(f) $1.8462 - j5.2308$.
(g) $234 - j52$.
(h) $-0.8966 + j0.7586$.

P.8.2 Angles are given in degrees:
(a) $v_1 = 5.831e^{j59.036}$.
(b) $v_2 = 2.236e^{j116.565}$.
(c) $v_3 = 4.472e^{-j116.565}$.
(d) $v_4 = 4.123e^{-j14.036}$.
(e) $v_5 = 5e^{-j143.13}$.
(f) $v_6 = 3e^{j90}$.
(g) $v_7 = 1e^{j180}$.
(h) $v_8 = 5e^{-j90}$.

P.8.3 (a) $v_1 = 2.5 + j4.33$.
(b) $v_2 = -1 + j1.73205$.
(c) $v_3 = -3.464 - j2$.

(d) $v_4 = 2.8191 - j1.0261$.
(e) $v_5 = -4.33 + j2.5$.
(f) $v_6 = 0 + j4$.
(g) $v_7 = -5 + j0$.
(h) $v_8 = 0 - j4$.

P.8.4 Angles are given in degrees:
(a) Re $3e^{j(2t + 40)}$.
(b) Re $2e^{j(2t + 150)}$.
(c) Re $4e^{j(4t - 30)}$.
(d) Re $5e^{j(\omega t - 90)}$.
(e) Re $4e^{j(\omega t - 70)}$.
(f) Re $2e^{j(377t - 40)}$.

P.8.5 Angles are given in degrees:
(a) $5 \cos(2t + 3.13)$.
(b) $1.414 \cos(\omega t - 45)$.
(c) $4.927 \cos(\omega t + 27.98)$.
(d) $3.884 \cos(\omega t - 70.473)$.
(e) $4.76 \cos(\omega t + 36.457)$.
(f) $2.879 \cos(\omega t + 40)$.

(g) 4.472 cos (ωt + 116.565).

(h) 1.732 cos (ωt + 30).

(i) 2.993 cos (ωt + 69.21).

(j) 1.414 cos (ωt − 45).

P.8.6 $Z = 1 + 2j\omega$, $Z(j1) = 1 + j2$, $Z(j2) = 1 + j4$, $Z(j3) = 1 + j6$, $Y(j1) = \dfrac{1}{5} - j\dfrac{2}{5}$, $Y(j2) = \dfrac{1}{17} - j\dfrac{4}{17}$,

$Y(j3) = \dfrac{1}{37} - j\dfrac{6}{37}$.

P.8.7 $Z = \dfrac{1 + 6j\omega}{3j\omega}$, $Z(j1) = 2 - j0.333$,

$Z(j2) = 2 - j0.167$, $Z(j3) = 2 - j0.111$,

$Y(j1) = 0.486 + j0.081$, $Y(j2) = 0.497 + j0.041$,

$Y(j3) = 0.498 + j0.028$.

P.8.8 $Y = 2 + j3\omega$, $Y(j1) = 2 + 3j$,

$Y(j2) = 2 + j6$, $Y(j3) = 2 + j9$,

$Z(j1) = 0.154 - j0.231$, $Z(j2) = 0.05 - j0.15$,

$Z(j3) = 0.024 - j0.106$.

P.8.9 $Y = \dfrac{2 + j3\omega}{j6\omega}$, $Y_1 = 0.5 - j\dfrac{1}{3}$,

$Y(j2) = 0.5 - j\dfrac{1}{6}$, $Y(j3) = 0.5 - j\dfrac{1}{9}$,

$Z(j1) = 1.385 + j0.923$, $Z(j2) = 1.8 + j0.6$,

$Z(j3) = 1.906 + j0.424$.

P.8.10 $2 + j\dfrac{5}{6}$.

P.8.11 $Z(j5) = 0.024 - j0.204$.

P.8.12 $i(t) = 5.6568 \cos (2t + 75°)$.

P.8.13 $v_2(t) = 0.149 \cos (3t - 128.435°)$.

P.8.14 $V_2 = (2s + 3)J$, $V_2(j1) = 3 + j2$,

$V_2(j2) = 3 + j4$, $V_2(j3) = 3 + j6$.

P.8.15 $V_2 = \dfrac{2}{2s^2 + 5s + 2}$, $V_2(j1) = 0.4e^{-j90°}$,

$V_2(j2) = 0.171e^{-j120.964°}$,

$V_2(j3) = 0.091e^{-j136.848°}$.

P.8.16 $V_2 = \dfrac{2}{8s + 6}$, $V_2(j1) = 0.2e^{-j53.13°}$,

$V_2(j2) = 0.177e^{-j69.444°}$.

P.8.17 $V_{out} = \dfrac{-1}{3s + 6}$, $V_{out}(j1) = 0.149e^{j153.435°}$.

P.8.18 $I_2 = \dfrac{2}{9s^2 + 3s}$, $I_2(j1) = 0.211e^{-j161.57°}$.

P.8.19 $V_2 = \dfrac{6s^2}{6s^2 + 19s + 5}$, $V_2(j1) = 0.315e^{j86.987°}$,

$V_2(j2) = 0.565e^{j63.435°}$.

P.8.20 $Z = 1 + 2s$.

(a) For $e(t) = \cos \omega t$:

$I = EY = \dfrac{1}{1 + j2\omega}$, $S(j\omega) = \dfrac{1}{2 - j4\omega}$,

$S(j1) = 0.1 + j0.2$, $S(j2) = 0.0294 + j0.1176$.

(b) For $i(t) = \cos \omega t$:

$V = ZI = 1 + j2\omega$,

$S(j\omega) = 0.5 + j\omega$, $S(j1) = 0.5 + j1$,

$S(j2) = 0.5 + j2$.

P.8.21 $Y = \dfrac{2 + j3\omega}{j6\omega}$.

(a) For $e(t) = \cos \omega t$:

$I = \dfrac{2 + j3\omega}{j6\omega}$, $S(j\omega) = \dfrac{2 - j3\omega}{-j12\omega}$, $S(j1) = \dfrac{1}{4} + j\dfrac{1}{6}$,

$S(j2) = \dfrac{1}{4} + j\dfrac{1}{12}$.

(b) For $i(t) = \cos \omega t$:

$V = \dfrac{j6\omega}{2 + j3\omega}$, $S(j1) = \dfrac{9}{13} + j\dfrac{6}{13}$, $S(j2) = \dfrac{9}{10} + j\dfrac{3}{10}$.

P.8.22 $Z = \dfrac{(1 - 1.5\omega^2) + j6\omega}{j3\omega}$.

(a) For $e(t) = \cos \omega t$: $S(j\omega) = \dfrac{j1.5\omega}{(1 - 1.5\omega^2) + j6\omega}$,

$S(j1) = \dfrac{36}{145} - j\dfrac{3}{145}$, $S(j2) = \dfrac{36}{169} - j\dfrac{15}{169}$.

(b) For $i(t) = \cos \omega t$: $S(j\omega) = \dfrac{(1 - 1.5\omega^2) + j6\omega}{j6\omega}$,

$S(j1) = 1 - j\dfrac{1}{12}$, $S(j2) = 1 + j\dfrac{5}{12}$.

P.8.23 $Z = \dfrac{2 + j\omega}{(2 - \omega^2) + j2.5\omega}$.

(a) For $e(t) = \cos \omega t$: $S(j\omega) = \dfrac{(2 - \omega^2) - j2.5\omega}{4 - j2\omega}$,

$S(j1) = 0.45 - j0.4$, $S(j2) = 0.375 - j0.875$.

(b) For $i(t) = \cos \omega t$: $S = \dfrac{1 + 0.5j\omega}{(2 - \omega^2) + j2.5\omega}$,

$S(j1) = 0.31 - j0.2758$, $S(j2) = 0.1034 - j0.2414$.

P.8.24 $V_1 = \dfrac{4s+8}{2s^2+5s+2}$, $I_{RL} = \dfrac{4}{2s^2+5s+2}$.

$S(j\omega) = \dfrac{8(2+j\omega)}{(2-2\omega^2)^2+25\omega^2}$, $S(j2) = \dfrac{4}{34} + j\dfrac{4}{34}$.

P.8.25 $V_1 = 2s+2$, $S(j\omega) = 2 + 2j\omega$,
$S(j2) = 2 + j4$.

P.8.26 $I_1 = \dfrac{8s^2+12s}{2s^2+5s+2}$,

$S(j2) = \dfrac{-32-j24}{-6-j10} = 3.1765 - j1.2941.$

P.8.27 $I_1 = \dfrac{s+4}{10s+20}$,

$S(j2) = \dfrac{2-j}{10-j10} = \dfrac{3}{20} + j\dfrac{1}{20}.$

SOLUTIONS CHAPTER 9

The symbol *c.c.t.*, appearing in some solutions, denotes complex conjugate terms which should follow those given.

P.9.1 (a) $F(s) = \dfrac{3}{s} - \dfrac{2}{s+4}$; $f(t) = (3 - 2e^{-4t})u(t)$.

(b) $F(s) = \dfrac{2}{s} + \dfrac{1}{s+1}$; $f(t) = (2 + e^{-t})u(t)$.

(c) $F(s) = \dfrac{1}{s} + \dfrac{2}{s+1} + \dfrac{3}{s+2}$;
$v(t) = (1 + 2e^{-t} + 3e^{-2t})u(t)$.

(d) $F(s) = 2 - \dfrac{4}{s+2}$; $f(t) = 2\delta(t) - 4e^{-2t}u(t)$.

(e) $F(s) = \dfrac{2}{s+1} + \dfrac{4}{s^2+1}$;
$f(t) = (2e^{-t} + 4\sin t)u(t)$.

(f) $F(s) = 2 - \dfrac{5}{s+3}$; $f(t) = 2\delta(t) - 5e^{-3t}u(t)$.

(g) $F(s) = \dfrac{s+1}{(s+1)^2+1}$; $f(t) = (e^{-t}\cos t)u(t)$.

(h) $F(s) = \dfrac{8}{s+1} + \dfrac{-3.5+j3}{s+1+j} + c.c.t.$;
$f(t) = [8e^{-t} + e^{-t}(-7\cos t + 6\sin t)]u(t)$.

(i) $F(s) = \dfrac{-2+j3}{s+j} + \dfrac{2-j0.5}{s+1+j} + c.c.t.$;
$f(t) = [-4\cos t + 6\sin t + e^{-t}(4\cos t - \sin t)]u(t)$.

(j) $F(s) = \dfrac{0.2}{s} + \dfrac{0.4+j0.45}{s+1+j2} + c.c.t.$;
$f(t) = [0.2 + e^{-t}(0.8\cos 2t + 0.9\sin 2t)]u(t)$.

(k) $F(s) = \dfrac{0.3145}{s} + \dfrac{0.1875}{s+2} + \dfrac{0.1875}{s+4} + \dfrac{0.3125}{s+6}$;
$f(t) = (0.3125 + 0.1875e^{-2t} + 0.1875e^{-4t} + 0.3125e^{-6t})u(t)$.

(l) $F(s) = \dfrac{1.5}{s+1} - \dfrac{3}{s+2} + \dfrac{2.5}{s+3}$;
$f(t) = (1.5e^{-t} - 3e^{-2t} + 2.5e^{-3t})u(t)$.

(m) $F(s) = \dfrac{0.6}{s+1} + \dfrac{-2.8+j1.35}{s+2+j2} + c.c.t.$;
$f(t) = [0.6e^{-t} + e^{-2t}(-5.6\cos 2t + 2.7\sin 2t)]u(t)$.

(n) $F(s) = \dfrac{1}{s} + \dfrac{2}{s+j} + c.c.t.$;
$f(t) = (1 + 4\cos t)u(t)$.

P.9.2 (a) $F(s) = \dfrac{3}{s} - \dfrac{2}{s+4}$, $f(t) = (3 - 2e^{-4t})u(t)$,

$sF(s) = 1 + \dfrac{8}{s+4}$; $f(t) = \delta(t) + 8e^{-4t}u(t)$.

(b) $F(s) = \dfrac{2}{s+1} - \dfrac{2}{s+2}$;
$f(t) = (2e^{-t} - 2e^{-2t})u(t)$,
$sF(s) = \dfrac{-2}{s+1} + \dfrac{4}{s+2}$;
$f'(t) = (-2e^{-t} + 4e^{-2t})u(t)$.

(c) $F(s) = \dfrac{0.5}{s+1} - \dfrac{1}{s+2} + \dfrac{0.5}{s+3}$,
$f(t) = (0.5e^{-t} - e^{-2t} + 0.5e^{-3t})u(t)$;
$sF(s) = \dfrac{-0.5}{s+1} + \dfrac{2}{s+2} - \dfrac{1.5}{s+3}$,
$f'(t) = (-0.5e^{-t} + 2e^{-2t} - 1.5e^{-3t})u(t)$.

(d) $F(s) = \dfrac{0.5}{s} + \dfrac{-0.25+j0.25}{s+1+j} + c.c.t.$;
$f(t) = [0.5t + e^{-t}(-0.5\cos t + 0.5\sin t)]u(t)$;
$sF(s) = \dfrac{s+1}{(s+1)^2+1}$; $f'(t) = e^{-t}\cos t\, u(t)$.

P.9.3

(a) $f(0) = 1$; $f(\infty) = 3$.
(b) $f(0) = 3$; $f(\infty) = 0$.
(c) $f(0) = 6$; $f(\infty) = 0$.
(d) $f(0) = \delta(t)$; $f(\infty) = 0$.
(e) $f(0) = 2$; $f(\infty) = 0$.
(f) $f(0) = \delta(t)$; $f(\infty) = 0$.
(g) $f(0) = 1$; $f(\infty) = 0$.
(h) $f(0) = 1$; $f(\infty) = 0$.
(i) $f(0) = 0$; $f(\infty) = 0$.

(j) $f(0) = 1$; $f(\infty) = \dfrac{1}{5}$.

(k) $f(0) = 1$; $f(\infty) = \dfrac{15}{48}$.

(l) $f(0) = 1$; $f(\infty) = 0$.
(m) $f(0) = \delta(t)$; $f(\infty) = 0$.
(n) $f(0) = 5$; $f(\infty) = 1$.

P.9.4

(d) $F(s) = \dfrac{2}{s+2}$; $f(t) = 2e^{-2t}u(t)$.

(e) $F(s) = \dfrac{6}{s} - \dfrac{2}{s+1} + \dfrac{-2}{s-j} + c.c.t.$;
$f(t) = (6 - 2e^{-t} - 4\cos t)u(t)$.

(f) $F(s) = \dfrac{1}{3s} + \dfrac{5}{3}\dfrac{1}{s+3}$; $f(t) = \left(\dfrac{1}{3} + \dfrac{5}{3}e^{-3t}\right)u(t)$.

(g) $F(s) = \dfrac{0.5}{s} + \dfrac{-0.5 - j0.25}{s+1+j} + c.c.t.$;
$f(t) = (0.5 - e^{-t}\cos t - e^{-t}\sin t)u(t)$.

(h) $F(s) = \dfrac{7.5}{s} - \dfrac{8}{s+1} + \dfrac{0.25 - j3.25}{s+1+j} + c.c.t.$;
$f(t) = [7.5 - 8e^{-t} + e^{-t}(0.5\cos t - 6.5\sin t)]u(t)$.

(i) $F(s) = \dfrac{7.5}{s} + \dfrac{-3 - j2}{s+j} + \dfrac{-0.75 - j1.25}{s+1+j}$
$+ c.c.t.$;
$f(t) = [7.5 - 6\cos t - 4\sin t$
$+ e^{-t}(-1.5\cos t + 2.5\sin t)]u(t)$.

(l) $F(s) = \dfrac{5}{6}\dfrac{1}{s} - \dfrac{1.5}{s+1} + \dfrac{1.5}{s+2} - \dfrac{5}{6}\dfrac{1}{s+3}$;
$f(t) = \left(\dfrac{5}{6} - 1.5e^{-t} + 1.5e^{-2t} - \dfrac{5}{6}e^{-3t}\right)u(t)$.

(m) $F(s) = \dfrac{0.875}{s} - \dfrac{0.6}{s+1} + \dfrac{0.3625 - j1.0375}{s+2+j2}$
$+ c.c.t.$;
$f(t) = [0.875 - 0.6e^{-t}$
$+ e^{-2t}(0.725\cos t - 2.075\sin 2t)]u(t)$.

P.9.5

(a) $F(s) = \dfrac{3}{s^2} - \dfrac{1}{s} + \dfrac{1}{2}\dfrac{1}{s+4}$;
$f(t) = \left(3t - \dfrac{1}{2} + \dfrac{1}{2}e^{-4t}\right)u(t)$.

(b) $F(s) = \dfrac{2}{s^2} + \dfrac{1}{s} - \dfrac{1}{s+1}$;
$f(t) = (2t + 1 - e^{-t})u(t)$.

(c) $F(s) = \dfrac{1}{s^2} + \dfrac{7}{2}\dfrac{1}{s} - \dfrac{2}{s+1} - \dfrac{3}{2}\dfrac{1}{s+2}$;
$f(t) = \left(t + \dfrac{7}{2} - 2e^{-t} - \dfrac{3}{2}e^{-2t}\right)u(t)$.

(j) $F(s) = \dfrac{0.2}{s^2} + \dfrac{0.52}{s} + \dfrac{-0.26 + j0.07}{s+1+j2} + c.c.t.$;
$f(t) = [0.2t + 0.52 + e^{-t}(-0.52\cos 2t$
$+ 0.14\sin 2t)]u(t)$.

(n) $f(s) = \dfrac{1}{s^2} + \dfrac{4}{s^2+1}$; $f(t) = (t + 4\sin t)u(t)$.

P.9.6

(a) $F(s) = \dfrac{1}{(s+1)^3} - \dfrac{2}{(s+1)^2} + \dfrac{3}{s+1} - \dfrac{1}{(s+2)^2}$
$- \dfrac{3}{s+2}$;
$f(t) = \left(\dfrac{t^2}{2}e^{-t} - 2te^{-t} + 3e^{-t} - te^{-2t} - 3e^{-2t}\right)u(t)$.

(b) $F(s) = \dfrac{2}{(s+1)^3} - \dfrac{4}{(s+1)^2} + \dfrac{5}{s+1} - \dfrac{5}{s+2}$;
$f(t) = (t^2 - 4t + 5)e^{-t}u(t) - 5e^{-2t}u(t)$.

(c) $F(s) = \dfrac{2}{(s+1)^2} - \dfrac{4}{s+1} + \dfrac{9}{(s+1)^2} + \dfrac{4}{s+2}$;
$f(t) = (2te^{-t} - 4e^{-t} + 9te^{-2t} + 4e^{-2t})u(t)$.

SOLUTIONS CHAPTER 10

All solutions are given in the Laplace domain by their partial fraction expansions. For complex roots the results are in the form

$$\frac{A + jB}{s + c + jd} + c.c.t.$$

where c.c.t. stands for *complex conjugate term*. In some cases the inverse into time domain is not given, to save space. The reader can use formula 13 in Table 9.2.1 for such inversions. The decompositions were obtained by the program from Appendix B.

P.10.1 $V_2 = \dfrac{sCJ}{sC(G_1 + G_2) + G_1 G_2} = \dfrac{sJ}{3s + 2}$.

(a) For $j(t) = \delta(t)$: $V_2 = \dfrac{1}{3} - \dfrac{2}{9} \dfrac{1}{s + \dfrac{2}{3}}$;

$v_2(t) = \dfrac{1}{3}\delta(t) - \dfrac{2}{9}e^{-2t/3}u(t)$.

(b) For $j(t) = u(t)$: $V_2 = \dfrac{1}{3\left(s + \dfrac{2}{3}\right)}$,

$v_2(t) = \dfrac{1}{3}e^{-2t/3}u(t)$.

P.10.2 $V_2 = \dfrac{J}{sLG_1 G_2 + G_1 G_2} = \dfrac{J}{4s + 3}$.

(a) For $j(t) = \delta(t)$: $v_2(t) = \dfrac{1}{4}e^{-3t/4}u(t)$.

(b) For $j(t) = u(t)$: $V_2 = \dfrac{1}{3s} - \dfrac{1}{3}\dfrac{1}{s + \dfrac{2}{3}}$,

$v_2(t) = \dfrac{1}{3}(1 - e^{-3t/4})u(t)$.

P.10.3 $V_2 = \dfrac{(s - 1)J}{4s + 2}$.

(a) For $j(t) = \delta(t)$: $V_2 = \dfrac{1}{4} - \dfrac{3}{8}\dfrac{1}{s + 0.5}$,

$v_2(t) = \dfrac{1}{4}\delta(t) - \dfrac{3}{8}e^{-0.5t}u(t)$.

(b) For $j(t) = u(t)$: $V_2 = -\dfrac{1}{2s} + \dfrac{3}{4}\dfrac{1}{s + 0.5}$,

$v_2(t) = \left(-\dfrac{1}{2} + \dfrac{3}{4}e^{-0.5t}\right)u(t)$.

(c) For $j(t) = \sin 2t$:

$V_2 = \dfrac{-0.17646}{s + 0.5} + \dfrac{0.08824 + j0.10294}{s + j2} + c.c.t.$;

$v_2(t) = [-0.17647e^{-0.5t} + 2(0.0824 \cos 2t + 0.1029 \sin 2t]u(t)$.

(d) For $j(t) = \cos 2t$:

$V_2 = \dfrac{0.04413}{s + 0.5} + \dfrac{0.10294 - j0.08824}{s + j2} + c.c.t.$

P.10.4 $V_2 = \dfrac{1 - s}{2s + 4}J$.

(a) For $j(t) = \delta(t)$: $v_2(t) = -0.5\delta(t) + 1.5e^{-2t}u(t)$.

(b) For $j(t) = u(t)$: $V_2 = \dfrac{0.25}{s} - \dfrac{0.75}{s + 2}$,

$v_2(t) = (0.25 - 0.75e^{-2t})u(t)$.

(c) For $j(t) = \sin 2t$:

$V_2 = \dfrac{0.375}{s + 2} + \dfrac{-0.1875 + j0.0625}{s + j} + c.c.t.$

(d) For $j(t) = \cos 2t$:

$V_2 = -\dfrac{0.375}{s + 2} + \dfrac{-0.0625 + j0.1875}{s + j} + c.c.t.$

P.10.5 $V_2 = \dfrac{sJ}{3s + 10}$.

(a) For $j(t) = \delta(t)$: $v_2(t) = \dfrac{1}{3}\delta(t) - \dfrac{10}{9}e^{-10t/3}u(t)$.

(b) For $j(t) = u(t)$: $v_2(t) = \dfrac{1}{3}e^{-10t/3}u(t)$.

(c) For $j(t) = \sin 2t$:

$V_2 = \dfrac{-0.14706}{s + \dfrac{10}{3}} + \dfrac{0.07353 + j0.04412}{s + j2} + c.c.t.$

(d) For $j(t) = \cos t$:

$V_2 = \dfrac{0.2451}{s + \dfrac{10}{3}} + \dfrac{0.04412 - j0.07353}{s + j2} + c.c.t.$

P.10.6 $V_{out} = \dfrac{2 - 4s}{2s + 5} E.$

(a) For $e(t) = \delta(t)$: $V_{out} = -2 + \dfrac{6}{s + 2.5}$,

$v_{out}(t) = -2\delta(t) + 6e^{-2.5t}u(t).$

(b) For $e(t) = u(t)$: $V_{out} = \dfrac{2}{5s} - \dfrac{12}{5(s + 2.5)}$,

$v_{out}(t) = \left(\dfrac{2}{5} - \dfrac{12}{5} e^{-2.5t}\right)u(t).$

(c) For $e(t) = \sin 2t$:

$V_{out} = \dfrac{1.17073}{s + 2.5} + \dfrac{-0.58537 - j0.26829}{s + j2} + c.c.t.$

P.10.7 $V_{out} = \dfrac{sE}{s + 2} - \dfrac{J_0}{s + 2}$, $J_0 = 3.$

(a) For $e(t) = \delta(t)$: $V_{out} = 1 - \dfrac{5}{s + 2}$,

$v_{out}(t) = \delta(t) - 5e^{-2t}u(t).$

(b) For $e(t) = u(t)$: $V_{out} = \dfrac{-2}{s + 2}$,

$v_{out}(t) = -2e^{-2t}u(t).$

(c) For $e(t) = \sin 2t$:

$V_{out} = \dfrac{-3.5}{s + 2} + \dfrac{0.25 + j0.25}{s + j2} + c.c.t.$

P.10.8 $V_{out} = \dfrac{2E + 3}{s + 2}.$

(a) For $e(t) = \delta(t)$: $V_{out} = \dfrac{5}{s + 2}$, $v_{out}(t) = 5e^{-2t}u(t).$

(b) For $e(t) = u(t)$: $V_{out} = \dfrac{1}{s} + \dfrac{2}{s + 2}$,

$v_{out}(t) = (1 + 2e^{-2t})u(t).$

(c) For $e(t) = \sin 2t$:

$V_{out} = \dfrac{3.5}{s + 2} + \dfrac{-0.25 + j0.25}{s + j2} + c.c.t.$

P.10.9 $V_{out} = \dfrac{sJ - 4}{3s + 2}.$

(a) For $j(t) = \delta(t)$: $V_{out} = \dfrac{1}{3s} - \dfrac{14}{9} \dfrac{1}{s + \dfrac{2}{3}}$,

$v_{out}(t) = \left(\dfrac{1}{3} - \dfrac{14}{9} e^{-2t/3}\right)u(t).$

(b) For $j(t) = u(t)$: $V_{out} = \dfrac{-1}{s + \dfrac{2}{3}}$,

$V_{out}(t) = -e^{-2t/3}u(t).$

(c) For $j(t) = \sin 2t$:

$V_{out} = \dfrac{-1.43333}{s + \dfrac{2}{3}} + \dfrac{0.05 + j0.15}{s + j2} + c.c.t.$

P.10.10 $V_{out} = \dfrac{J + 6}{4s + 3}.$

(a) For $j(t) = \delta(t)$: $V_{out} = \dfrac{7}{4} \dfrac{1}{s + \dfrac{3}{4}}$,

$v_{out}(t) = \dfrac{7}{4} e^{-3t/4}u(t).$

(b) For $j(t) = u(t)$: $V_{out} = \dfrac{1}{3s} + \dfrac{7}{6} \dfrac{1}{s + \dfrac{3}{4}}$,

$v_{out}(t) = \left(\dfrac{1}{3} + \dfrac{7}{6} e^{-3t/4}\right)u(t).$

P.10.11 $V_{out} = \dfrac{s - 2}{5s + 2} J - \dfrac{4E_0}{5s + 2}$, $E_0 = 3.$

(a) For $j(t) = \delta(t)$: $V_{out} = \dfrac{1}{5} - \dfrac{72}{25} \dfrac{1}{s + \dfrac{2}{5}}$,

$v_{out}(t) = \dfrac{1}{5} \delta(t) - \dfrac{72}{25} e^{-0.4t}u(t).$

(b) For $j(t) = u(t)$: $V_{out} = \dfrac{-1}{s} - \dfrac{1.2}{s + 0.4}$,

$v_{out}(t) = (-1 - 1.2e^{-0.4t})u(t).$

P.10.12 $V_{out} = \dfrac{12 + J(1 - 2s)}{2s + 5}.$

(a) For $j(t) = \delta(t)$: $V_{out} = -1 + \dfrac{9}{s + 2.5}$,

$v_{out}(t) = -\delta(t) + 9e^{-2.5t}u(t).$

(b) For $j(t) = u(t)$: $V_{out} = \dfrac{1}{5s} + \dfrac{24}{5}\dfrac{1}{s + 2.5}$,

$v_{out}(t) = \dfrac{1}{5}(1 + 24e^{-2.5t})u(t).$

P.10.13 $V_{out} = \dfrac{6 + E(1 - 2s)}{s + 2.5}.$

(a) For $e(t) = \delta(t)$: $V_{out} = -2 + \dfrac{12}{s + 2.5}$,

$v_{out}(t) = -2\delta(t) + 12e^{-2.5t}u(t).$

(b) For $e(t) = u(t)$: $V_{out} = \dfrac{0.4}{s} + \dfrac{3.6}{s + 2.5}$,

$v_{out}(t) = (0.4 + 3.6e^{-2.5t})u(t).$

P.10.14 $V_{out} = \dfrac{(2s - 2)E - 24}{5s + 1}.$

For $e(t) = \sin 3t$:

$V_2 = \dfrac{-4.95929}{s + 0.2} + \dfrac{0.07965 + j0.19469}{s + j3} + c.c.t.$

P.10.15 $V_{out} = \dfrac{J + 2}{2s^2 + 3s + 3}.$

(a) For $j(t) = 0$: $V_{out} = \dfrac{0 + j0.5164}{s + 0.75 + j0.96825} + c.c.t.;$

$v_{out}(t) = (2e^{-0.75t}0.5 \sin 0.96825t)u(t).$

(b) For $j(t) = u(t)$:

$V_{out} = \dfrac{0.3333}{s} + \dfrac{0.16667 + j0.38730}{s + 0.75 + j0.96825} + c.c.t.$

P.10.16 $V_{out} = \dfrac{(3s + 3)E + 12}{s + 6}.$

(a) For $e(t) = 0$: $v_{out} = 12e^{-6t}u(t).$

(b) For $e(t) = \delta(t)$: $V_{out} = 3 - \dfrac{3}{s + 6}$,

$V_{out}(t) = 3\delta(t) - 3e^{-6t}u(t).$

P.10.17 $V_{out} = \dfrac{(s - 0.5)J + 4s + 4}{2s^2 + 3.5s}.$

(a) For $j(t) = o$: $V_{out} = \dfrac{8}{7s} + \dfrac{6}{7}\dfrac{1}{s + \dfrac{7}{4}}$,

$v_{out}(t) = \left(\dfrac{8}{7} + \dfrac{6}{7}e^{-1.75t}\right)u(t).$

(b) For $j(t) = \delta(t)$: $V_{out} = \dfrac{1}{s} + \dfrac{1.5}{s + 1.75}$,

$v_{out}(t) = (1 + 1.5e^{-1.75t})u(t).$

P.10.18 $V_{out} = \dfrac{s^3}{2s^3 + 3s^2 + 2s + 1} J.$

(a) For $j(t) = \delta(t)$:

$V_{out} = 0.5 + \dfrac{0.5}{s + 1} + \dfrac{-0.125 + j0.04725}{s + 0.25 + j0.66144} + c.c.t.$

(b) For $j(t) = u(t)$:

$V_{out} = \dfrac{0.5}{s + 1} + \dfrac{0 - j0.18898}{s + 0.25 + j0.66144} + c.c.t.$

P.10.19 $V_{out} = \dfrac{-6s - 3}{10s^2 + 7s + 1} E.$

For $e(t) = u(t)$: $V_{out} = \dfrac{-3}{s} + \dfrac{3}{s + 0.2}$,

$v_{out}(t) = (-3 + 3e^{-0.2t})u(t).$

P.10.20 $V_{out} = \dfrac{4s + 3.5 + 0.125E}{s^2 + s + 0.125}.$

(a) For $e(t) = 0$: $V_{out} = \dfrac{0.412132}{s + 0.14645} - \dfrac{0.12132}{s + 0.85355}.$

(b) For $e(t) = u(t)$:

$10.19 V_{out} = \dfrac{4.2981}{s + 0.14645} - \dfrac{0.2981}{s + 0.85355}.$

P.10.21 $V_{out} = \dfrac{4s - 5 + sE}{s^2 + 2s + 1.25}.$

(a) For $e(t) = 0$: $V_{out} = \dfrac{1 + j2}{s + 1 + j0.5} + c.c.t.$

(b) For $e(t) = u(t)$: $V_{out} = \dfrac{2 - j8}{s + 1 + j0.5} + c.c.t.$

P.10.22 $V_{out} = \dfrac{8s + 28}{2s^2 + 9s + 6} = \dfrac{3.74078}{s + 0.81386}$

$\qquad\qquad\qquad + \dfrac{0.25922}{s + 0.25922}.$

P.10.23 $V_{out} = \dfrac{s^2 E}{3s^2 + 2s + 1}.$

(a) For $e(t) = 0$:

$V = \dfrac{-0.3333 - j0.4714}{s + 0.33333 + j0.4714} + c.c.t.$

(b) For $e(t) = u(t)$:

$V_{out} = \dfrac{-0.166666 + j0.82496}{s + 0.33333 + j0.4714} + c.c.t.$

P.10.24 $V_{out} = \dfrac{-16 + (6s + 4)E}{9s + 2}.$

(a) For $e(t) = 0$: $V_{out} - \dfrac{16}{9}\dfrac{1}{s + \dfrac{2}{9}}.$

(b) For $e(t) = u(t)$: $V_{out} = 0.666667 - \dfrac{1.48148}{s + \dfrac{2}{9}}.$

P.10.25 $V_{out} = \dfrac{10s + 3 + sJ}{4s^2 + 5s + 1}.$

(a) For $j(t) = 0$. $V_{out} = \dfrac{0.166667}{s + 0.25} + \dfrac{2.33333}{s + 1}.$

(b) For $j(t) = \delta(t)$: $V_{out} = \dfrac{0.083333}{s + 0.25} + \dfrac{2.66667}{s + 1}.$

P.10.26 $V_{out} = \dfrac{(2s + 5)E_0 + (4s + 6)E}{2s^2 + 11s + 11}.$

(a) For $e(t) = 0$, $E_0 = 2$:

$V_{out} = \dfrac{0.82592}{s + 1.31386} + \dfrac{1.17408}{s + 4.18614}.$

(b) For $e(t) = u(t)$, $E_0 = 0$:

$V_{out} = \dfrac{0.54545}{2} - \dfrac{0.09865}{s + 1.31386} - \dfrac{0.4468}{s + 4.18614}.$

(c) For $e(t) = u(t)$, $E_0 = 2$:

$V_{out} = \dfrac{0.54545}{s} + \dfrac{0.72727}{s + 1.31386} + \dfrac{0.72727}{s + 4.18614}.$

P.10.27 The dc source creates $E_0 = 1$ V and $J_0 = 1$ A.

$V_{out} = \dfrac{-6}{4s^2 + 6s + 3} = \dfrac{0 + j1.73205}{s + 0.75 + 0.43301} + c.c.t.$

P.10.28 For the first case $V_{out} = 2/s - 2/(s + 0.5)$ and $v_{out}(t) = (2 - 2e^{-0.5t})u(t)$ for $t \leq 1$. At $t = 1$ this gives $E_0 = 0.78694$ V.

For the second case with no source but with initial condition $V_{out} = E_0/(s + 1/6)$; $v_{out}(t) = E_0 e^{-(t-1)/6}u(t-1)$ for $t > 1$.

P.10.29 For the first case $V_{out} = (2/3)e^{-2t/3}u(t)$. The inductor current is $I_L = 1/s - [1/(s + 2/3)]$ and thus $i_L(t) = (1 - e^{-2t/3})u(t)$. This makes the initial current at switching $J_0 = 1 - e^{-2/3}$.

After switching $V = -2J_0/(s + 2)$ and $v_{out}(t) = -2J_0 e^{-2(t-1)}u(t-1)$ for $t > 1$.

P.10.30 Only the first capacitor is charged to $V = 1/s - 1/(s + 0.5)$, thus $v(t) = (1 - e^{-0.5t})u(t)$. At $t = 1$ the initial voltage is $E_0 = 1 - e^{-0.5} = 0.3945$ V. There is no output across the second capacitor up to $t = 1$.

After switching the output voltage is

$$V = E_0\left(\dfrac{1}{3s} - \dfrac{1}{3}\dfrac{1}{s + \dfrac{2}{3}}\right)$$

or

$$v_{out}(t) = \dfrac{E_0}{3}(1 - e^{-2(t-1)/3})u(t-1)$$

for $t > 1$.

P.10.31 In the first network the current is $I = 2/s - 2/(s + 1.5)$. The current through the inductor at $t = 1$ is $J_0 = 2(1 - e^{-1.5})$. Since the output resistor is $R = 1\,\Omega$, the output voltage is the same function as the current. After switching the output voltage is $V_{out} = J_0/(s + 1.5)$ and we have

$$v_{out}(t) = J_0 e^{-1.5(t-1)}u(t-1)$$

610   Solutions to Problems

SOLUTIONS CHAPTER 11

We do not give solutions for plotting results since such problems can be displayed by running the program from Appendix B.

To see exact equivalents to Bode plots run frequency domain analysis with logarithmic scale and in dB. Give the initial and final frequency values as powers of 10.

P.11.8a $D = s^2 + 3s + 1$
$$= (s + 0.38197)(s + 2.61803).$$

$$T_V = \frac{V_2}{E} = \frac{s^2}{(s + 0.38197)(s + 2.61803)}.$$

$$Y_{TR} = \frac{I_{out}}{E} = \frac{s^2}{(s + 0.38197)(s + 2.61803)}.$$

$$Y_{in} = \frac{I_1}{E} = \frac{2s(s + 1)}{(s + 0.38197)(s + 2.61803)}.$$

P.11.8b $D = s^2 + 3s + 1$
$$= (s + 0.38197)(s + 2.61803).$$

$$Z_{TR} = \frac{V_2}{E} = \frac{1}{(s + 0.38197)(s + 2.61803)}.$$

$$Y_{TR} = \frac{I_{out}}{E} = \frac{1}{(s + 0.38197)(s + 2.61803)}.$$

$$Y_{in} = \frac{I_1}{E} = \frac{s + 2}{(s + 0.38197)(s + 2.61803)}.$$

P.11.8c $D = s^2 + 3s + 1$
$$= (s + 0.38197)(s + 2.61803).$$

$$Y_{in} = \frac{I_1}{E} = \frac{s(s + 2)}{(s + 0.38197)(s + 2.61803)}.$$

$$Y_{TR} = \frac{I_2}{E} = \frac{s}{(s + 0.38197)(s + 2.61803)}.$$

$$T_V = \frac{V_{out}}{E} = \frac{1}{(s + 0.38197)(s + 2.61803)}.$$

P.11.8d $D = s^2 + 3s + 1$
$$= (s + 0.38197)(s + 2.61803).$$

$$Y_{in} = \frac{I_1}{E} = \frac{2(s + 0.5)}{(s + 0.38197)(s + 2.61803)}.$$

$$Y_{TR} = \frac{I_{out}}{E} = \frac{s}{(s + 0.38197)(s + 2.61803)}.$$

$$T_V = \frac{V_{out}}{E} = \frac{s^2}{(s + 0.38197)(s + 2.61803)}.$$

SOLUTIONS CHAPTER 12

P.12.1a $T = \dfrac{s^2 C_1 C_2}{s^2 C_1 C_2 + s(C_1 G_2 + C_2 G_2) + G_1 G_2}.$
After simplification $T = s^2 C^2/(s^2 C^2 + 2sCG + G^2)$. For this case $\omega_0 = G/C$ and $Q = 1/2$. The network has a double real pole at $p_{1,2} = -G/C$.

P.12.1b $T = \dfrac{G_1 G_2}{s^2 C_1 C_2 + sC_2(G_1 + G_2) + G_1 G_2}.$
After simplification $T = G^2/(s^2 C^2 + 2sCG + G^2)$, $\omega_0 = G/C$, $Q = 1/2$, $p_{1,2} = -G/C$.

P.12.1c
$$T = \frac{sC_2 G_1}{s^2 C_1 C_2 + s(G_2 C_2 + G_1 C_2 + G_2 C_1) + G_1 G_2 + G_2 G_3}.$$
After simplification

$$T = \frac{sCG}{s^2 C^2 + 3sCG + 2G^2} = \frac{G}{C}\frac{s}{\left(s + \dfrac{2G}{C}\right)\left(s + \dfrac{G}{C}\right)}$$

$\omega_0 = (G/C)\sqrt{2}$, $Q = \sqrt{2}/3$.

P.12.2a $T = \dfrac{sC + G}{G}.$

P.12.2b $T = \dfrac{s(C_1 + C_2) + G}{sC_2 + G}.$

P.12.2c $T = \dfrac{-sC_1 - G_1}{G_2}.$

P.12.2d $T = \dfrac{-G_2}{sC + G_1}$.

P.12.2e $T = \dfrac{sC_1G_3 - G_1G_2}{sC_1G_3 + G_1G_3}$.

P.12.3 $T = \dfrac{G_1G_2}{(sC_1 + G_1)(sC_2 + G_2)}$.

After simplification

$$T = \dfrac{G^2}{C^2\left(s + \dfrac{G}{C}\right)^2}$$

The network has a double pole. If the amplifier is short-circuited and the same resistors and capacitors used, then

$$T = \dfrac{G^2}{s^2C^2 + 3sCG + 2G^2}$$

P.12.4 $T = \dfrac{s^2C_1C_2}{s^2C_1C_2 + sG_2(C_1 + C_2) + G_1G_2}$.

P.12.5 $T = \dfrac{-sC_1G_1}{s^2C_1C_2 + sG_2(C_1 + C_2) + G_1G_2}$.

P.12.6 $T = \dfrac{1}{s^2 + 4s + 2}$

$$= \dfrac{1}{(s + 0.585786)(s + 3.41421)}.$$

P.12.7

$$T = \dfrac{s^2C_1C_2}{s^2C_1C_2 + s(G_2C_1 + G_1C_2 + G_2C_2 - G_1C_2A) + G_1G_2}.$$

After simplification

$$T = \dfrac{s^2C^2}{s^2C^2 + sCG(3 - A) + G^2}$$

$\omega_0 = G/C$, $Q = 1/(3 - A)$. To get $Q = 2$ we must select $A = 2.5$.

P.12.8 $T = \dfrac{A}{s^3 + s^2(5 - A) + s(6 - 2A) + 1}$.

P.12.9 $T = \dfrac{6A}{2s^3 + s^2(15 - A) + s(23 - 5A) + 6}$.

P.12.10 $Z_{in} = \dfrac{sG_2C_4}{G_1G_3G_5}$.

P.12.11 $T = \dfrac{s^2C^2 - sCG_1 + G^2}{s^2C^2 + sCG_1 + G^2}$.

$\omega_0 = G/C$ and $Q = G/G_1$.

P.12.12 $T = \dfrac{s^2 + s(2G_1 - G_2) + G_1}{(1 + G_2)(s^2 + 2sG_1 + G_1)}$.

Selecting $G_2 = 4G_1$ results in

$$T = \dfrac{s^2 - 2sG_1 + G_1}{(1 + G_2)(s^2 + 2sG_1 + G_1)}$$

which is an all-pass. For $G_1 = 1$ it has a double pole at $p_{1,2} = -1$ and a double zero $z_{1,2} = +1$.

P.12.13 $T = \dfrac{G_1 + G_2}{s^2G_2 + s(2G_2 - G_1) + G_2}$.

P.12.14 $T = \dfrac{-sG_1}{s^2 + s(G_1 + G_2) + G_1G_2}$.

P.12.15 $T = \dfrac{-sC_1 - 1}{s^2C_1C_2 + 3sC_2 + 1}$.

P.12.16 $T = \dfrac{s^2C_1C_2 + s(2C_2 - C_1) + 1}{2s^2C_1C_2 + 4sC_2 + 2}$.

P.12.17 Denote $M = (G_5 + G_6)/(G_3 + G_4)$. Then

$$T = \dfrac{s^2C_1C_2G_3M}{s^2C_1C_2G_6 + sC_2G_1G_4M + G_1G_2G_5}$$

P.12.18

$$T = \dfrac{G_1G_2A}{s^2C_1C_2 + s[C_1G_2(1 - A) + C_2G_1 + C_2G_2] + G_1G_2}.$$

After simplification

$$T = \dfrac{G^2}{C^2}\dfrac{A}{s^2 + s\dfrac{G}{C}(3 - A) + \dfrac{G^2}{C^2}}$$

For the first network $G/C = \sqrt{2}$ and $A = 3 - \sqrt{2} = 1.5858$. For the second network $G/C = \sqrt{5}$ and $A = 3 - (2/\sqrt{5}) = 2.10557$.

SOLUTIONS CHAPTER 13

P.13.1

$$
Z = \begin{bmatrix} \dfrac{s^2LC(G_1+G_2)+s(LG_1G_2+C)+G_2}{sC(G_1+G_2)+G_1G_2} & \dfrac{s^2LC(G_1+G_2)+s(LG_1G_2+C)}{sC(G_1+G_2)+G_1G_2} \\[4mm] \dfrac{s^2LC(G_1+G_2)+s(LG_1G_2+C)}{sC(G_1+G_2)+G_1G_2} & \dfrac{s^2LC(G_1+G_2)+s(LG_1G_2+C)+G_1}{sC(G_1+G_2)+G_1G_2} \end{bmatrix}
$$

P.13.2

$$
Z = \begin{bmatrix} \dfrac{s^2+2s+2}{s^2+2s} & \dfrac{2s+2}{s^2+2s} \\[4mm] \dfrac{2s+2}{s^2+2s} & \dfrac{s^2+2s+2}{s^2+2s} \end{bmatrix}
$$

P.13.3

$$
T = \begin{bmatrix} \dfrac{s^2+2s+2}{2s+2} & \dfrac{s^2+2s+2}{2s+2} \\[4mm] \dfrac{s^2+2s}{2s+2} & \dfrac{s^2+2s+2}{2s+2} \end{bmatrix}
$$

P.13.4

$$
Z = \begin{bmatrix} \dfrac{(Z_1+Z_3)(Z_2+Z_4)}{Z_1+Z_2+Z_3+Z_4} & \dfrac{Z_2Z_3-Z_1Z_4}{Z_1+Z_2+Z_3+Z_4} \\[4mm] \dfrac{Z_2Z_3-Z_1Z_4}{Z_1+Z_2+Z_3+Z_4} & \dfrac{(Z_1+Z_2)(Z_3+Z_4)}{Z_1+Z_2+Z_3+Z_4} \end{bmatrix}
$$

P.13.5

$$
T = \begin{bmatrix} \dfrac{Z_2+Z_1}{Z_2-Z_1} & \dfrac{2Z_1Z_2}{Z_2-Z_1} \\[4mm] \dfrac{2}{Z_2-Z_1} & \dfrac{Z_2+Z_1}{Z_2-Z_1} \end{bmatrix}
$$

P.13.6

$$
T = \begin{bmatrix} \dfrac{(1-K)G_2}{KG_1} & \dfrac{1}{KG_1} \\[4mm] \dfrac{G_2}{K} & \dfrac{1+K}{K} \end{bmatrix}
$$

P.13.7

$$
T = \begin{bmatrix} \dfrac{(1-K)G_2+G_3}{KG_1} & \dfrac{1}{KG_1} \\[4mm] \dfrac{G_2+G_3(1+K)}{K} & \dfrac{1+K}{K} \end{bmatrix}
$$

P.13.8

$$H = \begin{bmatrix} \dfrac{1}{G_1(1+K)} & \dfrac{-G_2K}{G_1(1+K)} \\[3mm] \dfrac{-K}{1+K} & \dfrac{G_2}{1+K} \end{bmatrix}$$

P.13.9

$$T = \begin{bmatrix} \dfrac{sC_2 + G}{sC_2 - g} & \dfrac{1}{sC_2 - g} \\[3mm] \dfrac{s^2C_1C_2 + s(C_1G + C_2G + C_2g)}{sC_2 - g} & \dfrac{s(C_1 + C_2)}{sC_2 - g} \end{bmatrix}$$

P.13.10

$$Z = \begin{bmatrix} \dfrac{s^2LC_2 + 1}{s^3LC_1C_2 + s(C_1 + C_2)} & \dfrac{1}{s^3LC_1C_2 + s(C_1 + C_2)} \\[3mm] \dfrac{1}{s^3LC_1C_2 + s(C_1 + C_2)} & \dfrac{s^2LC_1 + 1}{s^3LC_1C_2 + s(C_1 + C_2)} \end{bmatrix}$$

$$T = \begin{bmatrix} s^2LC_2 + 1 & sL \\ s^3LC_1C_2 + s(C_1 + C_2) & s^2LC_1 + 1 \end{bmatrix}$$

P.13.11

$$T = \begin{bmatrix} 1 & 1 \\ 0 & 1 \end{bmatrix} \begin{bmatrix} s^2 + 1 & s \\ s^3 + 2s & s^2 + 1 \end{bmatrix} \begin{bmatrix} 1 & 0 \\ 1 & 1 \end{bmatrix} = \begin{bmatrix} s^3 + 2s^2 + 3s + 2 & s^2 + s + 1 \\ s^3 + s^2 + 2s + 1 & s^2 + 1 \end{bmatrix}$$

P.13.12

$$Z = \begin{bmatrix} R_B + R_E & R_E \\ R_E - \alpha R_C & R_C(1 - \alpha) + R_E \end{bmatrix}$$

P.13.13

$$H = \begin{bmatrix} R_B + \dfrac{R_C R_E}{R_C(1 - \alpha) + R_E} & \dfrac{R_E}{R_C(1 - \alpha) + R_E} \\[3mm] \dfrac{\alpha R_C - R_E}{R_C(1 - \alpha) + R_E} & \dfrac{1}{R_C(1 - \alpha) + R_E} \end{bmatrix}$$

P.13.14

$$T = \begin{bmatrix} \dfrac{R_B + R_E}{R_E + \alpha R_C} & \dfrac{R_B R_C(1 - \alpha) + R_E(R_B + R_C)}{R_E - \alpha R_C} \\[3mm] \dfrac{1}{R_E - \alpha R_C} & \dfrac{R_C(1 - \alpha) + R_E}{R_E - \alpha R_C} \end{bmatrix}$$

P.13.15

$$Z = \begin{bmatrix} R_B + R_E & R_B \\ R_B + \alpha R_C & R_B + R_C \end{bmatrix}$$

$$H = \begin{bmatrix} R_E + \dfrac{R_B R_C(1 - \alpha)}{R_B + R_C} & \dfrac{R_B}{R_B + R_E} \\[3mm] \dfrac{-R_B - \alpha R_C}{R_B + R_C} & \dfrac{1}{R_B + R_C} \end{bmatrix}$$

P.13.16

$$T = \begin{bmatrix} \dfrac{sC + G}{sC} & \dfrac{2sC + G}{s^2 C^2} \\[3mm] G & \dfrac{sC + G}{sC} \end{bmatrix}$$

SOLUTIONS CHAPTER 14

P.14.1 $T_V = \dfrac{sMR_2}{s^2(L_1 L_2 - M^2) + s(L_1 R_2 + L_2 R_1) + R_1 R_2}$

$$= \dfrac{0.5s}{1.75s^2 + 3s + 1}$$

P.14.2 $T_V = \dfrac{1.75s^2 - 0.5s}{1.75s^2 + 6s + 1}$

P.14.3 $T_V = \dfrac{0.5s + 2}{0.75s^3 + 4s^2 + 6s + 3}$

P.14.4 $T_V = \dfrac{0.96s^2 + 1.8s + 1}{0.96s^2 + 2.6s + 1}$

P.14.5 $T_V = \dfrac{3s}{7s^2 + 18s + 2}$

P.14.6 $T_V = \dfrac{2s^2 + 2}{3s^3 + 6s^2 + 4s + 2}$

P.14.7 $T_V = \dfrac{5s^3 + 4s}{5s^3 + 2s^2 + 7s + 1}$

P.14.8 $Z_{TR} = \dfrac{s}{3s^4 + 4s^3 + 5s^2 + 2s + 1}$

P.14.9 $Y_{TR} = \dfrac{0.75s^2 + 2s + 1}{1.5s^2 + s}$

P.14.10 $T_V = \dfrac{3s + 2}{s(3s^3 + 10s^2 + 8s + 2)}$

P.14.11 $T_V = \dfrac{7s^2 + 14s}{7s^2 + 20s + 11}$

P.14.12 $T_V = \dfrac{73s^2 + 1200s}{144s^2 + 2600s + 10000}$

P.14.13 $T_V = \dfrac{16s^3 - 4s}{48s^3 + 96s^2 + 20s + 8}$

SOLUTIONS CHAPTER 15

P.15.1 $V_{GS} = E \dfrac{R_2}{R_1 + R_2} = \dfrac{E}{1.02}$,

$\dfrac{V_{out}}{E} = -\dfrac{g_m R_2 R_3}{R_1 + R_2} = -\dfrac{5}{1.02} = 4.902$

P.15.2 $V_G = \dfrac{E}{1.02}$, $V_S = \dfrac{E}{2.04}$,

$\dfrac{V_{out}}{E} = -\dfrac{2.5}{2.04} = 1.2255$

P.15.3–P.15.7 Require solutions by NETOPT.

P.15.8 Leads to quadratic equation $v^2 + 10v - 20 = 0$ with the solution $v = 1.7082$.

P.15.9 Leads to iterative solution of

$$\Delta x^k = \frac{-(e^{40v} + 10v - 21)}{40e^{40v} + 10}$$

$$x^{k+1} = x^k + \Delta x^k$$

P.15.10 Leads to iterative solution of

$$\begin{bmatrix} 1 + 2(v_1 - v_2) & -2(v_1 - v_2) \\ -2(v_1 - v_2) & 2(v_1 - v_2) + 2 \end{bmatrix} \begin{bmatrix} \Delta v_1^k \\ \Delta v_2^k \end{bmatrix}$$
$$= \begin{bmatrix} -v_1 - (v_1 - v_2)^2 + 2 \\ (v_1 - v_2)^2 - 2v_2 \end{bmatrix}$$
$$\begin{bmatrix} v_1^{k+1} \\ v_2^{k+1} \end{bmatrix} = \begin{bmatrix} v_1^k \\ v_2^k \end{bmatrix} + \begin{bmatrix} \Delta v_1^k \\ \Delta v_2^k \end{bmatrix}$$

P.15.11 Leads to iterative solution of

$$\begin{bmatrix} 2x_1 & -1 \\ 2 & -1 \end{bmatrix} \begin{bmatrix} \Delta x_1^k \\ \Delta x_2^k \end{bmatrix} = \begin{bmatrix} -x_1^2 + x_2 + 2 \\ -2x_1 + x_2 - 3 \end{bmatrix}$$
$$\begin{bmatrix} x_1^{k+1} \\ x_2^{k+1} \end{bmatrix} = \begin{bmatrix} x_1^k \\ x_2^k \end{bmatrix} + \begin{bmatrix} \Delta x_1^k \\ \Delta x_2^k \end{bmatrix}$$

P.15.12 Leads to iterative solution of

$$\begin{bmatrix} 1 & -2x_2 \\ 1 & 3 \end{bmatrix} \begin{bmatrix} \Delta x_1^k \\ \Delta x_2^k \end{bmatrix} = \begin{bmatrix} -x_1 + x_2^2 \\ -x_1 - 3x_2 + 5 \end{bmatrix}$$
$$\begin{bmatrix} x_1^{k+1} \\ x_2^{k+1} \end{bmatrix} = \begin{bmatrix} x_1^k \\ x_2^k \end{bmatrix} + \begin{bmatrix} \Delta x_1^k \\ \Delta x_2^k \end{bmatrix}$$

SOLUTIONS CHAPTER 16

P.16.1 $Y = \dfrac{sC + G}{s^2 LC + s(CR + LG) + 1 + RG}$,

$Q = \dfrac{\sqrt{LC(RG + 1)}}{CR + LG}$, $\omega_0 = \left[\dfrac{RG + 1}{LC} \right]^{-1/2}$

$S_R^{\omega_0} = S_G^{\omega_0} = \dfrac{1}{2} \dfrac{RG}{RG + 1}$, $S_L^{\omega_0} = S_C^{\omega_0} = -\dfrac{1}{2}$

$S_R^Q = \dfrac{RG}{2(RG + 1)} - \dfrac{RC}{RC + LG}$

$S_G^Q = \dfrac{RG}{2(RG + 1)} - \dfrac{LG}{RC + LG}$

$S_L^Q = \dfrac{1}{2} - \dfrac{LG}{CR + LG}$, $S_C^Q = \dfrac{1}{2} - \dfrac{RC}{RC + LG}$

Numerically $Q = 0.70711$, $\omega_0 = 1.414$, $S_R^{\omega_0} = S_G^{\omega_0} = 1/4$, $S_L^{\omega_0} = S_C^{\omega_0} = -1/2$, $S_R^Q = S_G^Q = -1/4$, $S_L^Q = S_C^Q = 0$.

P.16.2 $F = \dfrac{s(G + g)}{s^2 G + s(1 - gG) + G}$

For $Q = 4$ we need $g = 3/4 = 0.75$. For this choice $\omega_0 = 1$ and $Q = G/(1 - gG)$. Then $S_G^{\omega_0} = S_g^{\omega_0} = 0$, $S_G^Q = 4$, $S_g^Q = 3$.

P.16.3 $F = \dfrac{G - g}{s^2 + 2sG + Gg}$

For $G = 1$ and $Q = 1$ we need $g = 4$. Then $\omega_0 = 2$ and $S_G^{\omega_0} = S_g^{\omega_0} = 1/2$, $S_G^Q = -1/2$, $S_g^Q = 1/2$.

P.16.4 $F = \dfrac{1-g}{s^2 + 2s + g}$

For $g = 4$, $S_g^F = 4/3 - 4/(s^2 + 2s + 4)$. For $\omega = \omega_0 = 2$ we have $F(j2) = 3/4 = 0.75$, $S_g^F(j2) = 4/3 + j$ and $S_g^{|F|}(j2) = \text{Re } S_g^F = 4/3$.

P.16.5 NETOPT finds that the maximum of $S_g^{|F|} = 1.6666$ occurs at $\omega = 2.82828$. At this frequency $|F(j\omega)| = 0.433$.

P.16.6 $F = \dfrac{s^2 A}{s^2 + s(2G_2 + G_1 - G_1 A) + G_1 G_2}$

To obtain $Q = 4$ and $\omega_0 = 3$ with $G_1 = G_2 = G$, we need $A = 2.75$ and $G = 3$.

$$S_A^F = S_A^N - S_A^D = 1 + \frac{3sA}{s^2 + 3s(3 - A) + 9}$$

$S_A^F(j2) = 1.9083 + j3.0275$ and $S_A^{|F|}(j2) = \text{Re } S_A^F = 1.9083$.

P.16.7

$$F = \frac{sG_1 A}{s^2 + s(2G_2 + G_1 + G_3 - G_3 A) + G_2(G_1 + G_3)}$$

A possible choice is $G_1 = G_2 = G_3 = G = 1$ and $A = 2$:

$$F = \frac{sA}{s^2 + s(4 - A) + 2}$$

Then

$$S_A^F = S_A^N - S_A^D = 1 + \frac{2s}{s^2 + s(4 - A) + 2}$$

$S_A^F(j3) = 1.4235 - j0.4941$ and $S_A^{|F|}(j3) = 1.4235$. $Q = A/(4 - A)$ and $S_A^Q = 1$. Since the denominator is $P = s^2 + s(4 - A) + 2$, we have $dP/dA = -s$, $dP/ds = 2s + 2$, the pole is $-1 - j1$. Using Eq. (16.6.4) $d(\text{pole})/dA = 0.5 + j0.5$ and the normalized sensitivity is $S_A^{pole} = 0 - j$.

P.16.8 $F = \dfrac{sC_1 G_3 - G_1 G_2}{sC_1 G_3 + G_1 G_3}$

Element selection is $G_1 = G_2 = G_3 = 1\,\text{S}$ and $C_1 = 1\,\text{F}$. The sensitivities at $\omega = 3$ are

$S_{C_1}^F = 0 - j0.6$, $S_{C_1}^{|F|} = 0$,

$S_{G_1}^F = 0 + j0.6$, $S_{G_1}^{|F|} = 0$,

$S_{G_2}^F = 0.1 + j0.3$, $S_{G_2}^{|F|} = 0.1$,

$S_{G_3}^F = -0.1 - j0.3$, $S_{G_3}^{|F|} = -0.1$,

$zero = \dfrac{G_1 G_2}{C_1 G_3}$,

$S_{C_1}^{zero} = S_{G_3}^{zero} = -1$, $S_{G_1}^{zero} = S_{G_2}^{zero} = 1$,

$pole = -\dfrac{G_1}{C_1}$,

$S_{C_1}^{pole} = -1$, $S_{G_1}^{pole} = 1$, $S_{G_2}^{pole} = S_{G_3}^{pole} = 0$.

P.16.9 $F = \dfrac{6A}{s^3 + s^2(9 - A) + s(18 - 5A) + 6}$

for $A = 2$

$$S_A^F = 1 - \frac{2s^2 + 10s}{s^3 + 7s^2 + 8s + 6} \quad \text{for } s = j3$$

$S_A^F(j3) = 1.2873 - j0.5414$ and $S_A^{|F|}(j3) = 1.2873$.

P.16.10 $F = \dfrac{s^2 + s(2G_1 - G_2) + G_1}{(1 + G_2)(s^2 + 2sG_1 + G_1)}$

$$S_{G_1}^F = S_{G_1}^N - S_{G_1}^D = \frac{2sG_1 + G_1}{s^2 + s(2G_1 - G_2) + G_1}$$

$$- \frac{2sG_1 + G_1}{s^2 + 2sG_1 + G_1}$$

For $G_2 = 4G_1$ and $s = j3$, $S_{G_1}^F(j3) = -0.72 + j0.12$, $S_{G_1}^{|F|}(j3) = -0.72$.

$$S_{G_2}^F = \frac{-sG_2}{s^2 + s(2G_1 - G_2) + G_1} - \frac{G_2}{1 + G_2}$$

For the same choice of frequency and elements, $S_{G_2}^F(j3) = -0.08 + j0.96$ and $S_{G_2}^{|F|}(j3) = -0.08$.

SOLUTIONS CHAPTER 17

P.17.1

$$
\begin{bmatrix}
G_1 & -G_1 & 0 & 1 & 0 \\
-G_1 & G_1 + G_2 + sC & -G_2 & 0 & 0 \\
0 & -G_2 & G_2 & 0 & 1 \\
1 & 0 & 0 & 0 & 0 \\
0 & 0 & 1 & 0 & 0
\end{bmatrix}
\begin{bmatrix}
V_1 \\ V_2 \\ V_3 \\ I_E \\ I_{sc}
\end{bmatrix}
=
\begin{bmatrix}
0 \\ 0 \\ 0 \\ E \\ 0
\end{bmatrix}
$$

P.17.2

$$
\begin{bmatrix}
G_1 & -G_1 & 0 & 1 & 0 \\
-G_1 & G_1 + sC_1 & 0 & 0 & 1 \\
0 & 0 & sC_2 + G_2 & 0 & -1 \\
1 & 0 & 0 & 0 & 0 \\
0 & 1 & -1 & 0 & -sL
\end{bmatrix}
\begin{bmatrix}
V_1 \\ V_2 \\ V_3 \\ I_E \\ I_L
\end{bmatrix}
=
\begin{bmatrix}
0 \\ 0 \\ 0 \\ E \\ 0
\end{bmatrix}
$$

P.17.3

$$
\begin{bmatrix}
G + sC_1 & -G \\
-G + g & G + sC_2
\end{bmatrix}
\begin{bmatrix}
V_1 \\ V_2
\end{bmatrix}
=
\begin{bmatrix}
J \\ 0
\end{bmatrix}
$$

P.17.4

$$
\begin{bmatrix}
G + sC & -sC & -G & 0 & 0 \\
-sC & sC & 0 & 0 & 1 \\
-G & 0 & G & 1 & 0 \\
0 & 0 & 1 & 0 & 0 \\
0 & 1 & 0 & -r & 0
\end{bmatrix}
\begin{bmatrix}
V_1 \\ V_2 \\ V_3 \\ I_1 \\ I_2
\end{bmatrix}
=
\begin{bmatrix}
J \\ 0 \\ 0 \\ 0 \\ 0
\end{bmatrix}
$$

P.17.5

$$
\begin{bmatrix}
G_1 & -G_1 & 0 & 1 & 0 \\
-G_1 & G_1 + sC_1 + sC_3 & -sC_3 & 0 & 1 \\
0 & -sC_3 & G_2 + sC_2 + sC_3 & 0 & -1 \\
1 & 0 & 0 & 0 & 0 \\
0 & 1 & -1 & 0 & -sL
\end{bmatrix}
\begin{bmatrix}
V_1 \\ V_2 \\ V_3 \\ I_E \\ I_L
\end{bmatrix}
=
\begin{bmatrix}
0 \\ 0 \\ 0 \\ E \\ 0
\end{bmatrix}
$$

P.17.6

$$
\begin{bmatrix}
G_1 + sC & -sC & 0 \\
-sC - g & sC + G_2 & 1 \\
0 & 1 & -sL
\end{bmatrix}
\begin{bmatrix}
V_1 \\ V_2 \\ I_L
\end{bmatrix}
=
\begin{bmatrix}
J \\ 0 \\ 0
\end{bmatrix}
$$

P.17.7

$$
\begin{bmatrix}
0 & 0 & 1 & 0 \\
0 & G & 0 & 1 \\
1 & 0 & 0 & 0 \\
1 & -1+B & 0 & 0
\end{bmatrix}
\begin{bmatrix}
V_1 \\
V_2 \\
I_E \\
I_O
\end{bmatrix}
=
\begin{bmatrix}
0 \\
0 \\
E \\
0
\end{bmatrix}
$$

P.17.8

$$
\begin{bmatrix}
G & -G & 0 & 0 & 1 & 0 & 0 & 0 \\
-G & G & 0 & 0 & 0 & 0 & 1 & 0 \\
0 & 0 & sC & 0 & 0 & 1 & -1 & 0 \\
0 & 0 & 0 & 0 & 0 & -1 & 0 & 1 \\
1 & 0 & 0 & 0 & 0 & 0 & 0 & 0 \\
0 & 0 & 1 & -1 & 0 & -sL & 0 & 0 \\
0 & 1 & -1 & 0 & 0 & 0 & 0 & 0 \\
0 & 0 & 0 & 1 & 0 & 0 & -r & 0
\end{bmatrix}
\begin{bmatrix}
V_1 \\
V_2 \\
V_3 \\
V_4 \\
I_E \\
I_L \\
I_1 \\
I_2
\end{bmatrix}
=
\begin{bmatrix}
0 \\
0 \\
0 \\
0 \\
E \\
0 \\
0 \\
0
\end{bmatrix}
$$

P.17.9

$$
\begin{bmatrix}
G_1 & -G_1 & 0 & 0 & 1 & 0 \\
-G_1 & (G_1+G_2+G_3+sC_1) & -G_2 & -G_3 & 0 & 0 \\
0 & -G_2 & G_2+sC_2 & 0 & 0 & 0 \\
0 & -G_3 & 0 & G_3 & 0 & 1 \\
1 & 0 & 0 & 0 & 0 & 0 \\
0 & 0 & A & -1 & 0 & 0
\end{bmatrix}
\begin{bmatrix}
V_1 \\
V_2 \\
V_3 \\
V_4 \\
I_E \\
I_A
\end{bmatrix}
=
\begin{bmatrix}
0 \\
0 \\
0 \\
0 \\
E \\
0
\end{bmatrix}
$$

P.17.10

$$
\begin{bmatrix}
G_1+G_3 & -G_1 & 0 & -G_3 & 0 & 1 & 0 & 0 \\
-G_1 & (G_1+G_2+G_5) & -G_2 & 0 & -G_5 & 0 & 0 & 0 \\
0 & -G_2 & (G_2+G_4) & -G_4 & 0 & 0 & 1 & 0 \\
-G_3 & 0 & -G_4 & (G_3+G_4+G_6) & -G_6 & 0 & 0 & 0 \\
0 & -G_5 & 0 & -G_6 & (G_5+G_6) & 0 & 0 & 1 \\
1 & 0 & 0 & 0 & 0 & 0 & 0 & 0 \\
0 & -A_1 & 1 & 0 & 0 & 0 & 0 & 0 \\
0 & 0 & 0 & -A_2 & 1 & 0 & 0 & 0
\end{bmatrix}
\begin{bmatrix}
V_1 \\
V_2 \\
V_3 \\
V_4 \\
V_5 \\
I_E \\
I_{A1} \\
I_{A2}
\end{bmatrix}
=
\begin{bmatrix}
0 \\
0 \\
0 \\
0 \\
0 \\
E \\
0 \\
0
\end{bmatrix}
$$

P.17.11

$$\begin{bmatrix} G_1 & -G_1 & 0 & 0 & 1 & 0 & 0 \\ -G_1 & G_1 + sC & 0 & -sC & 0 & 1 & 0 \\ 0 & 0 & G_2 & 0 & 0 & -1 & \alpha \\ 0 & -sC & 0 & sC & 0 & 0 & 1 \\ 1 & 0 & 0 & 0 & 0 & 0 & 0 \\ 0 & 1 & -1 & 0 & 0 & -sL & 0 \\ 0 & 0 & 0 & 1 & 0 & 0 & 0 \end{bmatrix} \begin{bmatrix} V_1 \\ V_2 \\ V_3 \\ V_4 \\ I_E \\ I_L \\ I_{cc} \end{bmatrix} = \begin{bmatrix} 0 \\ 0 \\ 0 \\ 0 \\ E \\ 0 \\ 0 \end{bmatrix}$$

P.17.12

$$\begin{bmatrix} sC_1 + sC_2 & -sC_2 & 0 \\ -sC_2 & sC_2 + sC_3 & 1 \\ 0 & 1 & -R \end{bmatrix} \begin{bmatrix} V_1 \\ V_2 \\ I_R \end{bmatrix} = \begin{bmatrix} J \\ 0 \\ 0 \end{bmatrix}$$

P.17.13

$$\begin{bmatrix} 0 & 0 & 1 & 1 & 0 \\ 0 & 0 & 0 & -1 & 1 \\ 1 & 0 & -R_1 & 0 & 0 \\ 1 & -1 & 0 & -R_2 & 0 \\ 0 & 1 & 0 & 0 & -R_3 \end{bmatrix} \begin{bmatrix} V_1 \\ V_2 \\ I_{R1} \\ I_{R2} \\ I_{R3} \end{bmatrix} = \begin{bmatrix} J \\ 0 \\ 0 \\ 0 \\ 0 \end{bmatrix}$$

P.17.14

$$\begin{bmatrix} sC & -sC & 1 & 1 & 0 \\ -sC & sC & 0 & 0 & 1 \\ 1 & 0 & -sL & 0 & 0 \\ 1 & 0 & 0 & -R_1 & 0 \\ 0 & 1 & 0 & 0 & -R_2 \end{bmatrix} \begin{bmatrix} V_1 \\ V_2 \\ I_L \\ I_{R1} \\ I_{R2} \end{bmatrix} = \begin{bmatrix} J \\ 0 \\ 0 \\ 0 \\ 0 \end{bmatrix}$$

P.17.15

$$\begin{bmatrix} sC_1 + G_1 & -G_1 & 0 & 0 & 0 \\ -G_1 + g & G_1 + G_2 - g & -G_2 & 1 & 0 \\ 0 & -G_2 & G_2 & 0 & 1 \\ 0 & 1 & 0 & -sL & 0 \\ 0 & 0 & 1 & 0 & 0 \end{bmatrix} \begin{bmatrix} V_1 \\ V_2 \\ V_3 \\ I_L \\ I_{sc} \end{bmatrix} = \begin{bmatrix} J \\ 0 \\ 0 \\ 0 \\ 0 \end{bmatrix}$$

P.17.16

$$
\begin{bmatrix}
G_1 & 0 & 0 & 0 & 1 & 0 & 0 \\
0 & 0 & 0 & 0 & -1 & 0 & 1 \\
0 & 0 & sC + G_2 & -G_2 & 0 & 0 & -1 + \alpha \\
0 & 0 & -G_2 & G_2 & 0 & 1 & 0 \\
1 & -1 & 0 & 0 & -sL & 0 & 0 \\
0 & 0 & 0 & 1 & 0 & 0 & 0 \\
0 & 1 & -1 & 0 & 0 & 0 & 0
\end{bmatrix}
\begin{bmatrix}
V_1 \\ V_2 \\ V_3 \\ V_4 \\ I_L \\ I_{sc} \\ I_{cc}
\end{bmatrix}
=
\begin{bmatrix}
J \\ 0 \\ 0 \\ 0 \\ 0 \\ 0 \\ 0
\end{bmatrix}
$$

P.17.17

$$
\begin{bmatrix}
G_1 + sC_1 + sC_2 & -sC_2 & 0 & 0 & 0 & 1 & 0 \\
-sC_2 & sC_2 + sC_4 + G_2 & -sC_4 & 0 & 0 & 0 & 1 \\
0 & -sC_4 & sC_4 & 0 & 1 & 0 & 0 \\
0 & 0 & 0 & sC_3 & 0 & -1 & -1 \\
0 & 0 & 1 & 0 & 0 & 0 & 0 \\
1 & 0 & 0 & -1 & 0 & -sL_1 & sM \\
0 & 1 & 0 & -1 & 0 & sM & -sL_2
\end{bmatrix}
\begin{bmatrix}
V_1 \\ V_2 \\ V_3 \\ V_4 \\ I_E \\ I_{L1} \\ I_{L2}
\end{bmatrix}
=
\begin{bmatrix}
J \\ 0 \\ 0 \\ 0 \\ E \\ 0 \\ 0
\end{bmatrix}
$$

P.17.18

$$
\begin{bmatrix}
G_1 & -G_1 & 0 & 0 & 1 & 0 & 0 \\
-G_1 & G_1 + sC_1 & -sC_1 & 0 & 0 & 1 & 0 \\
0 & -sC_1 & sC_1 + G_2 & 0 & 0 & 0 & 1 \\
0 & 0 & 0 & sC_2 & 0 & -1 & -1 \\
1 & 0 & 0 & 0 & 0 & 0 & 0 \\
0 & 1 & 0 & -1 & 0 & -sL_1 & -sM \\
0 & 0 & 1 & -1 & 0 & -sM & -sL_2
\end{bmatrix}
\begin{bmatrix}
V_1 \\ V_2 \\ V_3 \\ V_4 \\ I_E \\ I_{L1} \\ I_{L2}
\end{bmatrix}
=
\begin{bmatrix}
0 \\ 0 \\ J \\ 0 \\ E \\ 0 \\ 0
\end{bmatrix}
$$

SOLUTIONS CHAPTER 19

P.19.1 The function is $f(t) = (A/T)t$. The expansion is

$$
f(t) = A\left[\frac{1}{2} - \frac{1}{\pi}\sum_{n=1}^{\infty}\frac{1}{n}\sin n\omega t\right]
$$

P.19.2 The function is $f(t) = \sin t$, the period is $T = 2\pi$ but the integration bounds are from 0 to $T/2$. The expansion is

$$
f(t) = \frac{2}{\pi} + \frac{4}{\pi}\sum_{n=1}^{\infty}\frac{\cos nt}{1 - 4n^2}
$$

P.19.3 The function is $f(t) = \sin t$ with $T = \pi$. Thus the fundamental frequency is $\omega = 2$. The expansion is

$$f(t) + \frac{4}{\pi}\left[1 + \sum_{n=1}^{\infty} \frac{2}{1 - 4n^2} \cos 2nt\right]$$

P.19.4 From the Laplace transform $F(s) = 1/(s + a)$. The spectrum is $S(\omega) = 1/(a + j\omega)$ and $|S(\omega)| = [a^2 + \omega^2]^{-1/2}$.

P.19.5 The Laplace transform is

$$F(s) = \frac{s + a}{(s + a)^3 + \omega^2}$$

Substitute $s = j\omega$.

Index

DISKS AVAILABLE

The three programs discussed in this text are available on two 5-1/4 inch disks prepared for the IBM PC (or equivalent) for a fee of $12.00 to cover shipping and handling.

To obtain the disks, please complete the form below and return it, with a check or money order, to:

Van Nostrand Reinhold
Order Fulfillment
P.O. Box 668
Florence, KY 41042-9979
(1-800-926-2665)

A larger version of the program (capable of dealing with larger problems), and a manual for its use, are available from the author through the University of Waterloo.

VAN NOSTRAND REINHOLD
Order Fulfillment
P.O. Box 668
Florence, KY 41042-9979

Please send me the 5-1/4 inch disks, containing three executable programs, for *Basic Network Theory with Computer Applications* by Jiri Vlach

ISBN:0-442-01200-4

☐I have enclosed my check or money order for $12.00
☐Please bill my credit card (see below)

Name _____ Title _____

Company _____

Address _____

City _____ State _____ Zip _____

Phone _____

Credit card number _____ Expiration date _____

Signature _____ Date _____

Master, Visa, American Express credit Cards accepted